Science
Intermediate Level

Authors
Joan Wagner & Alan Fiero

Topical Review Books Co.
P.O. Box 328
Onsted, MI 49265
Phone: 1-800-847-0854; Fax: 1-800-847-0851
E-MAIL: Sales@TopicalRBC.com
www.topicalrbc.com

DEDICATION

This book is dedicated to all of our past and present students and to all students who use our book as a resource to improve their knowledge of the natural world. It is our students who have inspired us to do better as teachers.

This book is also dedicated to all of our colleagues and science educators who strive to inculcate their students with knowledge and understanding about the natural world so that their students can become better decision makers about themselves and the high-tech, information society in which they live and will some day work.

This book was first published for articulation with the New York State Science Learning Standards. We have revised our book for *Topical Review Book Company* to more be in alignment with NGSS. The book's deep roots in the nature and learning of science facilitated our adaption to NGSS because a 3D approach to the teaching and learning of science permeates the book. Lastly, we thank Eugene B. Fairbanks and Wayne Garnsey for all artwork and graphics, which appeared in our original book.

ABOUT THE AUTHORS

Dr Alan Fiero taught science for 44 years, with 32 years at the middle school level. His doctorate from the University at Albany is in teaching students in middle school to use metacognitive skills to improve their learning. Dr. Fiero and his students worked for over 15 years doing active research and restoration in the rare Pine Bush Ecosystem. This work with his students won many awards and grants. Students girdled over 50,000 overpopulating aspen trees and raised endangered Karner blue butterflies so that they could be returned to their natural habitat. He created a student run butterfly house called "Butterfly Station" at their school. Upon retiring he worked three years at an elementary school creating and teaching a NGSS aligned science program. Alan spends his free time raising bees, chickens and making maple syrup.

Ms. Joan Wagner has taught middle and high school science for 34 years and has been the recipient of a number of teaching awards. She received a degree in biology from Syracuse University and her masters of science education from Teachers College, Columbia University. She is one of the authors of the New York State Intermediate Level Core Curriculum Guide and has also worked on the Intermediate Level assessments. She is a past president of the Science Teachers Association of New York State and presently sits on the Board of Trustees for the Dudley Observatory and the Eastern Section of STANYS. Joan has written the *Learn Science* series for Pearson Publishing and recently completed a Middle Level Science book called "Tales of Science," also published by *Topical Review Book Company*. She continues to be the Director of the Greater Capital Region Science and Engineering Fair (33 years), a regional fair for the Regeneron International Science and Engineering Fair, STANYS Science Congress and the Thermo Fisher Scientific Junior Innovators Challenge. Joan enjoys skiing, hiking and playing tennis.

TABLE OF CONTENTS

TO THE TEACHER

The Next Generation Science Standards (NGSS) provides a three-dimensional approach to the teaching and learning of science. Shown on the Mobius Loop NGSS logo are Crosscutting 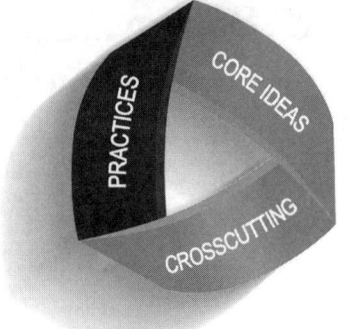 Concepts, Science and Engineering Practices and the Disciplinary Core Ideas. "Crosscutting" concepts such as cause and effect permeate the four main domains of NGSS: Physical Science, Life Science, Earth and Space Science and Engineering & Design. "Practices" emphasize inquiry. Students learn how to investigate and analyze the natural world. "Core Ideas" encompass the age appropriate science needed for the development of science literacy. There is a caveat important for students to understand in their journey to develop science literacy. Scientific knowledge is not a collection of immutable facts. There are times when observations cannot be explained by present scientific concepts. For example, it is Einstein's theory of gravity that is used for GPS because time is relative and it goes faster on satellites. If we used Newton's theory, we would get lost.

The 3D approach to learning should help students develop the critical thinking skills necessary to navigate an increasingly complex world. The Performance expectations in Unit 7 provides a number of 3D learning assessments.

This book does not cover all of the engineering standards since that is better taught through a technology course. However, they are addressed as they play an important role in our understanding and advancements in science. We highly recommend all intermediate level teachers work with math and technology teachers to coordinate curriculum and team together on projects when feasible.

TO THE STUDENT

Understanding the natural world is an important part of your education. This book should strengthen your knowledge of important core science ideas and help you to develop critical thinking skills needed to make informed decisions. The book is designed to help you with middle level assessments.

The first section of this book called "Assessing and Understanding" and the last section of the book called "General Science Skills" are all about the practices that are used in all areas of science and engineering. It also provides many crosscutting concepts explained above.

All units are followed by assessments useful to both students and teachers as a way of measuring a student's strengths and weaknesses in 3D learning. You should appreciate that the great strides made in science is due to the work of great minds. A famous quote by Sir Isaac Newton (1675) is "If I have seen further it is by standing on the shoulders of Giants."

ASSESSING YOUR UNDERSTANDING

The questions at the end of each chapter test your understanding of core ideas in science, your ability to analyze and interpret data, use models, explain phenomena and propose solutions to problems. It will also ask you to look for patterns and cause and effect relationships in the natural world.

ASSESSING YOUR WRITING FOR SHORT ANSWERS

When writing answers to questions strive to create answers that clearly show you are using scientific thinking. Refer to the "Credits" list below. Note: papers that meet the criteria at the top of the list are the best papers, deserving of excellent grades. Papers written to the criteria in the middle of the list receive partial credit, while papers near the bottom of the credits list receive few or no points.

FULL CREDIT - EXCELLENT
YOUR RESPONSE IS COMPLETE AND CORRECT IF IT...

- DEMONSTRATES A THOROUGH UNDERSTANDING OF THE SCIENTIFIC CONCEPTS IN THE TASK

- INDICATES THAT YOU HAVE RESPONDED CORRECTLY, USING SCIENTIFICALLY SOUND REASONING

- CONTAINS CLEAR, COMPLETE EXPLANATIONS

PARTIAL CREDIT - ACCEPTABLE
YOUR RESPONSE IS MOSTLY CORRECT IF IT...

- INDICATES THAT YOU HAVE DEMONSTRATED ONLY A PARTIAL UNDERSTANDING OF THE SCIENTIFIC CONCEPTS IN THE TASK

- ADDRESSES SOME ELEMENTS OF THE TASK CORRECTLY BUT MAY BE INCOMPLETE OR CONTAIN INCORRECT REASONING

NO CREDIT - TOTALLY UNACCEPTABLE
YOUR RESPONSE IS UNACCEPTABLE IF IT...

- INDICATES A CORRECT RESPONSE HAS BEEN ARRIVED AT USING AN INCORRECT PROCEDURE

- NO RESPONSE IS GIVEN

Let us look at a sample question:

A COMPARISON OF STRUCTURES USED FOR SIMILAR FUNCTIONS

ONE-CELLED ORGANISM	MULTICELLULAR ORGANISM
mitochondria	lungs
cytoplasm	heart
cell membrane	kidney

Pick one pair of structures from the chart above and describe how their functions are alike and different.

_____

_____

_____

_____

Below are three sample answers for this question. Compare your written answer above with the sample answers below.

> **SAMPLE ANSWER 1** (Excellent)
> Both mitochondria and lungs help the organism to release the energy stored in food. A mitochondria is different from a lung because it is where the oxygen is actually used to release energy. Lungs function to get the oxygen to the blood so it can be brought to the cells.
>
> **SAMPLE ANSWER 2** (Good)
> Mitochondria and lungs both help respiration. They are different, because one does its job by itself and one is part of a system.
>
> **SAMPLE ANSWER 3** (No Credit)
> Mitochondria and lungs both help circulation. They are different because only one has materials traveling through the cell membrane.

Look back at the previous page and review the criteria for rating a response answer. Based on the criteria and the sample answers above, what grade would you give your answer? Why?

_____

_____

Rewrite your answer. Keeping in mind the criteria for rating a response to a question. Try to improve your answer and therefore your rating.

_____

_____

_____

_____

ASSESSING YOUR WRITING FOR SHORT ESSAYS

Another typical question you will be asked to answer will be used to check if you can apply thinking and problem solving skills to what you have already learned. In order to answer a short essay question you must show a clear understanding of the science concepts being tested. Most extended short essay questions will contain more than one part, so you must be sure to address all aspects of the question. In many cases you will be analyzing, comparing and/or contrasting issues and events.

FULL CREDIT - EXCELLENT
YOUR RESPONSE IS COMPLETE AND CORRECT IF IT...

- SHOWS YOU HAVE A CLEAR UNDERSTANDING OF THE SCIENTIFIC CONCEPTS IN THE TASK
- ADDRESSES ALL ASPECTS OF THE TASK

PARTIAL CREDIT - VERY GOOD
YOUR RESPONSE IS MOSTLY CORRECT IF IT...

- SHOWS YOU UNDERSTAND AND/OR HAVE BEEN ABLE TO DEFINE THE SCIENTIFIC CONCEPTS IN THE TASK
- ADDRESSES MOST ASPECTS OF THE TASK AND INCLUDES ACCURATE FACTS, EXAMPLES, AND DETAILS, BUT MAY BE SOMEWHAT UNEVEN IN TREATMENT

PARTIAL CREDIT - GOOD
YOUR RESPONSE IS PARTIALLY CORRECT IF IT...

- PRESENTS AN ACCEPTABLE DEFINITION OF THE SCIENTIFIC CONCEPTS IN THE TASK
- FAILS TO ADDRESS ALL ASPECTS OF THE TASK
- MINIMAL FACTUAL ERRORS ARE PRESENT

PARTIAL CREDIT - POOR
YOUR RESPONSE IS ONLY PARTIALLY CORRECT IF IT...

- ATTEMPTS TO ADDRESS THE SCIENTIFIC CONCEPTS IN THE TASK, BUT USES VAGUE AND/OR INACCURATE INFORMATION
- RESPONSE LACKS FOCUS

PARTIAL CREDIT - VERY POOR
YOUR RESPONSE IS ONLY A LITTLE CORRECT IF IT...

- SHOWS LIMITED KNOWLEDGE OR UNDERSTANDING OF THE SCIENTIFIC CONCEPTS IN THE TASK
- DETAILS ARE WEAK

NO CREDIT - TOTALLY UNACCEPTABLE
YOUR RESPONSE IS NOT CORRECT AT ALL IF IT...

- FAILS TO ADDRESS THE TASK
- MAKES NO SENSE
- CONTAINS NO RESPONSE

Here is an example of a **short essay** question:

Many plants have roots, stems, leaves, and flowers. Each of these structures is adapted to a specific function. Choose TWO of these structures and describe their function and how they are adapted to carry out that function. Include a labeled diagram of ONE of the structures that illustrates how the structure is adapted.

Drawing

_____

_____

_____

_____

SAMPLE ANSWER 2

The roots of some plants are adapted to support the plant in soil. They may be long, thick, and travel far down into the soil to provide balance for the plant. The leaves of the plant allow it to make food through the process of photosynthesis. They are thin and flat to collect more sunlight.

leaves

stem

roots

Look back at the previous page and review the credits rating a **short essay** answer. Based on the credits list and the sample answer above, what grade would you give your answer? Why? [5]

_____

_____

Rewrite your answer. Keeping in mind the credits for rating a **short essay**, try to improve your answer and therefore your rating.

_____

_____

_____

_____

_____

_____

SCIENCE PRACTICES:
ANALYSIS, INQUIRY, & DESIGN

CONSTRUCTING EXPLANATIONS & DESIGNING SOLUTIONS
THE CENTRAL PURPOSE OF SCIENTIFIC INQUIRY IS TO DEVELOP EXPLANATIONS OF NATURAL PHENOMENA IN A CONTINUING, CREATIVE PROCESS.

WHAT IS SCIENCE?

The goal of science is to understand the natural world. Therefore, science is like a never ending mystery. In science, people uncover clues and seek explanations for why our world is like it is. Each answer brings us new knowledge and more questions.

Knowledge in science is not written in stone. As you make new observations and test new explanations your view of the natural world may change again and again. Science is therefore self-correcting. Ideas, explanations and even theories are constantly revised as you learn more. A basic part of understanding science is understanding that what you know changes as you learn more about our world. For example, for many years people believed that the Sun revolved around Earth. When closer observations were made they found that Earth revolves around the Sun.

1 Can you think of another idea about the natural world that has changed over time?

_____

_____

Science is a way of knowing. Building knowledge in science is characterized by observing, explaining, testing explanations, logical arguments, and critical review. "Doing science" is called **scientific inquiry**. Scientific inquiry refers to the many ways in which scientists study the natural world. It leads to an understanding of how the natural world works. Scientific inquiry has many components. This chapter will explain each component to help you build a better understanding of how to do science.

The central purpose of scientific inquiry is to develop explanations for natural occurrences. How does scientific inquiry begin? It begins by just observing the natural world. Each time you see, smell, taste, touch or hear anything in the natural world you may be on the road to doing science. What makes it science is *how* you observe the world. Many people look, but only a few observe. What type of observing allows scientists to begin to unravel the mysteries of the world around us?

2   What do you think was used to make the pencil you use? _____

_____

3   What observations could you make to prove this? _____

_____

4   What makes a good observer? _____

_____

Science and Engineering Practices describe how scientists investigate the world. They do this by...
• Asking Questions and Defining Problems
• Developing and Using Models
• Planning and Carrying Out Investigations
• Analyzing and Interpreting Data
• Using Mathematical and Computational Thinking
• Constructing Explanations and Designing Solutions
• Engaging in Argument from Evidence
• Obtaining, Evaluating, and Communicating Information

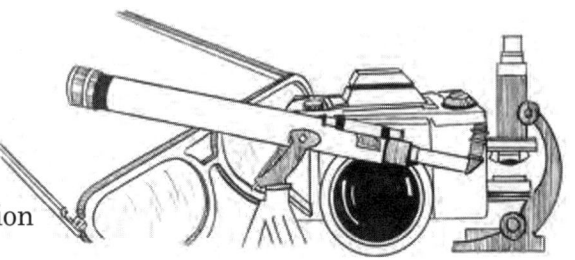

OBSERVING

"How do we come up with questions and problems in science? It usually begins by observing the world around us."

Observing something scientifically means using one or more of the senses to examine something carefully. Another important part of observing is to record what is observed. Good scientific observations also include **measurements** when possible. **Recording** quantities may help later on when observations are put to use.

A **checklist** for making good observations would include making note of color, size, shape, taste, smell, sounds, texture, actions, mass, length, volume, time or rate, and location. This list is for general observations. The specific question you study will direct you to which type of observations are most important.

A student was asked to scientifically observe a rock. The student looked at the rock carefully, picked it up, smelled it, and massed it. In his journal he wrote, "The rock was white with black speckles. It had no smell. It had a mass of 205 grams. It was 10 cm in height."

5   Did this student make a good scientific observation? Why or why not? _____

_____

_____

6   How could the student have made the observation better? _____

_____

_____

DESCRIPTIVE LINKING

Observe the object shown at the right. How many ways can you describe this apple? Starting with one descriptive word you can link together many descriptive words to make a better observation. For example, you could say:

oval
oval, smooth
oval, smooth, sweet smelling
oval, smooth, sweet smelling, sour tasting
oval, smooth, sweet smelling, sour tasting, mass is 353g
oval, smooth, sweet smelling, sour tasting, mass is 353g, makes a crunchy noise when bitten

Pick an object to observe. Record your observation. Read your observations to another person and see if he/she can guess what the object is. Safety Note: Do not touch, smell or taste the object unless directed by your teacher.

7   Name of Object: _____   Description: _____

_____

QUESTIONS PLEASE!

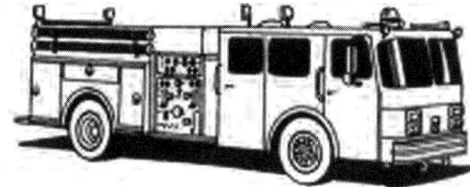

Observations are notations you make using your senses. Be sure not to confuse observations with things you did not sense. For example, if you hear a siren you may write in your journal that the siren was loud, 30 seconds long and high pitched. You might also write "I bet there was a fire!" However, this was not an observation.

8   "I bet there was a fire!" Why is this statement not an observation? _____

_____

Observations lead to questions. Science is based on our natural curiosity about the world around us. The human mind seeks explanations for why things occur. Science allows us to test our explanations. Science helps us to determine the answers to our questions. Since science seeks an understanding of the natural world, questions in science should be directed at this task. Usually you are looking at why something happens. Good scientific questions look to answer how one thing influences another. You want to understand the *how* and the *why* of natural occurrences.

Once a broad topic of study is determined, it is important to narrow down the study to a particular question of interest. It may be best to work with a group of students to brainstorm subtopics and possible questions. This is also a good time to look at reference materials you might find that have information about the topic so that you have some background in the area that you are studying.

Suppose you are interested in studying the motion of the Moon. You might identify some subtopics about the Moon's motion, such as gravitational pull, rotation, phases, or eclipses.

9   Take one of these subtopics and write three questions about it that you might be able to study.

Subtopic: _____

Question 1: _____

Question 2: _____

Question 3: _____

Constructing Explanations and Designing Solutions

Not all scientific knowledge is gained through experimenting. Some scientific questions may be answered by making more observations. A question could be answered by describing, measuring, or comparing.

As a scientist seeks to understand the natural world, they sometimes creates a possible explanation for their observations or question. The scientist does this so that they can test to see if the explanation is correct. This tentative, testable explanation of the observed relationship between factors is called a hypothesis. Some people call a hypothesis an educated guess. This is not a good definition. The hypothesis is much more than a mere guess. A hypothesis is "educated," in that you should know something about the topic before you try to explain it. It is a "guess" in that it is tentative and not yet proven. But most importantly, a hypothesis is an explanation. Science seeks explanations. The hypotheses is a scientist's first try at an explanation. A hypothesis must be testable. Testing if a hypothesis is correct will lead to understanding the natural world.

How do you get from a question to a hypothesis?

Writing a hypothesis is a creative part of science. You have to carefully examine what your observations show in order to think of a good explanation. Here are some hints for writing a hypothesis.

- What sort of information do you have? (Observations? Data from experiments?)
- What are the important properties of the objects or events being observed?
- Sort out events and objects. How do they relate?
- What things seem linked?
- What things do not seem logical?
- If any conditions varied, what were the consequences of them being varied?

Hypothesis Formation

You will use a specific format for writing a hypothesis to make it easier to evaluate it. Each hypothesis will have three parts:

- *if* (condition)
- *then* (predicted results)
- *because* (explanation)

Here are some examples of hypotheses.

If sufficiently cooled,
then a liquid will become a solid,
because its internal particles have slowed and changed their type of motion.

If given light,
then plants will grow,
because plants use light to make food.

If animals have long hair,
then they will survive better in cold climates,
because the hair keeps their body warm.

10 Write a hypothesis to explain why water runs faster through gravel than through sandy soil.

If _____ (What conditions are you going to set up?)

Then _____, (What do you predict will happen?)

Because _____. (Why?)

Rating a Hypothesis

A hypothesis must have several components. It must propose a relationship among the **variables** (factors in an investigation that may affect the results), so that the relationship can be tested. To make a better hypothesis, the quantities or qualities of the variables should be included in the hypothesis.

For example:

Good: *If* plants get light, *then* they will grow *because* they use light to make food.

Better: *If* plants get more sunlight, *then* they will grow faster, *because* they have more light to make food.

11 Make the following hypothesis better.

   Good: *If* you study, *then* you will receive a grade *because* you read the material.

   Better: _____

Developing and Using Models

Scientists and engineers use and construct models to help to understand an idea or explanation. A model can be a: diagram, drawing, physical replica, mathematical, or a computer simulation.

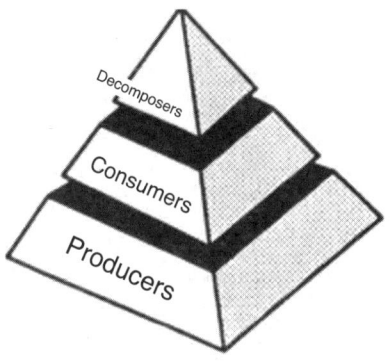

It is sometimes easier to understand the explanation your hypothesis is providing if you construct a mental **model**. For example, when explaining why there must be more producers (green plants) than herbivores (animals that eat green plants), scientists came up with a mental model called the **energy pyramid**.

The triangle shape shows by volume that there are more producers than consumers. It also shows by position that the producers are the base, or support, for the upper level consumers. Consumers are divided into three main groups – herbivores, carnivores, and omnivores – based on what they eat (see Unit 2 for more information on these groups and the decomposers).

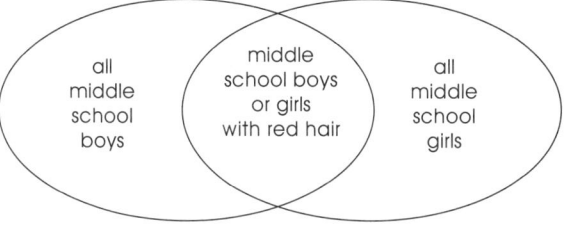

Models are used in science to help you get a better picture of things you cannot actually see. They are another way to express your ideas. Mental models often help others understand what you are trying to say. For example, in one type of model, a **Venn diagram**, all items in the same class or trait are in the same circle (field). Items of different classes or types are in different circles (fields). Items in more than one field are shown as overlapping. For example, Field 1 represents all middle school boys. Field 2 represents all middle school girls. If this model is being used to look at all people with red hair, the Venn diagram might look like the one at the right.

12 Construct a Venn diagram of what animals eat. Include herbivores (eat just plants), carnivores (eat just animals) and omnivores (eat both plants and animals). Start with two separate fields:

Plants   Animals

Sometimes an educated guess is *just* an educated guess. Not every statement in science is a hypothesis. It is important to distinguish among observations, inferences, predictions, and explanations.

Below is a diagram of a food web. This diagram was produced by observing what organisms used for food. The arrows represent where the energy is going. In one small diagram, the energy relationships among an entire community of organisms can be easily understood.

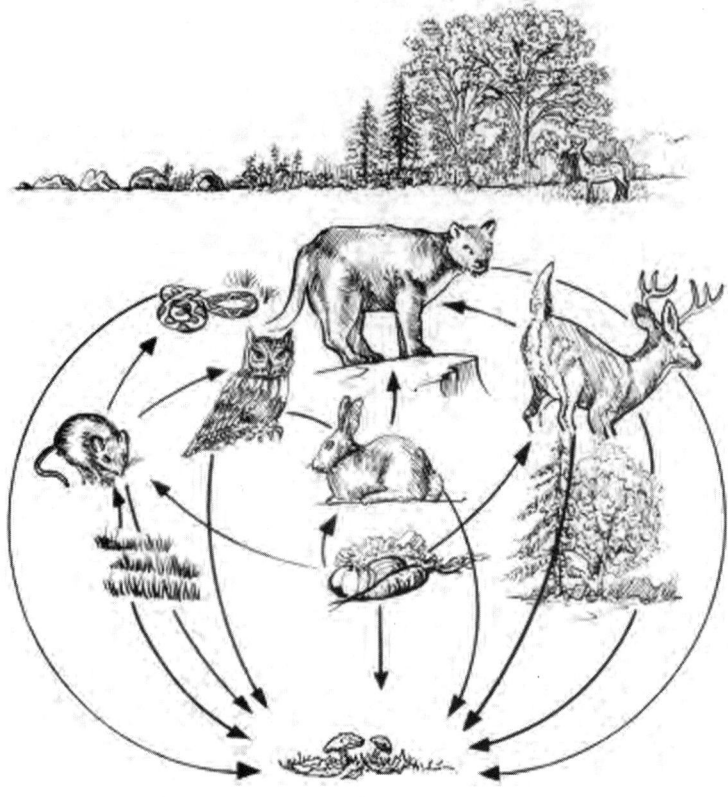

13 Choose two different types of plants to observe. Draw a diagram of each. Be sure to label all of the parts of the plants.

Plant A	Plant B

14 What differences between the plants can you distinguish by comparing the two diagrams?

_____

_____

_____

PLANNING AND CARRYING OUT INSTIGATIONS

OBSERVING

Observations come directly from your senses. They are the records of what the senses tell us about the object or event you are examining. For example, while walking on a nature trail, a student saw tracks.

15   What did the student observe? _____

_____

16   What sense was being used? _____

_____

INFERRING

Inferring is taking something you already know, and linking it to an observation. This is done in order to make a **conclusion** that explains the observation. For example, the student read that coyotes had returned to the area. The student inferred that the track had been made by a coyote. An **inference** is similar to a hypothesis, because both are untested explanations. Also, both help scientists make sense of their observations and make predictions. Yet, an inference is different from a hypothesis, because it explains specific cases. The hypothesis tries to explain a general relationships among variables.

© PhotoDisc

17   Did the student actually see the coyote make the track? _____

PREDICTING

Predicting is using an inference (explanation) to tell what will happen in the future. For example, the student predicted that the coyote would use this trail again.

18   What was the observation on which this prediction was based? _____

_____

19   What is the inference on which this prediction was based?_____

_____

HYPOTHESIZING

Hypothesizing is coming up with an explanation of why things occur in the natural world. For example, the student hypothesized that the coyote will use the trail again. The student believed coyotes travel on established trails rather than through the thick brush in the forest to find water. Hypothesis: *If* the coyotes returned, *then* they would use the trail again, *because* coyotes travel on established paths.

20   How could the student test this hypothesis? _____

_____

_____

BODY OF SCIENTIFIC KNOWLEDGE

THE PRODUCTS OF SCIENCE ARE EXPLANATIONS.

All of the work that scientists have ever done has contributed to our current understanding of the natural world. This is sometimes referred to as the "body" of scientific knowledge. When a scientist seeks to answer questions, they first examine what is already known about the topic. Through examining the body of scientific knowledge, a scientist may find the answer to their question or revise and refine a hypothesis. The scientist may also get ideas on how to test their own hypothesis. Researching might find the names of other scientists doing similar work so that they can collaborate (work together).

Checking what is already known is very important to the growth of scientific knowledge. Through checking, the scientist does not have to rediscover what is already known. The scientist can use what is already known to go on even further and learn even more.

The 17th-century English scientist **Sir Isaac Newton** when explaining how he had made such great advances in science said, "If I have seen further, it is by standing upon the shoulders of giants." He meant that he had used the work of other great scientists before him.

Sir Isaac Newton

© Wildside Press

21 Why is it important to find out what other scientists have already done before starting an investigation?

_____

_____

_____

THEORIES AND LAWS

What happens when a hypothesis is continually supported by evidence? When this happens, scientists sometimes agree that the hypothesis should become a theory. A **theory** is a carefully thought out explanation, logically reasoned, and based on well founded ideas about the nature of the world.

When a theory is continually supported by evidence, scientists may agree that the theory should become a law. A **law** is an idea or concept that is almost equal to a fact. Some scientists believe there should be no scientific laws. This is because science is constantly growing, and new evidence may lead to new ideas that contradict the law.

22 *a* How are hypotheses, theories, and laws alike? _____

_____

b How are hypotheses, theories, and laws different? _____

_____

OBTAINING, EVALUATING, AND COMMUNICATING INFORMATION

ENGAGING IN ARGUMENT FROM EVIDENCE

THE REVIEW CYCLE

Scientists communicate new findings through evidence-based knowledge. This form of knowledge gained through rigorous scientific research. In order for our knowledge of the natural world to grow, it is important that scientists share what they learn. Sharing ideas and findings is a major component of science. One way that scientists work together is to review each others' work. Peer review allows for an exchange of ideas. The questions that one scientist asks another may help clear up the thinking of both. If a scientist discovers an error in the work of another, incorrect data could be corrected or a flawed conclusion may be prevented. Whatever the case, reviewing by others helps to make the work better. Working with a group of your classmates can help you become better scientist.

HYPOTHESIS FOR THE REVIEW CYCLE

23 Write a hypothesis for the following observations: One year, there were gray and red squirrels in our trees. This year, there were only red squirrels.

Hypothesis: _____

_____

_____

24 *a* Construct a mental model (diagram) to explain your hypothesis.

b Present your hypothesis to your classmates or your parents.

c Explain how you came up with the hypothesis. Ask for questions. Record the comments that you get from others.

RETHINKING FOR THE REVIEW CYCLE

After you present your work to others, it is important to go back and reevaluate your work.

25 What were the strong points of your hypothesis? _____

_____

_____

26 What parts were unclear? _____

_____

_____

27 Do you need to rewrite your hypothesis? _____ If so, why? _____

_____

_____

28 Based on your discussions with others, what might a revised hypothesis be? _____

_____

ANALYZING AND INTERPRETING DATA
PLAN AN INVESTIGATION TO ANSWER QUESTIONS OR TEST SOLUTIONS TO A PROBLEM.
IDENTIFY CONTROLLED, DEPENDENT AND INDEPENDENT VARIABLES.

Once you have a hypothesis, you might think you should go right on and begin testing. In many cases, it is actually better to stop and take a look back. This is a good time to reflect on and improve your hypothesis. What further observations could you make that might help you develop a better explanation? Is there a better way to test your explanation? Is there further reading or other sources of information that can help guide you? For example, a student observed that not all plants grow at the same rate. She decided on the following hypothesis: *If* two different plants are given the same amount of light, *then* one will grow more than the other, *because* different plants need different amounts of light to grow.

29 What further observations should she make before she begins her investigation? _____

_____

30 What further information should she seek before she begins? _____

_____

DESIGNING A SCIENTIFIC INVESTIGATION

The basis of a scientific investigation is setting up conditions. These conditions should lead to results that will help you determine if your hypothesis is correct. You should set up the conditions so that you can predict what will happen if your hypothesis is indeed correct.

For example, a student noticed that most of the plants growing in the shaded area of her neighbor's garden had red or purple leaves, rather than bright green leaves. The student inferred that since her neighbor was a good gardener, she had planted them in the shade on purpose, knowing that they would grow well. The student hypothesized that if plants with red and purple leaves are placed in lower light conditions then they will grow better than if they are placed in the Sun, because their pigments work better with less light.

31 Describe the conditions you would set up to test the student's hypothesis. Based on her hypothesis, what prediction could you make about how the plants will grow in the conditions you set up?

Condition 1: _____

_____

Prediction 1: _____

_____

Condition 2: _____

_____

Prediction 2: _____

_____

Keys To A Successful Investigation

Understanding The Variables

One of the first things to consider in setting up an investigation is what factor (or factors) can affect the outcome of the investigation. Any factor (or condition) that can affect the outcome of an investigation is called a **variable**. If this factor varies or changes the investigation in some way, the variable can affect the results.

32 What are some variables that could affect the growth of the red and purple leafed plants? _____

_____

_____

Two types of variables are very important in any investigation. They are the **independent** (or manipulated) **variable** and the **dependent** (or responding) **variable**. In an experiment, one factor (*dependent* variable) is allowed to be manipulated (varied) by the action of a second factor (*independent* variable). The *dependent* variable is observed carefully for effects. If a change occurs in this variable, then it is truly a *dependent* variable, since it was changed by the action of the first *independent* variable.

33 In the student's investigation of green and purple plants:

   a   the amount of light both plants get would be the _____ variable, and

   b   the amount the plant grows would be the _____ variable.

Procedures are set up to change the independent variable. Observations then examine if this change affects the dependent variable. The hypothesis proposed was: if you vary the amount of light (independent variable), the plants would grow best (dependent variable) with less light.

34 In this investigation, what can vary? _____.

35 Make a drawing of how the investigation would look with the variables you chose above.

Controlling Other Variables

Many other factors could affect the growth of plants (such as the amount of water, temperature, size of pot, and type of soil). Since these factors vary in the way they influence the growth of the plant, these factors must be controlled (kept consistent). By controlling their effect on the investigation, they should affect *all* of the plants in the same way. Therefore, any differences in the plants' growth should be based on the independent variable, not on other factors.

A Controlled Experiment

The experimental group receives the independent (manipulated) variable. The control group acts as a baseline or frame of reference or does not receive the variable being tested at all.

EXPERIMENT EXAMPLE

A student wants to test which paper towel is the most absorbent. She hypothesizes that *if* five different brands of paper towels are placed in a dish of water, *then* the "Soaky" brand will absorb the most, *because* it has the most fibers per square cm.

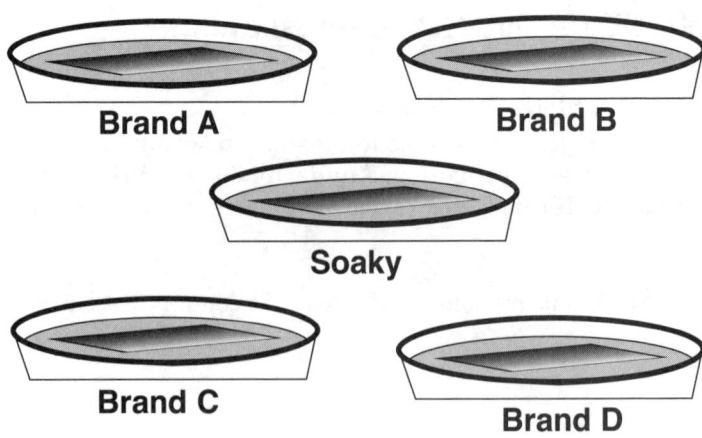

Brand A   **Brand B**

Soaky

Brand C   **Brand D**

She places a strip of paper towel 20 cm x 5 cm in each dish. She masses (measures of the quantity of matter) each paper towel. She leaves them in the dish for one minute and again masses the towel. Each cup is the same size and has the same amount of water.

36 What is the independent variable in this experiment? _____

_____

37 What is the dependent variable in this experiment? _____

_____

38 What variables are controlled in this experiment? _____

_____

Sometimes an experiment can be thought of as a race. The racers are the variations of the independent variable. In order to see who won, look at the dependent variable. For example, you are going to be in a 100-meter dash with a friend. The independent variable would be the runners (your friend and you). The dependent variable would be how fast you run (speed). What if you were forced to carry a book bag during the race, but your friend did not have to carry one? Would that be a fair race? What if your friend could wear track shoes and you could not, would that influence the results?

© PhotoDisc

In order to make it a "fair" race, the other variables, such as book bags and shoe type, must be controlled. That is, set up so that the race is fair. Their effect must not favor one racer or the other. That is what controlling an experiment is all about. It is setting up the variables in an experiment so that the variables that are not being tested do not influence the results.

A student is developing an experiment to see if plants grow better when more light is added. One plant gets three hours of light and the other gets six hours of light. On day one of his experiment, he gives both plants 100 ml of water. On day three, he gives both plants 200 ml of water.

39 Has the student controlled for the effect of the water? Why or why not? _____

_____

_____

_____

IDENTIFYING VARIABLES

SAMPLE EXPERIMENT 1

A student wanted to determine if temperature had an effect on the growth of bean plants. The student grew fifteen plants in fifteen pots containing the same type and amount of soil. Five plants were kept at 0°C, five plants were kept at 15°C, and five plants were kept at 25°C. The height of the plants was recorded after two weeks.

40 What is the student trying to explain? _____

_____

41 What is one variable that was controlled in this investigation? _____

_____

42 What is the dependent or responding variable? _____

_____

43 *a* What is the independent or manipulated variable? _____

b What is the control group? _____

c What are the experimental groups? _____

SAMPLE EXPERIMENT 2

A student wanted to find out if the volume of the water affects the amount of salt that can dissolve in that amount of water. Using four identical beakers, the student put 50 ml of water in one, 100 ml in the second, 150 ml in the third and 200 ml in the fourth. He kept the temperature during the experiment at 25°C. He then dissolved as much salt as he could into each of the beakers. He recorded the amount of salt dissolved for each volume of water.

44 What is the student trying to explain? _____

_____

45 Name a variable that is controlled for in this investigation. _____

_____

46 What is the dependent or responding variable? _____

_____

47 *a* What is the independent or manipulated variable? _____

b What is the control group? _____

c What are the experimental groups? _____

OPERATIONAL DEFINITIONS

An investigation requires you to observe and describe many things. Right from the beginning of the investigation, it is important that you make it clear what the important objects and events are. For example, you may use a simple word like "fast" to describe something. But what does "fast" really mean? It can mean different things to different people. What is fast to one person may be slow to another person. For scientists to communicate and understand each other, they must be sure that they agree on what terms for objects and events actually mean. Therefore, scientists will often define what they mean in a way so that others can agree with their definition of the terms. These type of definitions are called operational definitions. An **operational definition** tells you exactly what to observe and how to recognize the term being defined. An operational definition may tell you how to measure, how to create, or how to describe the items in the investigation. The key point is that different people agree with the definitions that are used in the investigation.

The results of the 100 meter dash at school today:

John, 10.8 sec   Jen, 11.7 sec   Bill, 12.2 sec   Alice, 11.0 sec

48 If fast is operationally defined as "running a 100 meter dash in 12 seconds or less," then which of

the runners are fast? _____

DESIGN
OBTAINING NEEDED OBSERVATIONS – COLLECTING DATA

As you plan your investigation, you must decide what observations will be needed and how you will obtain them. Here are some questions to ask yourself.

Who? – Who will collect the data? Can one person collect the data alone or is a team needed? If a team is needed, how will the jobs be split?

What? – What things will you need to observe? How will you describe your observations? What traits (characteristics, features, or properties) are you interested in? Will you be making comparisons?

How? – What measurements will you make? What tools or instruments will you need?

Where? – How will you record the data? How will you organize it? Can certain data be classified together?

When? – Is the timing of the observations important? Are the sequence of the events you are observing important?

A SIMPLE CONTROLLED EXPERIMENT

A student observed that after it rains, puddles form in some areas and not others. Upon closer examination of these areas, the student noticed that the soils in these areas seemed to be different. In the area without the puddles, the soil seemed loose. In the areas with puddles, the soil seemed more tightly packed. The student came up with a hypothesis to explain these observations.

> "*If* water is poured through loose and packed soil, *then* more water should flow through the loose soil, *because* loose soil has more spaces (pores) for the water to flow through."

Design a simple controlled experiment to test this hypothesis. The guiding questions on the next page can help you.

© PhotoDisc

49   *a*   What is the independent variable? _____

      b   How will you define (measure) it? _____

50   *a*   What is the dependent variable? _____

      b   How will you define (measure) it? _____

51   What other variables need to be controlled? _____

52   What things will you need to observe?_____

53   How will you take measurements? _____

54   Where will you record your data? _____

55   Who will collect the data? _____

56   *a*   When will you collect the data? _____

      b   What sequence of events must occur? _____

57   Use the answers to the "guiding questions" above to help you design a procedure for your investigation. A procedure lists, step by step, the activities that must occur during the experiment. Number each step and start each step on a new line. This will make the procedure easier to follow. Be sure not to leave any steps out. **A procedure** should be written so that another person could carry out your experiment without you being there. Do not take for granted that they will know what to do. It is always best to have someone else read your procedures to see if they could carry it out as it is written.

PROCEDURE

1 _____

2 _____

3 _____

4 _____

5 _____

6 _____

7 _____

8 _____

9 _____

10 _____

Using Mathematics & Computational Skills

MATHEMATICS AND COMPUTATION HELP US TO MAKE SENSE OF CAUSE AND EFFECT RELATIONSHIPS

Natural Variation

To test a hypothesis, you often compare results between a control or baseline condition, and a manipulated or experimental condition. For example, look at this experiment: Alan wanted to know if plants grow better (dependent variable) with more light (independent variable). He grew one plant under three lights and another under one light. After four weeks, the plant under three lights had grown 10 cm and the plant under one light had grown 7 cm.

If a difference in results is seen, as in this case, you might think that the manipulated variable (more lights) caused the difference. However, in science the scientist must be very sure before making conclusions about the results. The observer must be sure that it was the independent variable that caused the difference and not some other factor. In this experiment, Alan only used one plant in each condition. However, another student, Joan, repeated this experiment with five plants in each condition and got the following results:

JOAN'S EXPERIMENT:
AT START

JOAN'S EXPERIMENT:
AT START

ALL PLANTS
5 CM TALL

JOAN'S EXPERIMENT: AT END OF 4 WEEKS	Growth of plants with 3 lights	
	Plant 1	10 cm
	Plant 2	7 cm
	Plant 3	12 cm
	Plant 4	9 cm
	Plant 5	12 cm

JOAN'S EXPERIMENT: AT END OF 4 WEEKS	Growth of plants with 1 light	
	Plant 1	7 cm
	Plant 2	5 cm
	Plant 3	7 cm
	Plant 4	6 cm
	Plant 5	10 cm

Using more plants can provide more information and more evidence about whether the independent variable truly influenced the results. What if Alan just happened to have had the smallest plant (7 cm) under three lights and the largest plant (10 cm) under one light in his experiment? Using just these two plants, he might have concluded that having more light makes the plant grow less.

Repeating an investigation and also using a larger number of test cases are two ways of providing additional and often better data. More data helps you determine if the results that you are getting are *reliable* (can be repeated) and *valid* (really measuring what they are supposed to). When using more trials, an error in just one trial will not affect your **percentage of error** as greatly.

Another reason to duplicate (or repeat) your work and to use a larger number of cases is that it allows you to examine the natural variations in the variable you are studying. **Natural variation** is the difference due to just chance. For example, what if you were to measure the height of all the students in your class? You might find that the height varies 20 cm from the shortest to the tallest person. If you were going to do an experiment on what makes people grow taller, you would have to take into account that people naturally vary a certain amount. If your experiment was to show that something would make people grow taller, you would have to prove that the factor you were testing caused the increase in growth. Also, the growth produced would have to be greater than what you would expect from the natural differences of your subjects.

The following chart represents the scores from a class quiz. The students organized their data by how long they studied.

Studied *less* than **one hour**

student	score
John	10
Alex	6
Beth	5
Caity	7
Denis	3

Studied *more* than **three hours**

student	score
Bob	9
Eva	10
Michelle	10
Gary	9
Mike	10

58 Compare John's and Bob's scores on the test. What might you conclude about the effect of the amount of studying? _____

_____

59 Compare the scores of all of the students. What might you conclude about the effect of the amount of time of studying? _____

_____

60 Why is it important to repeat experiments? _____

_____

61 Why is it important to use a large number of trials in an experiment? _____

_____

QUALITATIVE & QUANTITATIVE RESULTS

When deciding on what type of data is to be collected, there is an important question to ask: Should the data be quantitative or qualitative? **Quantitative data**, as the name suggests, deals with quantities. It is numeric data, numbers. The data can be worked with mathematics. Any sort of data that has measurements such as grams, milliliters, and centimeters is quantitative. Quantitative data has an advantage over qualitative observations, because numbers are easier to agree upon. (For example, if something is measured at 1 cm long, it is 1 cm long no matter who measurers it.) Furthermore, you can do mathematical analysis on quantitative data (such as averages and percents).

Qualitative data refers to information that cannot be converted into numbers. It is descriptive data. Qualitative data is very useful in understanding complex situations and giving a broader view of important factors. One example is a description of an animal's behavior.

Both quantitative and qualitative data are important. The type of data needed in an investigation depends on the type of question or explanation being tested. Quantifying your data, making it so that it can be mathematically compared, may make it easier to analyze. However, you should not force data into a quantitative form if it causes you to lose important qualitative information.

62 Describe the relationship between weathering and the breaking down of rock.

a What quantitative data might you collect? _____

_____

_____

b What qualitative data might you collect? _____

_____

_____

ANALYZING & INTERPRETING DATA
DISPLAYING & COMPARING DATA

Charts, tables, and graphs (often the terms are interchangeable) help organize data so it can easily be compared. These records of data also are helpful in setting up comparison pictures (graphs). Two important characteristics of every chart, table, and graph are (1) a title that describes what is being shown and (2) labels with units for each variable.

The table below lists ten leading causes of death. The data has been organized (sorted) from greatest to least.

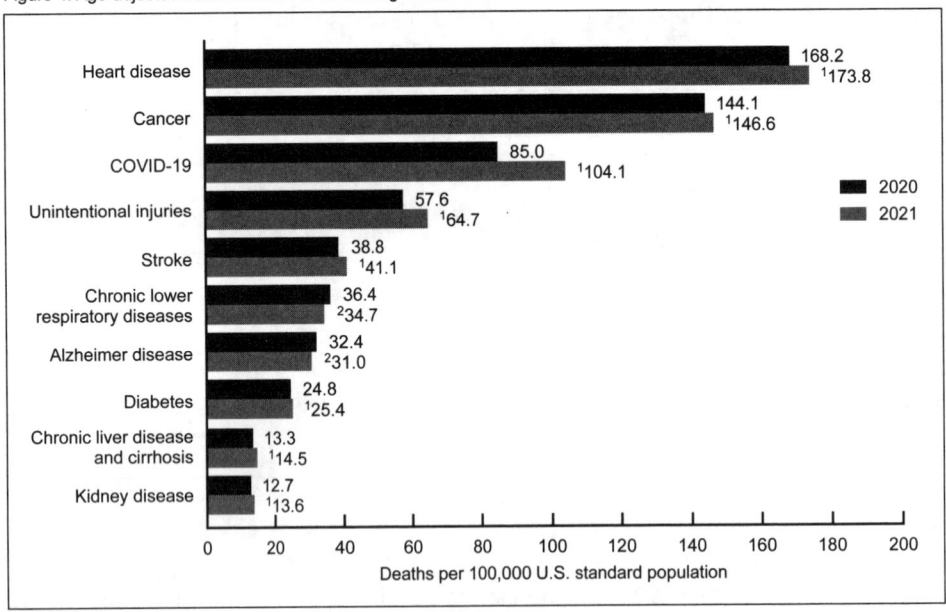

Figure 4. Age-adjusted death rate for the 10 leading causes of death in 2021: United States, 2020 and 2021

63 According to this table, what was the major cause of death in the United States?

_____

64 According to the table, what was the lowest cause of death in the United States?

_____

65 Compare the number of deaths from cancer and unintentional injuries. About how many times greater are deaths from cancer than those from unintentional injuries?

_____

Constructing A Data Table 1

A student was investigating the effect of surface area on the evaporation of water. She hypothesized that *if* pans with different surface areas were set out, *then* those with larger surface areas would evaporate more water, *because* evaporation increases with the amount of surface area exposed to the air. She did five trials and found the following results:

Pan A had a diameter of 10 cm² and evaporated 2 ml of water
Pan B had a diameter of 50 cm² and evaporated 10 ml of water
Pan C had a diameter of 20 cm² and evaporated 4 ml of water
Pan D had a diameter of 40 cm² and evaporated 80 ml of water
Pan E had a diameter of 70 cm² and evaporated 40 ml of water

66 Construct a table to better display these results.
Be sure to include a title for the table and labels and units for the columns.

Hint: There should be a column for both your dependent
and independent variable

Constructing A Data Table 2

A student noticed that her hot drinks seemed to cool more quickly in certain types of cups. She wanted to determine how well different cups could insulate and slow down the flow of heat energy. She hypothesized that *if* metal, Styrofoam, and pottery cups were filled with the same amount of water at the same temperature, *then* the water in the metal cup would stay warmer, *because* metal cups would reflect the heat energy back into the cup. She recorded the temperature change in each cup after 5 minutes. She repeated each trial three times. Here are her results:

Trial 1:   Pottery temperature change was 10°C
           Metal temperature change was 15°C
           Styrofoam temperature change was 15°C
Trial 2:   Pottery temperature change was 12°C
           Metal temperature change was 17°C
           Styrofoam temperature change was 10°C
Trial 3:   Pottery temperature change was 14°C
           Metal temperature change was 18°C
           Styrofoam temperature change was 12°C

67 Construct a data table to better display the results
of this experiment. You may need additional paper.

68 Do the results support the hypothesis? _____

69 Why or why not? _____

_____

70 What could be done to the data in the previous example to make it easier to understand? _____

_____

In experiments where multiple trials are conducted, one thing that can be done is to average the results. Averaging the results makes a much clearer picture of your results.

The chart at the right makes it easier to see that the metal cup was the least effective insulator.

EFFECT OF DIFFERENT MATERIALS ON TEMPERATURE CHANGE	
TYPE OF CUP (MATERIAL)	AVERAGE TEMPERATURE CHANGE (°C)
Styrofoam	10
Pottery	12
Metal	17

GRAPHING

"A PICTURE IS WORTH A THOUSAND WORDS."

This famous quote reflects the fact that a picture can show a tremendous amount of information. In this way, a graph is like a picture. It can display a large amount of data in a way that is easy to understand. It is visual and can often make patterns and relationships easier to recognize than data in a table.

TYPES OF GRAPHS

The most common types of graphs used in science are bar and line graphs. Bar graphs are used to compare definite quantities of different items. The bar graph, below right, illustrates the information from the chart on the left. The *relationship* between type and the amount of minerals in the Earth's crust is easier to "picture" in the bar graph.

MINERALS IN THE EARTH'S CRUST		
NAME	SYMBOL	% BY WT
oxygen	O	46.6
silicon	Si	27.7
aluminum	Al	8.1
iron	Fe	5.0
calcium	Ca	3.6
sodium	Na	2.8
potassium	K	2.6
magnesium	Mg	2.1
all others	-	1.5
		100.0

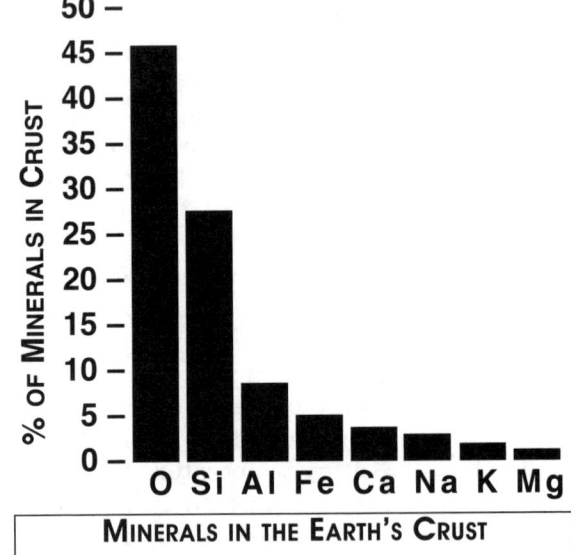

MINERALS IN THE EARTH'S CRUST

Line graphs are used to compare things when the data represents a continuous process. For example, the line graph at the right represents the effect of time on the growth of a plant. Although the plant's height was recorded only once a week, it is understood that the growth process was somewhat continuous and growing occurred between the times that data was collected.

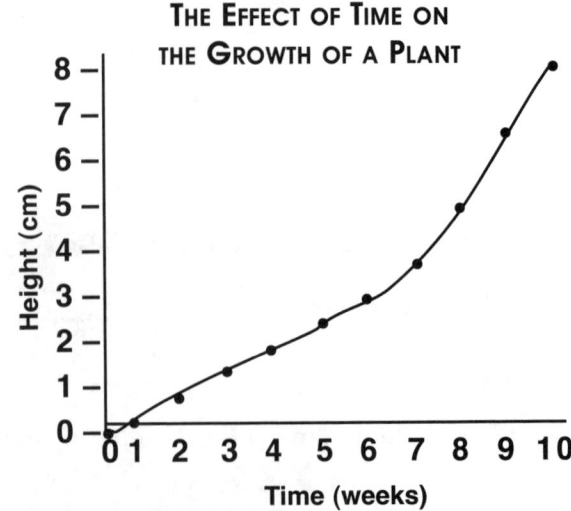

THE EFFECT OF TIME ON THE GROWTH OF A PLANT

CONSTRUCTING A LINE GRAPH

THE TITLE

The title of a graph is very important. It must tell the reader exactly what data the graph is showing. Creating a title for the graph is very easy when you know how and why the graph is being constructed. **Line graphs** are usually used to examine how the dependent (responding) variable relates to the independent (manipulated) variable. A line graph can show if there is a relationship between the variables. A line graph answers the question, does the changing of the independent (manipulated) variable affect the dependent (responding) variable? A title for most line graphs can therefore be written by using the following template.

RELATIONSHIP BETWEEN (INDEPENDENT VARIABLE) _____

AND (DEPENDENT VARIABLE) _____

71 Write a title for the graph at the right.

_____

72 What is the relationship between temperature and the number of yeast colonies present?

_____

_____

_____

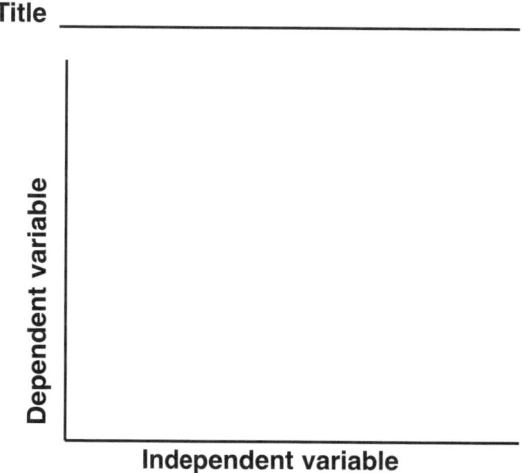

THE AXES

A graph consists of two axes; the **x-axis** (horizontal line) and the **y-axis** (vertical line). When using graph paper these should be drawn with a ruler right on the existing grid lines. The size of the graph depends on the scale of the graph. Two major concerns about size of a graph include (1) making it a size that will allow the reader to easily interpret what it represents and (2) large enough that all of the data will fit properly for comparison.

The axes of a graph must be labeled so that the reader will know what they represent. In order to make things more consistent, the x-axis should be labeled with the independent (manipulated) variable and the y-axis with the dependent (responding) variable. Since this format is used on all graphs, it is easier to understand what a graph is trying to show. Be sure to include units for both variables.

A student did an investigation trying to determine if the amount of sunlight affects the production of chlorophyll.

73 Title and label the graph at the right in a manner that would allow the student to graph his results.

Title _____

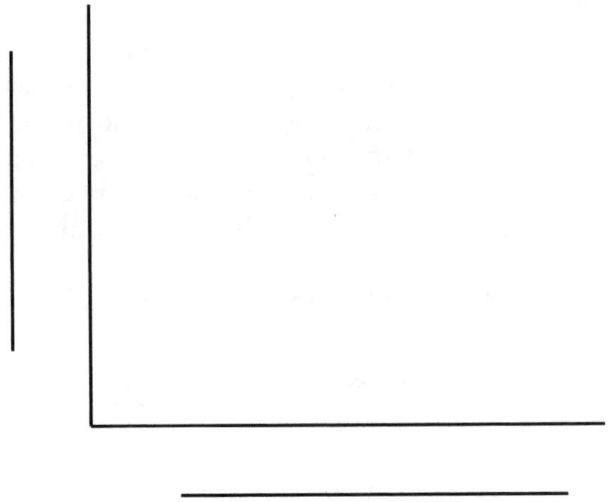

The units of measurement must always accompany the labels on the graph (see graph at the right). These are written in parentheses ( ).

THE EFFECT OF THE AMOUNT OF FERTILIZER
ON THE HEIGHT OF GRASS

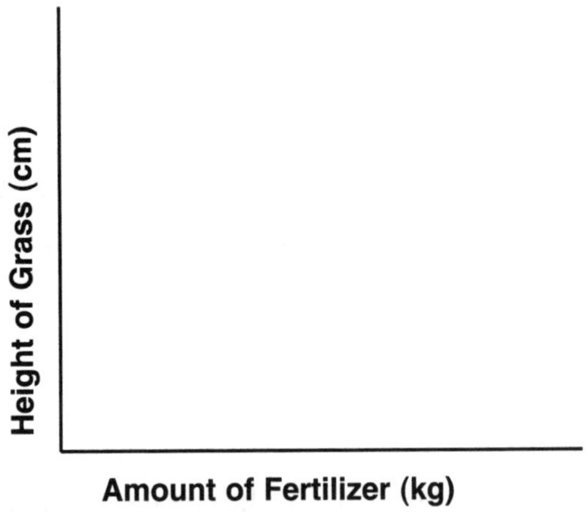

Height of Grass (cm)

Amount of Fertilizer (kg)

A student wanted to study the effect that the distance from the floor had on room temperature. The air temperature was measured in degrees Celsius and the height from the floor was measured in meters.

74 Fill in the title, labels, and units for a graph that could be made with this data.

Title _____

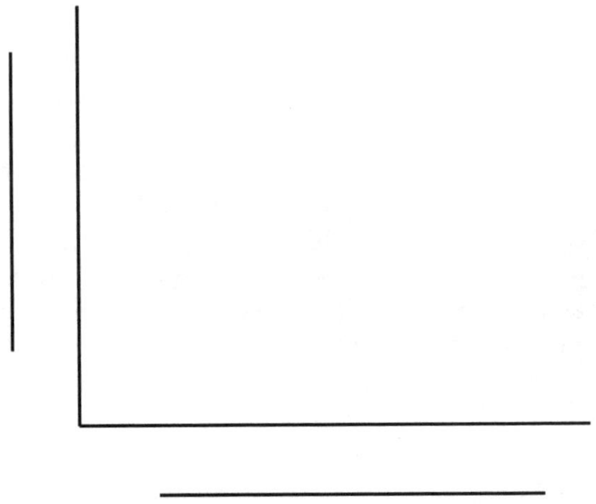

SCALING THE GRAPH

Scaling the graph means numbering the grid so that the data can be displayed for comparison. It is called scaling because you are actually making a scale, a set value per box, for both the x-axis and y-axis.

Begin by labeling the first line of the graph as **0**. This beginning point of the graph has a special name, the **origin**. (This has been done for you.)

75 Why is it necessary to begin both the x-axis and the

y-axis with a 0? _____

_____

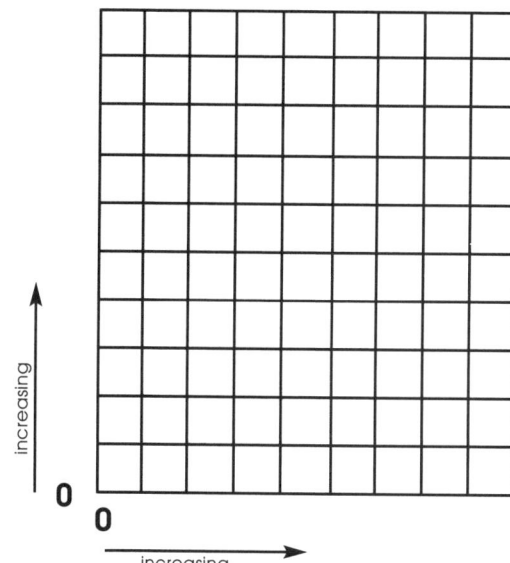

Notice that both lines have been labeled with 0s. One represents the x-axis, and one represents the y-axis. The values are written on the lines. This is because the line itself represents the value 0. Anything forward of the 0 on the x-axis represents more than 0. Anything above the 0 on the y-axis represents more than 0.

Start with a simple 1–to–1 scale. This means that each box represents 1 unit.

When a 1–to–1 scale is used, the graph may have to be gigantic for all the numbers to fit. In order to overcome this problem, change the scale of the graph. Instead of each axis line representing just 1 unit, you could make each axis line have a value of 2 or more units.

For example, if the data had values in the hundreds, each line of axis could be set at 100.

76 Fill in the missing values on the x-axis and y-axis on the graph at the right.

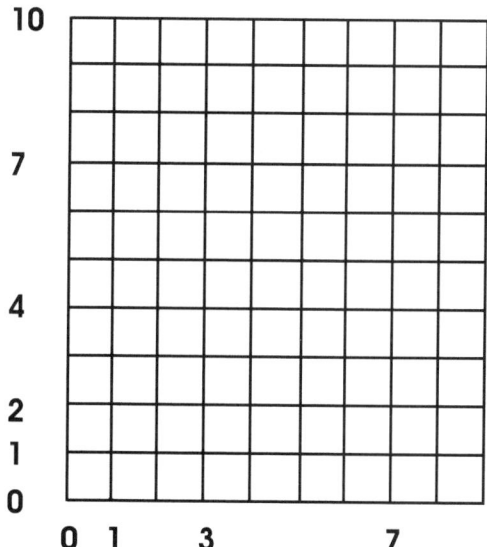

77 *a* What should you do if the data does not consist of

such low numbers? _____

_____

b What should you do if the data has values in the

hundreds or millions? _____

_____

78 For the graph at the right, fill in the missing values on the x-axis and y-axis.

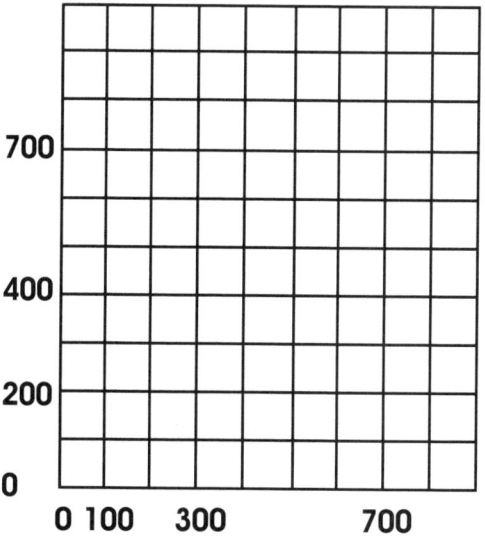

Each box could represent any number. However, it is important that each box has the same value along the axis. In this example at the right, each box has a value of 50.

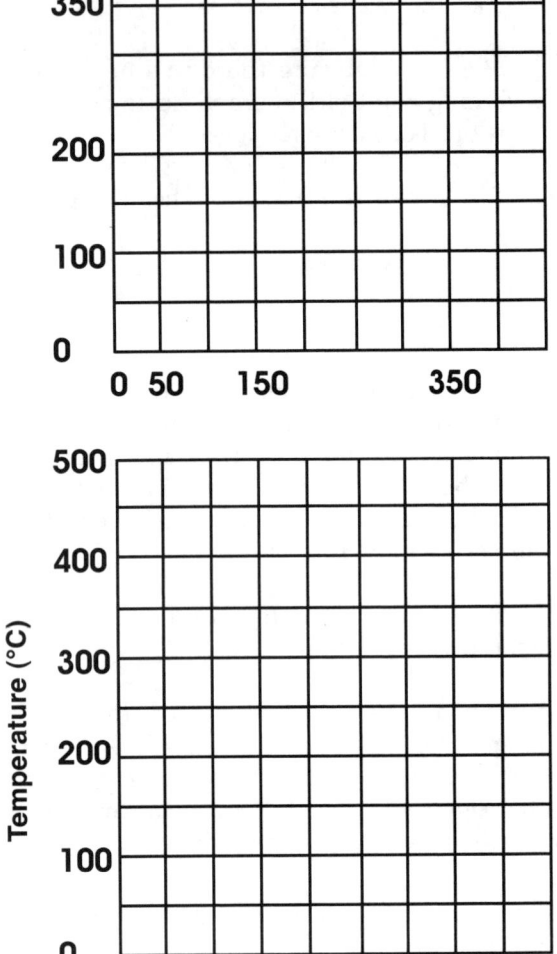

If the variables for both the x-axis and y-axis are the same, it might make it easier to interpret the graph if the same scale is used for both axes. For example, temperature can be both a dependent variable and an independent variable (see Data Table 2 on page 19).

If the axes stand for different types of information then the scales for the x-axis and y-axis are often different. Remember, the key to the scales is creating a graph that clearly displays the data.

79 On the graph at the right, create scales for the following axes:

Time must range from 0-100 minutes.

Temperature range is 0-200° C

a   the x-axis scale_____

b   the y-axis scale_____

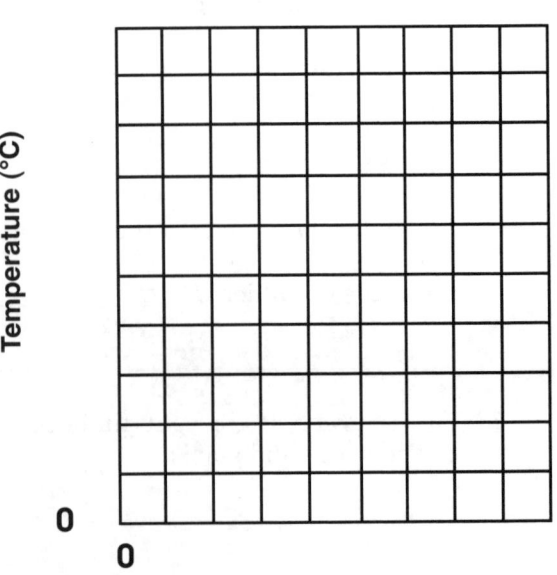

PLOTTING DATA POINTS

After the title is written and the axes are labeled and scaled, it is time to plot the graph. **Plotting** means to place the points on the graph that represent the information from the data.

An experiment was set up to determine the effect of light intensity on the rate of photosynthesis in a culture of the alga Chlorella. The results are shown in the data table at the right.

LIGHT INTENSITY (FOOT CANDLES)	BUBBLES OF OXYGEN (PER MINUTE)
200	8
500	22
750	29
1000	30
1250	30
1500	31
1750	30
2000	30

RELATIONSHIP OF LIGHT INTENSITY ON THE RATE OF PHOTOSYNTHESIS (IN A CULTURE OF ALGA *CHLORELLA*)

In order to plot a point, you need two pieces of information. One piece of information will tell you where the point is located on the x-axis. The other will tell you where the point is located on the y-axis.

For example, the first data point from this chart is a light intensity of 200 foot-candles with a rate of photosynthesis of 8 bubbles of oxygen per minute. Move your finger along the x-axis until you reach the line that represents 200 foot-candles. Then, move your finger up that line until you get to the line that represents 8 bubbles per minute. You should find a small dot at this point. (This point has already been plotted for you.)

Plot the second point, by moving your finger along the x-axis until you reach the line representing 500 foot-candles. Next, move up the that line. The number of bubbles per minute you are looking for is 22. Since there is no 22 line you must make use of our scale. Each line represents 4 bubbles per minute. So, a box also represents 4 bubbles per minute. Twenty-two bubbles per minute is represented by a spot about half way between the 20 and 24 line. You should find a point there.

80 Continue plotting the rest of the points for the graph below.

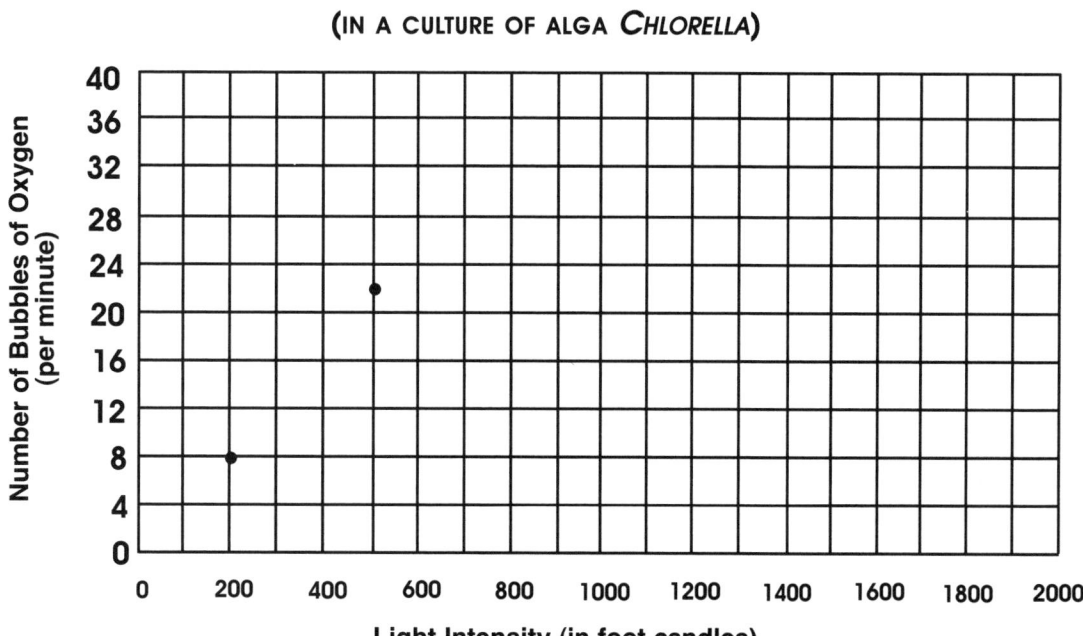

EFFECT OF LIGHT INTENSITY ON THE RATE OF PHOTOSYNTHESIS (IN A CULTURE OF ALGA *CHLORELLA*)

CONNECTING THE POINTS

Since a line graph is used to represent a *continuous* relationship, the points on a graph can be connected with a line. Remember, in doing this you are assuming what happens between the points is following the same pattern as the points. Making this assumption is called **interpolating**.

When connecting points on a line graph, remember that each point represents one piece of data. The entire collection of points represents a greater amount of data. Therefore the pattern that the points make provides more valid information than an individual point. One point may not represent the true relationship between the variables. That is why scientists often use the "line of best fit" when connecting points. Using the line of best fit means that instead of connecting the points such as in connect the dot puzzle, you draw a line that best 'fits' the pattern of the points. For example:

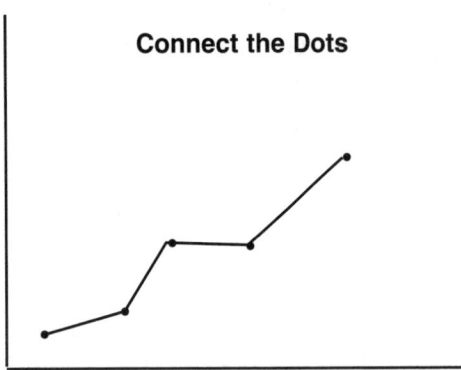

It is important to note that not all relationships will be represented as a straight line. Sometimes the relationship might appear as a curve or some other pattern. Sometimes, there is no relationship. For example, an investigation was performed to determine the effect of temperature on the action of the enzyme *catalase* (breaks down hydrogen peroxide, releasing oxygen). Six test tubes, each containing water and an equal concentration of hydrogen peroxide and *catalase*, were maintained at different temperatures. Observations made during the investigation were used to construct the following data table below.

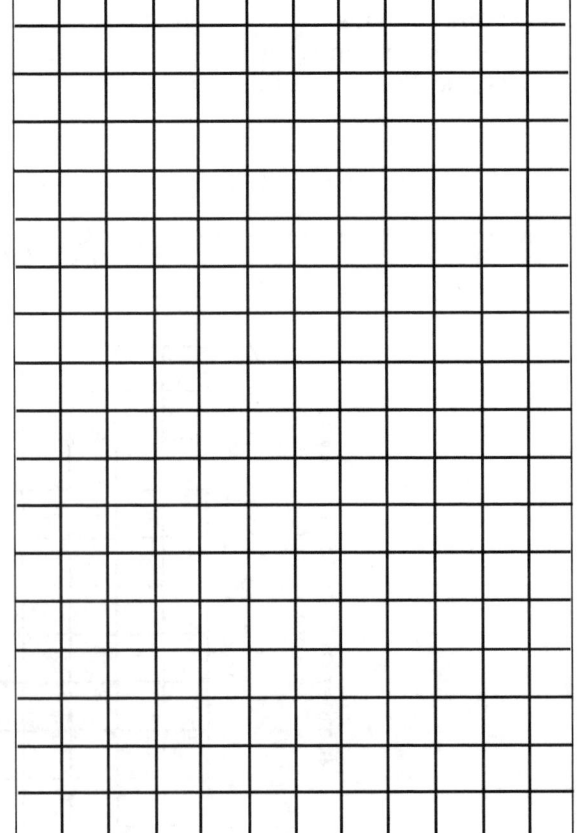

(TITLE)		
TEST TUBE	TEMPERATURE (°C)	BUBBLES OF OXYGEN (PER MINUTE)
1	0	3
2	10	22
3	20	40
4	30	58
5	40	71
6	50	2

81 Based on the data chart, what would the title for this chart and graph be? _____

_____

82 Draw and label the axes for a graph of this data. Be sure to include units.

83*a* Mark an appropriate scale on each of the axes.

83*b* Plot and connect the points.

CONSTRUCTING A BAR GRAPH

Bar graphs are used to compare an independent (manipulated) variable that has distinct and noncontinuous categories. Bar graphs are set up much as line graphs except that the x-axis is not a continuous scale. Instead, the x-axis is a set of distinct categories that represent the varying of some variable.

For example, the graph at the right represents the estimated fossil fuel reserves of the United States.

The three types of fuel represent three distinct categories of the variable (type of fuel). They do not represent a continuous flow of data and therefore a line graph would not be proper.

When constructing bar graphs, space the categories on the x-axis so that the graph is easy to interpret. Fill an entire column in the graphing grid up to the level on the y-axis that goes with that category. In this graph, the column for crude oil was shaded up to the 30-year level.

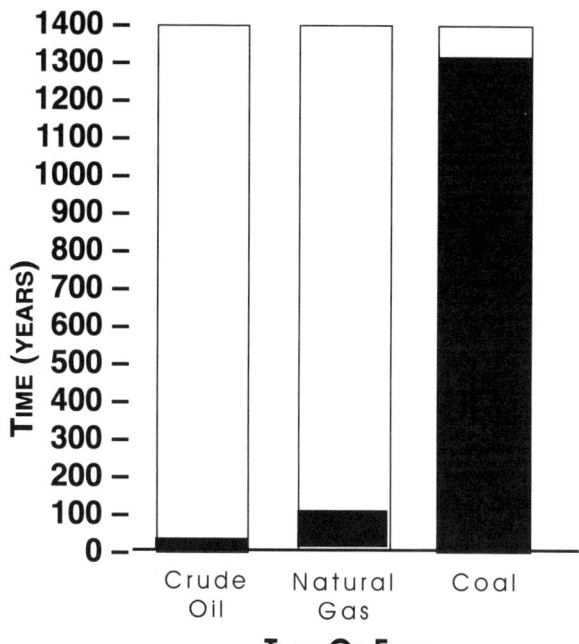

84 Complete the bar graph using the data from the following chart:

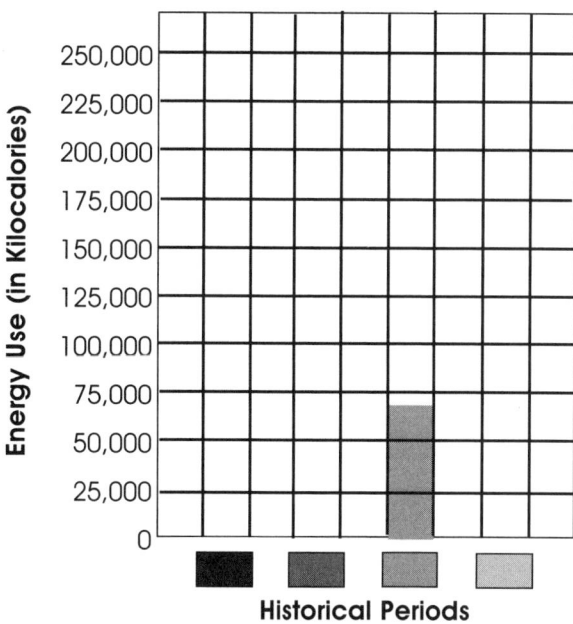

ENERGY USE HISTORICAL TIMELINE
Key to Graph
Source: *U.S. Bureau of Land Management*, 1999

HISTORICAL PERIOD	DAILY PER PERSON USE OF KILOCALORIES	SYMBOL
Primitive Society	2,000	
Early Agricultural Civilization	12,000	
19th Century	70,000	
20th Century	250,000	

Reviewing Graph Types

85 What type of graph should be used in each of the following cases?

a Effect of Time on the Growth Rate of Crystals _____

b Effect of Temperature on the Growth Rate of Crystals _____

c Growth Rate of Crystals of Different Substances _____

86 How do you know whether to use a bar or line graph? _____

_____

_____

INTERPRETATIONS

After the data has been collected and analyzed, the next step in an investigation is one of the most important in scientific inquiry. At this stage, a scientist will make some interpretations and conclusions based on the results of the investigation. In the interpretation section, the scientist discusses *what* happened. In the conclusion section, the scientist discusses *why* it happened.

INTERPRETING GRAPHS

A line graph shows the relationship between the independent (manipulated) variable and the dependent (responding) variable. The relationship will show up as a pattern. This pattern will show how changing the independent variable changes the dependent variable. If there is no relationship between the variables there should be no pattern. This means there is no regular effect of the independent variable on the dependent variable.

For example, a student hypothesized that a person's age would affect their sense of smell. Below is a graph of the results.

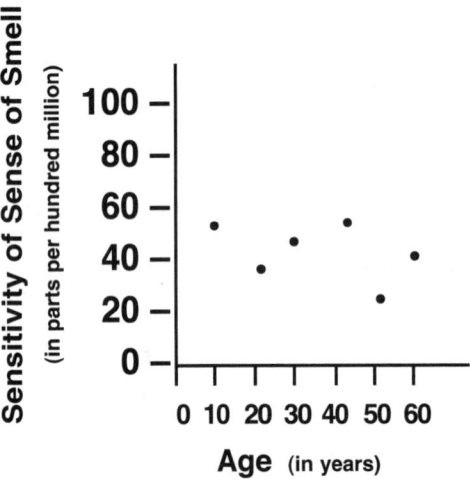

Notice there is no pattern, no relationship between age and ability to smell. You could not predict how good a person's sense of smell is by knowing their age.

Patterns allow you to make statements about how the independent variable relates to the dependent variable.

For example, below are three graphs of the relationship between time of heating and the temperature of a substance. Each shows a different relationship.

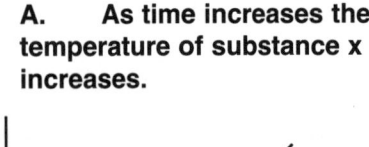

A.   As time increases the temperature of substance x increases.

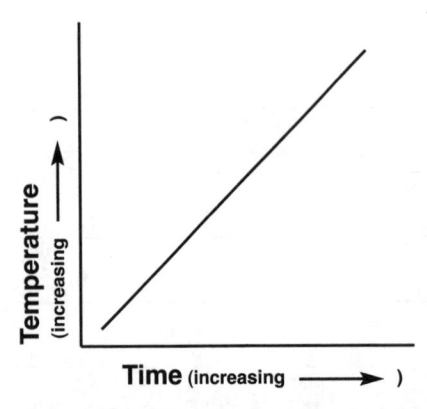

B.   As time increases the temperature of substance x decreases.

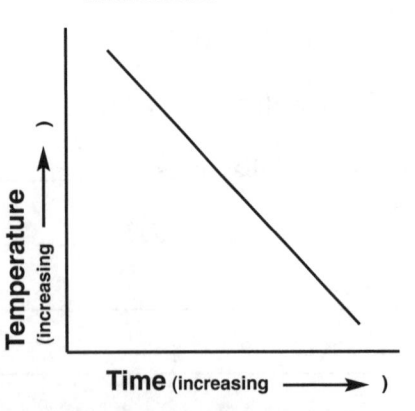

C.   As time increases the temperature of substance x remains the same.

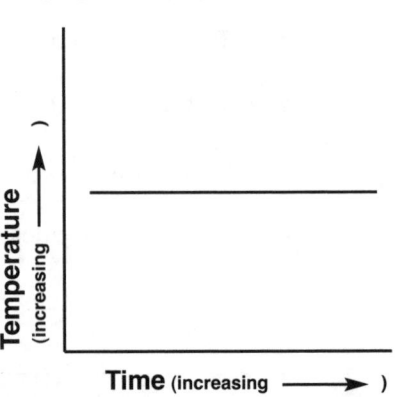

The relationships can be distinguished (shown) in the following two graphs.

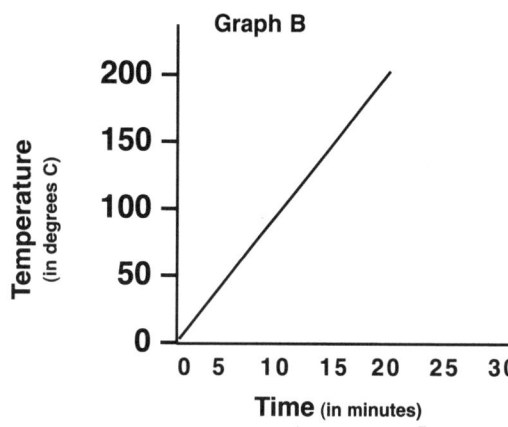

A quick look shows you that for both – as time increases the temperature increases. But the graphs are not the same. The *rate of change* is different in the graphs. Graphs make it easy to determine the rate of change. The rate of change can be determined by looking at the slope of the line on the graph. The **slope** of the line is how much the x-axis variable changes as compared to how much the y-axis variable changes.

The formula for slope is: Slope = $\dfrac{\text{change in y-axis } (\Delta y)}{\text{change in x-axis } (\Delta x)}$

where Δy is the *change* in the y-axis, (y final – y starting) and

where Δx is the *change* in the x-axis, (x final – x starting)

GRAPH A

In graph A, as the dependent variable (y-axis) changed from 0 to 100 degrees Celsius, the independent variable (x-axis) changed from 0 to 20 minutes. The slope of this line would be:

Slope = $\dfrac{100 - 0}{20 - 0}$ = $\dfrac{100 \text{ degrees C}}{20 \text{ minutes}}$ = 5 degrees Celsius per minute

GRAPH B

In graph B, as the dependent variable (y-axis) changed from 0 to 200 degrees Celsius, the independent variable (x-axis) changed from 0 to 20 minutes. The slope of this line would be:

Slope = $\dfrac{200 - 0}{20 - 0}$ = $\dfrac{200 \text{ degrees C}}{20 \text{ minutes}}$ = 10 degrees Celsius per minute

The rate of change in graph B is twice as fast as that of graph A. Slopes can be calculated for parts of a graph also. This is especially helpful when a graph represents one or more patterns or relationships.

In the following graph, several relationships must be described.

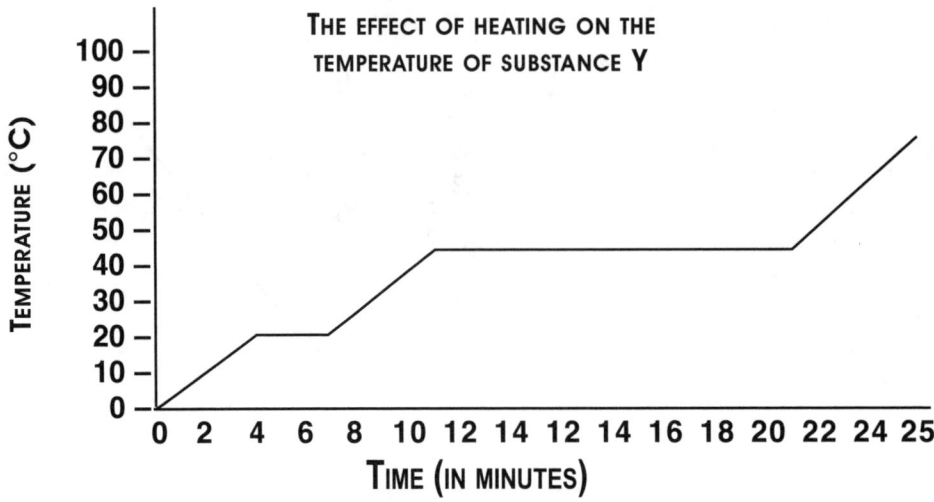

THE EFFECT OF HEATING ON THE
TEMPERATURE OF SUBSTANCE Y

From 0 to 4 minutes the temperature changed 20 degrees – a rate of 5 degrees per minute.

From 4 to 7 minutes the temperature remained the same.

From 7 to 11 minutes the temperature changed 24 degrees – a rate of 6 degrees per minute.

From 11 to 21 minutes the temperature remained the same.

From 21 to 25 minutes the temperature changed 30 degrees – a rate of 7.5 degrees per minute.

Notice that as each part of the graph is interpreted, the relationship between temperature and time is restated.

Be careful! Some students have the bad habit of saying things such as, "Between 5 and 7 minutes the line stayed flat." or "The line went up." The line has no meaning in and of itself, it is what the line represents that is important.

87 Describe the relationships shown
in the graph at the right.

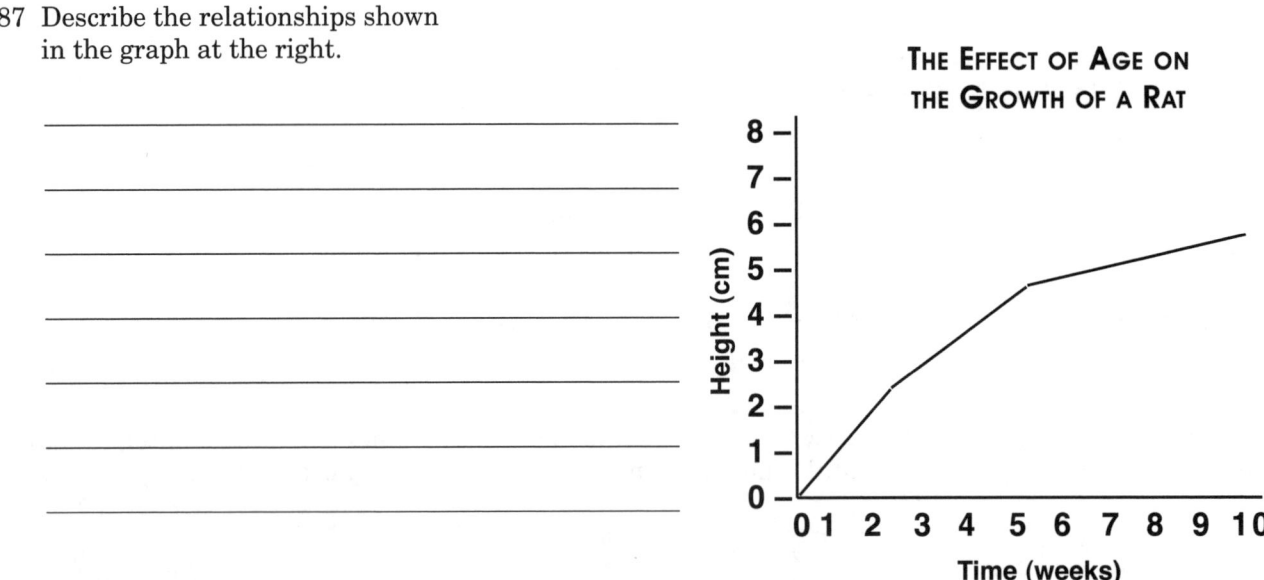

THE EFFECT OF AGE ON
THE GROWTH OF A RAT

_____

_____

_____

_____

_____

_____

_____

CONCLUSIONS

In the conclusion section of the investigation, several important topics must be addressed:

- evaluation of the original hypothesis, based on data
- explanation of results
- identification of sources of error
- new predictions and hypotheses
- recommendations for further study

Use the following investigation to practice writing conclusions.

Farmer McDonald constructed the following hypothesis. *If* grain plants receive more rain, *then* they should produce more grain, *because* water is essential to the growth of the plant. The table shows the average amount of grain in bushels per acre produced by a farm each year from 1994 to 2000. This table also shows the amount of rainfall received during the growing season of each year.

DATA TABLE

Year	Amount of Rain (cm)	Grain/acre (bushels)
1994	26	60
1995	14	50
1996	20	65
1997	18	60
1998	22	70
1999	30	20
2000	24	65

88 Based on the data, how is the independent

(manipulated) variable defined? _____

_____

_____

89 Based on the data how is the dependent (responding) variable defined? _____

_____

_____

90 *a* Construct a table using the table format below, that better displays the results.

b Construct a line graph, using the grid below, to analyze the data.

DATA TABLE

THE EFFECT OF RAIN ON THE PRODUCTION OF GRAIN

EVALUATION OF THE ORIGINAL HYPOTHESIS IN LIGHT OF THE DATA

Since an investigation is done to test a hypothesis, the first part of the conclusion should discuss whether or not the data supports the hypothesis. Restate your hypothesis so it is clear what you are discussing. You should use evidence from the data to make a statement about whether the data supports the hypothesis. When some of the data supports the hypothesis and some does not, sometimes a student wrongly tries to argue that the hypothesis is "sort of right." Remember, for a hypothesis to be true, the data must support it in *all* cases.

91 The results show that the amount of rainfall (did/did not) increase the amount of grain produced on McDonald's farm. The evidence that supports this interpretation is that when the farm received

_____, the amount of grain produced _____.

Based on the data from this experiment, the original hypothesis is not true. When the highest amounts of rain were received (24 cm, 26 cm, and 30 cm), the amount of grain produced actually *decreased*.

EXPLANATION OF RESULTS

In this section, scientists attempt to explain why the results occurred as they did. If the hypothesis was supported by the data, then the explanation comes from the hypothesis. Science is the search for explanations as to why things occur in the natural world. This is where those "whys" are answered. If any of the relationships studied seem to have a definite cause and effect (usually between the independent and dependent variables), it should be described. For example, if the hypothesis in the case of Farmer McDonald had proven to be true, the explanation discussed might have been: how water is essential for grain production. More water (cause) will allow more grain production (effect).

If the hypothesis is not supported by the data then other explanations may be offered in this section. These explanations should be supported by the data and the data should be used to defend your argument.

The results of this study might be explained by the idea that, although plants do need more water for more grain production, there is a limit to how much water a plant needs. Excess water may actually slow the grain production. As the rain increased from 14 to 22 cm, the production of grain increased from 50 to 70 bushels. The ideal amount of rain seemed to be 22 cm, because when the grain plants received more than this amount, 24-30 cm, the production of the grain plants decreased to 20 bushels.

IDENTIFICATION OF SOURCES OF ERROR

When investigations are conducted, there is always the possibility that errors will occur. In this section of the investigation, scientists reflect on possible errors that might be present. This self critique is very important. If you are trying to uncover the mysteries of the natural world, it is important that you do not get fooled by false data or conclusions. If there are any doubts about results, scientists must be honest and report them. Almost no experiment is perfect. You must take into account the possibility of errors and the limitations of the data when you make conclusions based on the data.

In this investigation, not much specific information was given about how the amount of rain or production of grain was measured. Possible errors could have occurred in measurement. What if the amount of rain was only checked weekly? Some of the rain could have evaporated before the amount was recorded.

In this part of the investigation, it is also proper to explain how the sources of errors could be reduced. Suggestions for improvements might help other scientists who wish to work in this area or in a similar area.

92 What could have been changed to reduce possible errors in this investigation?_____

_____

_____

_____

BEING A SKEPTIC

Scientists are skeptics – they have critical or doubting attitudes. Whenever looking at data or conclusions, it is important to examine them critically. Examine how the data was collected. Check if the experiment was properly controlled. Examine if the conclusions and explanations fit the data. Look for sources of error. You are building a foundation of knowledge from which your understanding of the natural world will grow. The scientist must be very sure that the foundation is based on firm, well supported research and thinking.

NEW PREDICTIONS & HYPOTHESES

Data from investigations can be thought of as new observations. New observations may lead to new explanations and predictions. Science is a never-ending cycle of observations, explanations, testing, and conclusions. Each component leads to the next as you cycle towards a better understanding of our natural world.

When making new hypotheses and predictions, be sure they are supported by data.

93 In the case of Farmer McDonald's investigation, what new hypotheses could be made? _____

_____

The data showed that plants produced more with up to 22 cm of rain. A new hypothesis could be if grain plants receive a certain amount of rain, they will produce more grain. Too much or too little rain will slow production.

94 What predictions could be made from this new hypothesis? _____

_____

When making a new hypothesis, scientists must be careful not to make too big a jump. Taking what happened in one case and presuming it will happen in different cases is called **generalizing**. If a scientist is trying to generalize conclusions then the investigation would have to be designed differently.

For example, the Farmer McDonald investigation dealt with only grain. Could he generalize that there was an optimal amount of water for each type of plant? Scientists must be careful to understand the limitations of their data. As you try to explain more and more of our natural world, you often try to find explanations that can be generalized instead of having different explanations for each similar case.

95 In order to generalize the results, how could you change the Farmer McDonald investigation? _____

_____

_____

RECOMMENDATIONS FOR FURTHER STUDY

An investigation usually ends with a discussion about ideas for further study. These ideas for research come out of the current study's findings. This shows how one discovery in science opens doors for further discoveries, again and again. Each investigation can open up new questions. Each investigation can help to pinpoint areas that need to be studied further. Some recommendations for further study may address the need to reduce errors. Other recommendations may address a new hypothesis based on the data.

96 What new questions or explanations can you think of based on the results from the Farmer McDonald

investigation? _____

_____

Interpret the data and write conclusions for the following investigation.

An investigation was conducted to study the effect of airborne sulfur dioxide (SO_2) on the growth of two tree species. The radial growth (increase in the radius of the trunk) of trees growing in an SO_2-free environment was compared with the radial growth of trees growing in environments with different concentrations of SO_2. Other environmental conditions were similar for both species. The average annual percent reduction in radial growth of the two tree species was recorded in the data table below.

The student hypothesized that *if* there was more sulfur dioxide (SO_2) in the air, *then* the growth of trees would be reduced, *because* SO_2 pollution would interfere with photosynthesis.

EFFECT OF SO_2 CONCENTRATIONS ON REDUCTION IN TREE GROWTH

Annual SO_2 Concentration (g/m^3)	Average Annual Percent Reduction in Radial Growth Scotch Pine	Average Annual Percent Reduction in Radial Growth Common Oak
0.00	0	0
0.05	8	18
0.10	52	35
0.15	97	50

97 What would you compare to analyze these results? _____

_____

98 Write your own interpretation of these results. _____

_____

99 Is the original hypothesis correct based on the data? _____ Why? _____

_____

100 Why does including two types of trees help to support the hypothesis? _____

_____

101 Explain the results. What happened? _____

_____

102 What are possible sources of error in this investigation? _____

_____

103 Make up a new hypothesis or new prediction based on the data. _____

_____

104 What recommendations can you make for further study in this area? _____

_____

SCIENCE PRACTICES REVIEW
ANALYSIS, INQUIRY, & DESIGN

Base your answers to questions 1, 2, and 3 on the diagram at the right, which shows what happens to incoming sunlight.

Construct a bar graph of what happens to incoming sunlight.

Include:

1   Title the graph and label the vertical axis. (1 point)

2   Scale the vertical axes. (1 point)

3   Plot the data. (1 point)

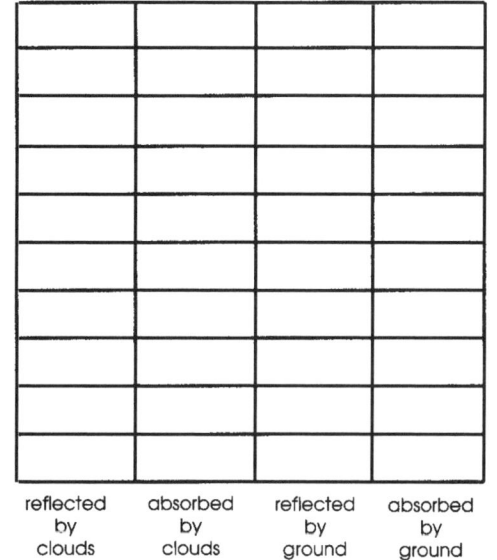

reflected by clouds   absorbed by clouds   reflected by ground   absorbed by ground

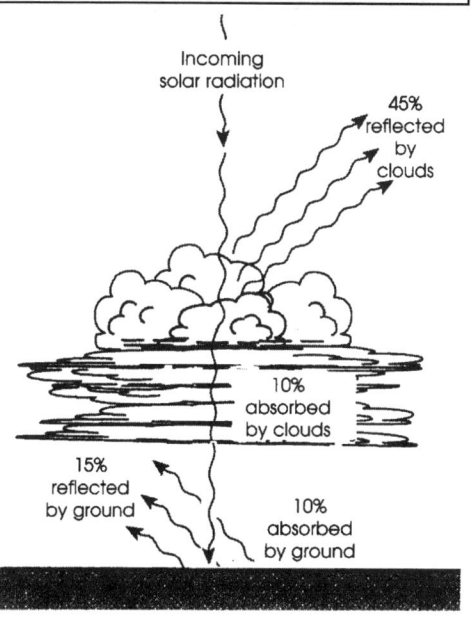

Incoming solar radiation

45% reflected by clouds

10% absorbed by clouds

15% reflected by ground

10% absorbed by ground

Base your answers to questions 4 and 5 on the diagram at the right, which shows an experiment on plant growth that is about to begin.

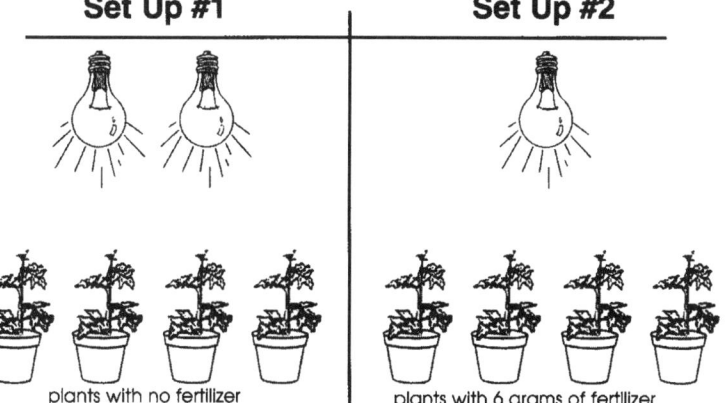

Set Up #1

Set Up #2

plants with no fertilizer

plants with 6 grams of fertilizer

4   Why would doing the experiment not be helpful in determining the effect of fertilizer on the growth of plants? (1 point) _____

_____

5   How could this error in experimental design be corrected? (1 point) _____

_____

Base your answers to questions 6 through 9 on the diagram below, which shows sets of equipment.

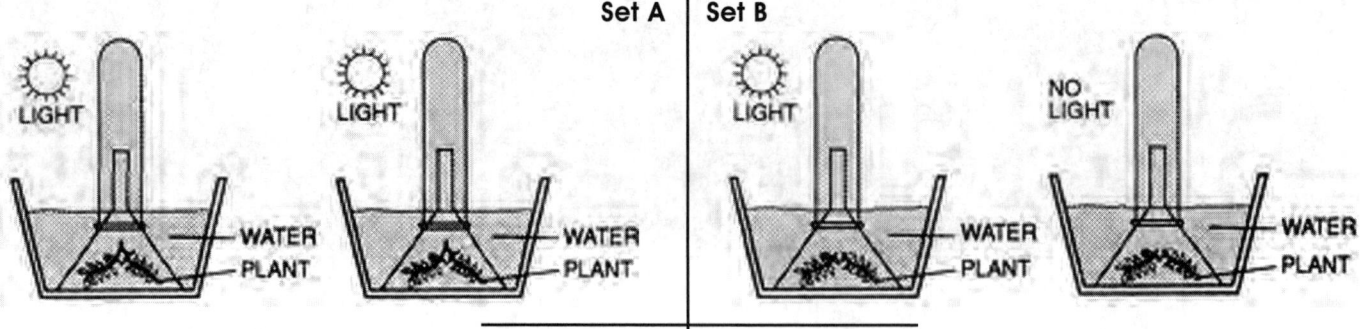

6   Which set of equipment, A or B, would you use to test the effect of light on the amount of gas produced

by plants? (1 point) _____

7   What is the independent (manipulated) variable in this investigation? (1 point) _____

_____

8   What is the dependent (responding) variable in this investigation? (1 point) _____

_____

9   List two variables that should be controlled in this investigation. (2 points) _____

_____

Base your answers to questions 10 and 11 on the diagram below, which shows the relationship between light intensity and plant growth.

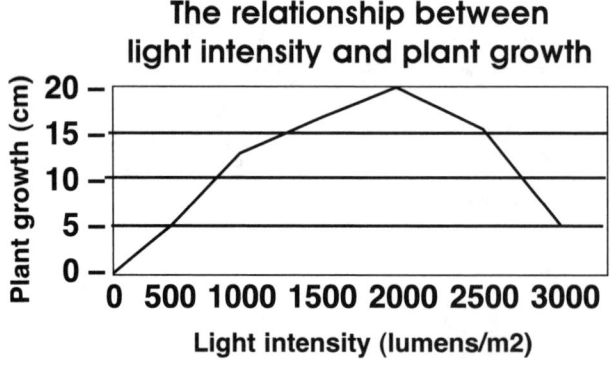

10 Describe the relationship between light intensity and plant growth shown in the graph. (2 points) _____

_____

_____

11 A scientist proposed the hypothesis that the greater the light intensity, the greater the plant growth.
    Based on the evidence shown in the graph, explain why the scientist was correct or incorrect. (1 point)

_____

_____

Base your answers to questions 12 and 13 on the diagram below, which shows the effect of seed position on the direction of root growth.

Diagram A

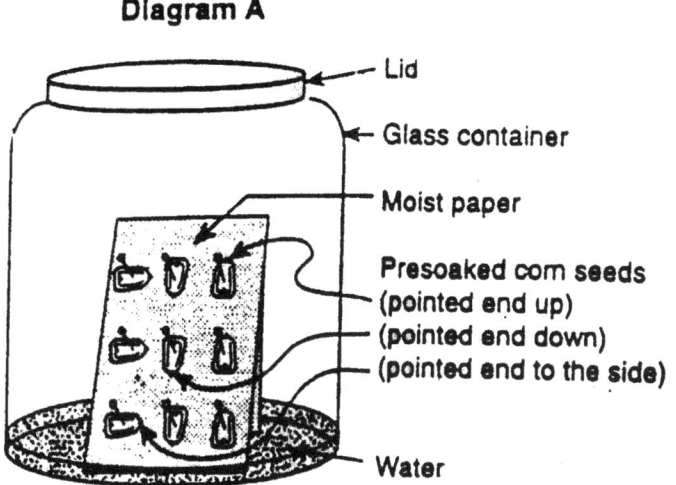

Lid

Glass container

Moist paper

Presoaked corn seeds
(pointed end up)
(pointed end down)
(pointed end to the side)

Water

Diagram B

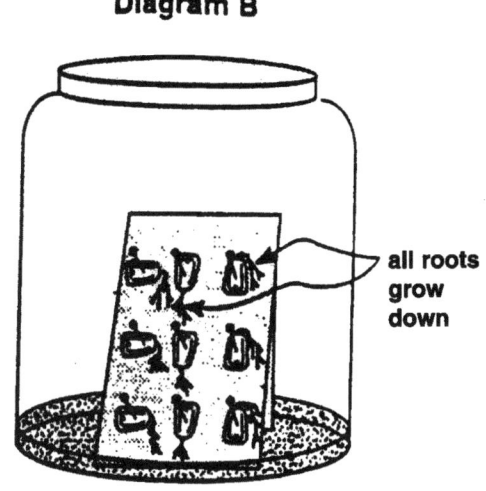

all roots
grow
down

12 A scientist proposed a hypothesis that the position of a corn seed would not affect the direction of root growth. Based on the results shown in the diagram, explain why this scientist was correct or incorrect. (1 point)

_____

_____

13 How should the experiment be changed if the scientist wanted to prove the hypothesis was true for all types of seeds? (1 point)

_____

_____

Base your answers to questions 14 and 15 on the diagram at the right, which shows a model of the human respiratory system.

Models help observers to understand how things in nature work. In the model at the right, the balloons represent lungs, because they both can fill with air. The rubber sheet represents the diaphragm, because it can be moved to affect the pressure in the system.

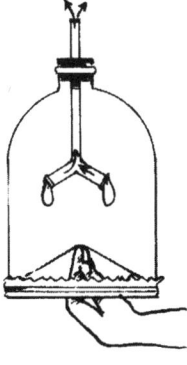
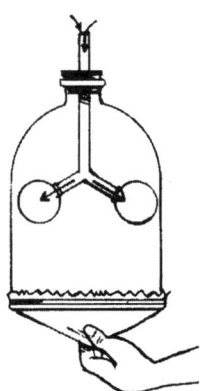

14 On the diagram at the right, label the parts that represent the lungs and the diaphram. (2 points)

15 How does this model illustrate how lungs inflate. (1 point) _____

_____

_____

Base your answers to questions 16 and 17 on the diagram at the right, which shows two insulated containers of water at different temperatures connected by a metal bar.

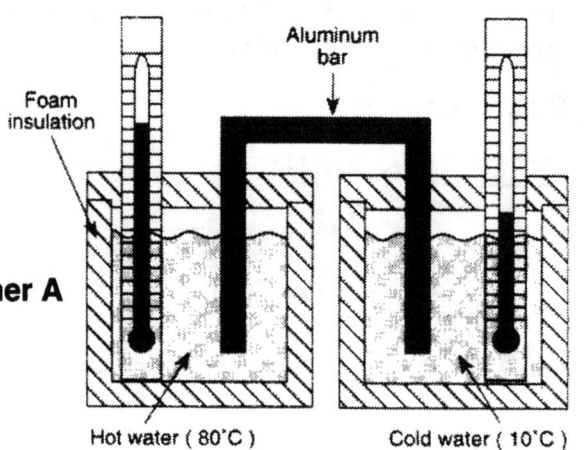

Aluminum bar

Foam insulation

Container A

Container B

Hot water (80°C)

Cold water (10°C)

16 Construct a hypothesis about the transfer of heat from Container A to Container B (1 point) _____

_____

_____

17 What variable would you manipulate to test your hypothesis? (1 point) _____

_____

Base your answers to questions 18 and 19 on the diagram below, which shows four sets of equipment that could be used in an experiment.

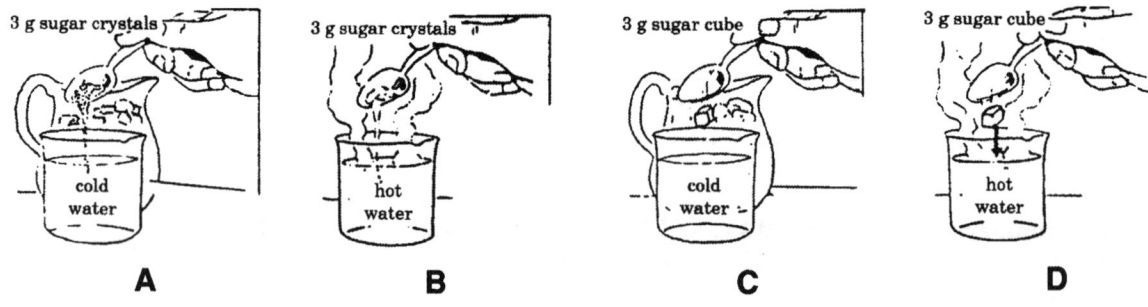

3 g sugar crystals

cold water

A

3 g sugar crystals

hot water

B

3 g sugar cube

cold water

C

3 g sugar cube

hot water

D

18 If you wanted to test the effect of water temperature on how quickly sugar dissolved, which two set ups could you use? (1 point)

_____ and _____   OR   _____ and _____

19 List two variables that should be controlled, if you wish to test the effect of water temperature on how

quickly sugar dissolves. (2 points) _____

_____

Base your answers to question 20 on the diagram at the right.

20 Propose a hypotheses for how the water droplets came to be on the outside of the pitcher. (2 points)

_____

_____

Base your answers to questions 21 through 25 on the table below left and on the diagram below right, which shows the results of attaching different masses to a spring.

DATA TABLE	
MASS	DISTANCE
100 g	3 cm
200 g	6 cm
300 g	9 cm
400 g	12 cm
500 g	15 cm
1000 g	30 cm

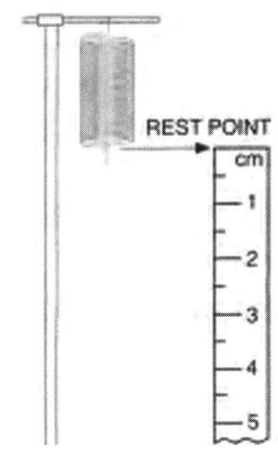
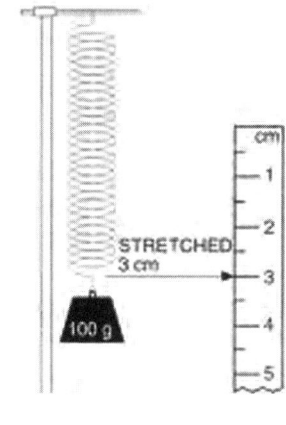

For questions 21 through 25, construct a line graph using the data table provided above.

Include the following:

21 Title (1 point)

22 labeled axes (1 point)

23 units (1 point)

24 scaled axes (1 point)

25 plot of points (1 point)

Base your answers to question 26 on the graph at the right, which shows the effect of time and heating on the temperature of a substance.

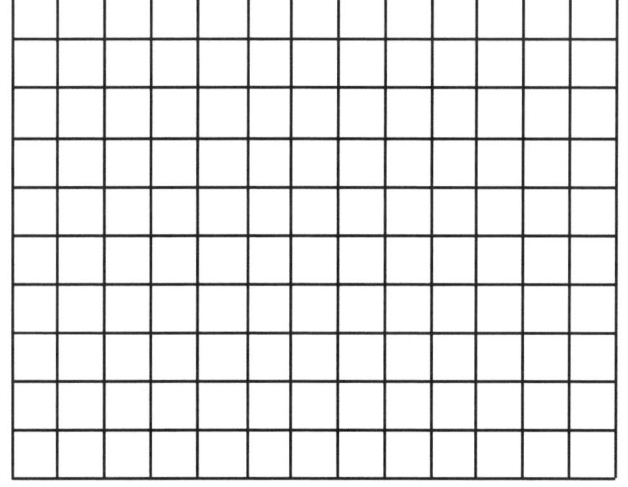

26 Explain the relationship between the time of heating and the temperature for this substance during the first 30 minutes (**A**) and the second 30 minutes (**B**) _____

_____

Base your answers to questions 27 on the diagram below, which shows the evolution of the modern horse.

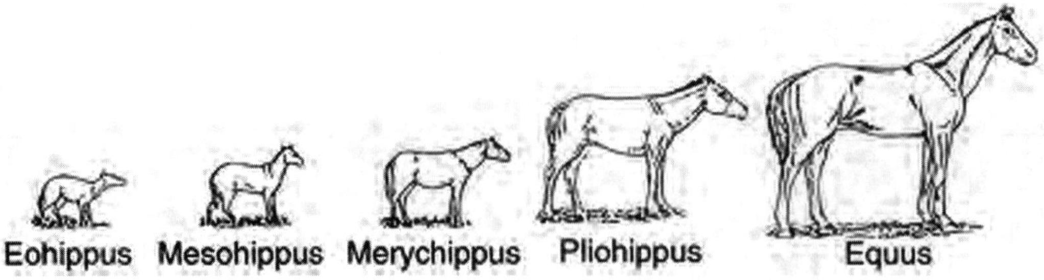

Eohippus   Mesohippus   Merychippus   Pliohippus   Equus

27 Twenty million years from now, what would you expect a horse to look like? Explain your answer. (2 points)

_____

_____

Base your answers to questions 28 through 30 on the diagram below – illustrating the process of erosion, which shows an investigation where students poured water on a stream table with three different materials.

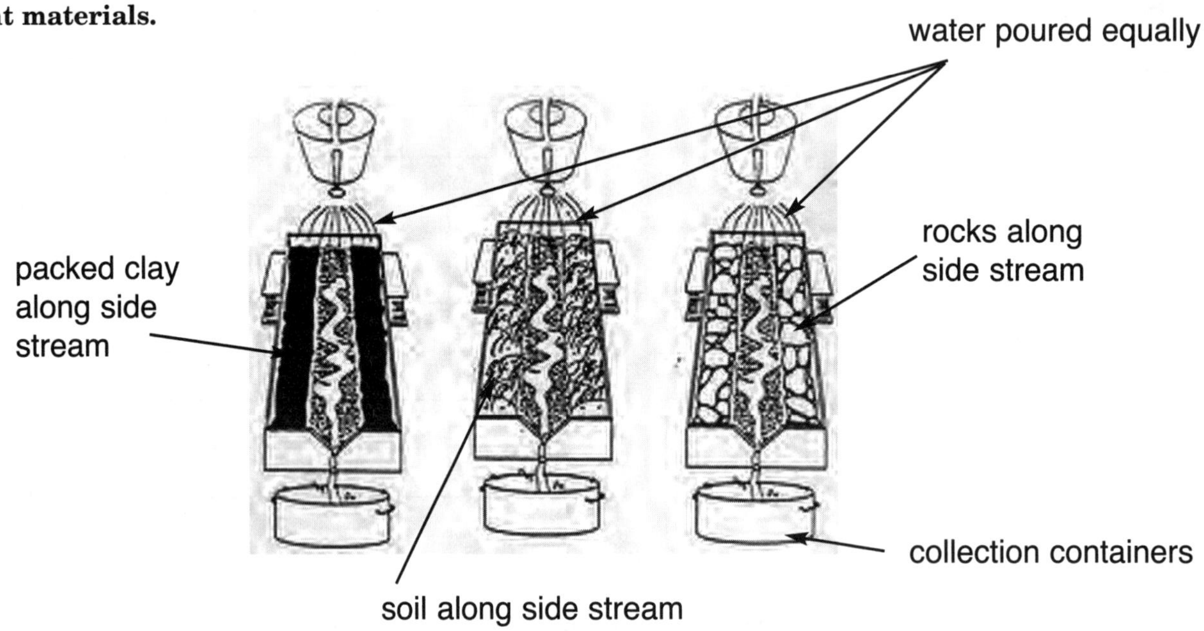

water poured equally

packed clay along side stream

rocks along side stream

collection containers

soil along side stream

28 What was the independent (manipulated) variable in this experiment? (1 point) _____

_____

_____

29 What was the dependent (responding) variable in this experiment? (1 point) _____

_____

_____

30 List two variables that were controlled in this experiment. (2 points) _____

_____

_____

LIFE SCIENCE

FROM MOLECULES TO ORGANISMS: STRUCTURES AND PROCESSES

DISCIPLINARY CORE IDEA:

LS1.A STRUCTURE AND FUNCTION:
LIVING THINGS ARE MADE OF CELLS, WHICH IS THE SMALLEST UNIT THAT IS SAID TO BE ALIVE.

DIFFERENT BUT ALIKE

Looking at all the forms of life on planet Earth, from fish to birds, from worms to lobsters, from ferns to redwoods, you might think that living things are quite different from one another. Actually, living things on planet Earth have much in common. What could a tree, worm, crayfish, squid, fish, frog, snake, bird, and bear all have in common? The answer becomes quite clear when you look at the basic structure from which they are all built.

1 List two ways all living things are alike.

_____

_____

2 List two ways living things are different from nonliving things.

_____

_____

OBSERVING - NOT JUST LOOKING!

When **microscopes** were first invented in the late 16th and early 17th centuries, one of the first things people looked at under microscopes was parts of living things. Little by little, the observations of many people came together. They built up into an explanation of what made up living things. In 1665, a British scientist, **Robert Hooke**, described his observations using a simple lens (magnifier) of thin slices of cork. He observed the empty cell walls of cork and it reminded him of little rooms or "cells." (As in jail cells). This is how cells got their name. In 1835, a German botanist, **Matthias Schleiden**, reported on his work with many different plants. He suggested that all plants were composed of **cells**. In 1836, a German zoologist, **Theodor Schwann**, reported on his

work with many different animals. He suggested that all animals consisted of cells. German scientist **Rudolf Virchow** (1855) recognized that all cells arise from preexisting cells by a process of division. As you can see, the work of many scientists may lead to the development of a theory.

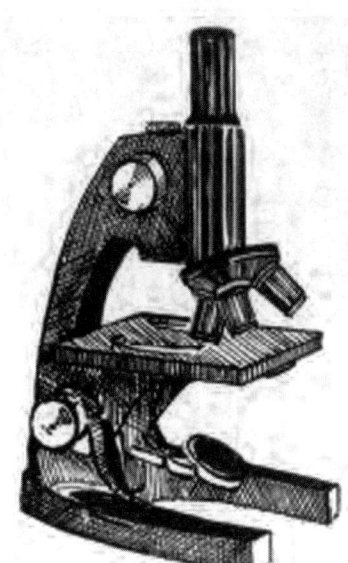

EXAMPLES OF HUMAN CELLS

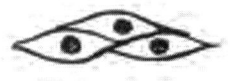 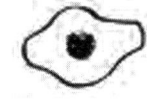 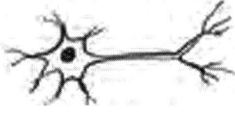 

SMOOTH MUSCLE   **SKIN**   **NERVE**   **HEART MUSCLE**

THE CELL THEORY

- **Living things are composed of cells.**
- **Cells are the basic unit of structure and function of living things.**
- **All cells come from other cells.**

This means single cell and multicellular organisms are all considered living things. While they may differ in how complex they are, all have the cell as the basic component where activities of life occur.

In science, it is often difficult to prove something is a **law** (true in every case). In order for an idea to reach the **theory** level (proven true in many cases), it takes many scientists and much research. The **cell theory** has at least one exception. **Viruses** may be considered alive, but they lack the typical cell structures.

3   How did the invention of the microscope make cell study possible? _____

_____

_____

4   Why are cells called "the basic unit of life?" _____

_____

_____

5   Why is the cell theory a *theory* and not a scientific law? _____

_____

_____

6   If life was found on another planet, what would it need if it were similar to life on Earth? _____

_____

Even Our Parts Have Parts!

Cells are composed of parts (known as **organelles**), each with a special function that help the cell survive.

STRUCTURE	FUNCTION
cell membrane	surrounds the cell and controls what enters and leaves
cytoplasm	area of the cell between the nucleus and membrane; watery material that contains organelles and makes up most of the cell
nucleus	information center for the cell; found in all cells; contains the chromosomes
chromosomes	located in the nucleus; contain most of the genes of a cell which are composed of DNA, the chemical that directs heredity
mitochondria	site of cellular respiration, where food is used to produce energy
ribosome	assists in building proteins
cell wall	protects and supports some types of cells such as plant cell
vacuole	storage for both nutrients and wastes
lysosome	contains enzymes that break down protein
chloroplasts	contain chlorophyll, a substance that can change light energy into chemical energy (food) for the cell

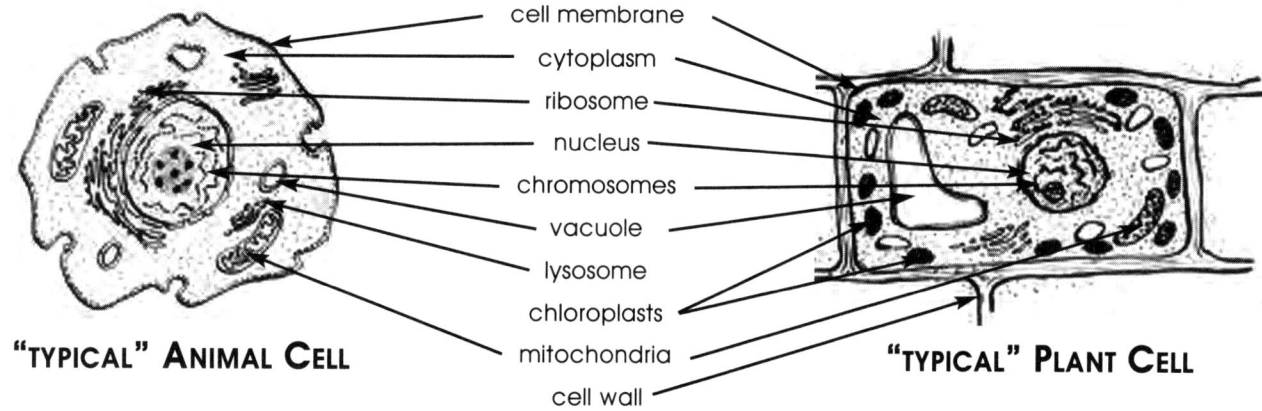

"TYPICAL" ANIMAL CELL

cell membrane
cytoplasm
ribosome
nucleus
chromosomes
vacuole
lysosome
chloroplasts
mitochondria
cell wall

"TYPICAL" PLANT CELL

In order for cells to survive, they must take in nutrients (through the cell membrane) which are converted to energy (in the mitochondria) or used to make materials that the cell or organism needs. Cells grow and split into more cells. This activity is determined and controlled by the chromosomes located in the nucleus.

THE CELL DOCTOR

You are the cell doctor.

7   List which cell structure is *not* working properly, based on each of the following observations.

Observation   **Cell Structure**

a   No food can enter the cell.   _____

b   The cell has divided incorrectly.   _____

c   The cell cannot function, it has no energy.   _____

d   The plant cell is wilting, falling over.   _____

e   The plant cell is not producing food.   _____

f   New cells are being produced but they are
      cancerous (unorganized cells not performing
      their functions).   _____

CELLS UNITE!

Organisms that consist of just one cell are called **unicellular**. This means that just one cell must carry out all of the life activities by itself.

Organisms that are **multicellular** consist of many cells. In these organisms the cells need to **specialize** (do fewer jobs and often do them better).

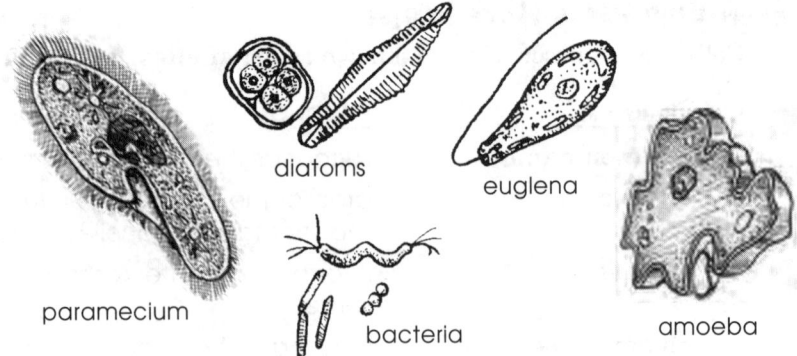

diatoms

euglena

paramecium

bacteria

amoeba

THE FIRST TOWN

What must life have been like for people before they lived together in towns? Each family living by itself would have to gather its own food, make its own clothes, make its own furniture, and care for its own sick.

In a town, you would have people doing different jobs. For example, you might have a shoemaker. A shoemaker would have the special tools to make shoes. Since the shoemaker specializes in making shoes and works on shoes most of the time, he would make shoes better.

However, this shoemaker cannot eat shoes! He must trade with other specialized workers, such as doctors, butchers, or farmers, to get what he needs to survive.

A multicellular organism is like a town. It has many individual specialized cells that work together, **interdependently**, so that the organism can function at a higher level of complexity than one cell working alone.

8   *a*   What advantages do unicellular organisms have over multicellular ones? _____

_____

_____

    b   What advantages do multicellular organisms have over single celled ones? _____

_____

_____

9   Why are the cells not all the same in multicellular organisms? _____

_____

LEVELS OF ORGANIZATION

In general, in a multicellular organism, cells are organized as cells, tissues, organs, organ systems, and organisms.

CELLS

In multicellular organisms, different cells perform different tasks. They specialize.

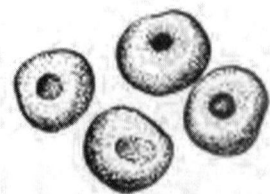

Red blood cells transport gases to and from other body cells.

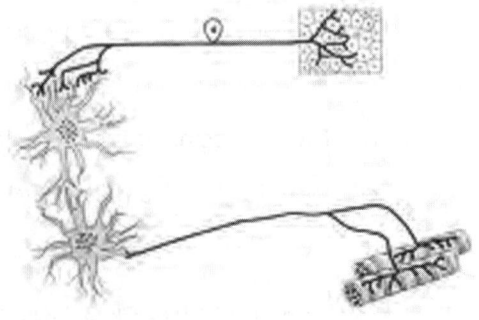

Nerve cells bring messages from one part of the body to another.

TISSUES

A **tissue** is a group of cells that performs the same function.

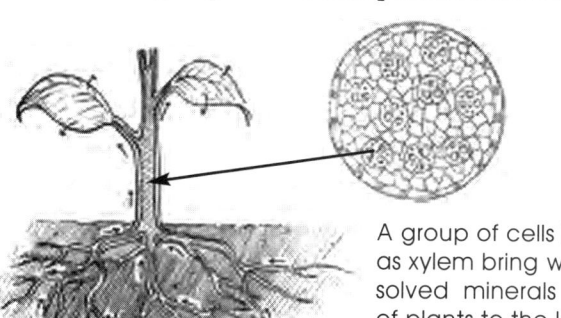

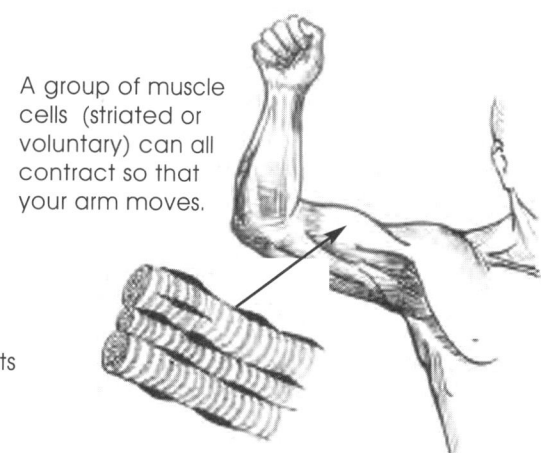

A group of muscle cells (striated or voluntary) can all contract so that your arm moves.

A group of cells in plants known as xylem bring water and dissolved minerals up from the roots of plants to the leaves.

Tissues are used to cover the body, connect other tissues and organs together, move parts of the body, carry messages to and from parts of the body, store nutrients, and do many other jobs.

10 What makes tissues different from cells? _____

_____

ORGANS

An **organ** is a group of tissues that work together to perform certain functions. In the 1800s, **Marie Francois Xavier Bichat** of France studied organs and helped give a new start to the study of diseases. He found that several tissues were common to many organs.

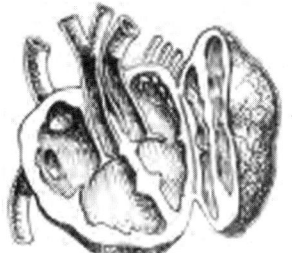

The heart is an organ composed of muscle, nervous, and blood tissues. They all work together to pump blood to your body.

A leaf is an organ composed of epidermal (protective outer cells), chlorophyll bearing cells, and phloem (transport cells) that work together to make food and distribute it to the plant.

11 How are organs different from just a group of tissues? _____

_____

ORGAN SYSTEMS

An **organ system** is a group of organs that work together to perform specific functions. For example, the circulatory system transports materials throughout an organism.

The circulatory system consists of the heart, arteries, veins, and capillaries. They work together to transport materials throughout the body.

12 *a* How does an organ system differ from just a group of organs? _____

_____

    b How is an organism different from just a group of organ systems? _____

_____

_____

PLANTS ARE ALSO ORGANISMS

Many plants consist of tissues and organs. These groups of tissues include **roots**, **stems**, and **leaves**. They are responsible for the plants life activities, called **metabolism**.

	FUNCTION	STRUCTURE
ROOTS	Anchors the plant. Stores nutrients. Absorbs water and minerals from soil. Transports water and minerals to the stems which transport to the leaves.	Root cells absorb water and minerals and pass them into tissue called xylem. **Xylem** carries water and minerals up through the plant. Another tissue, **phloem**, carries food products down the stem to the roots from the plants leaves.
STEMS	Supports the plant and holds leaves up. Stores food. Transports nutrients, minerals, and water between the leaves and the roots.	Stem cells include xylem and phloem which transport materials through the plant. The **epidermis** ("skin" of the plant) supports and protects the plant.
LEAVES	Site of **photosynthesis**, where light energy is converted into chemical energy during which oxygen and sugar are produced.	Many leaf cells contain chloroplasts where photosynthesis occurs. **Chloroplasts** contain chlorophyll the substance that can convert light energy into chemical energy in the form of sugar. Openings in the leaves, called **stomates**, allow gases to pass in and out of the leaves.

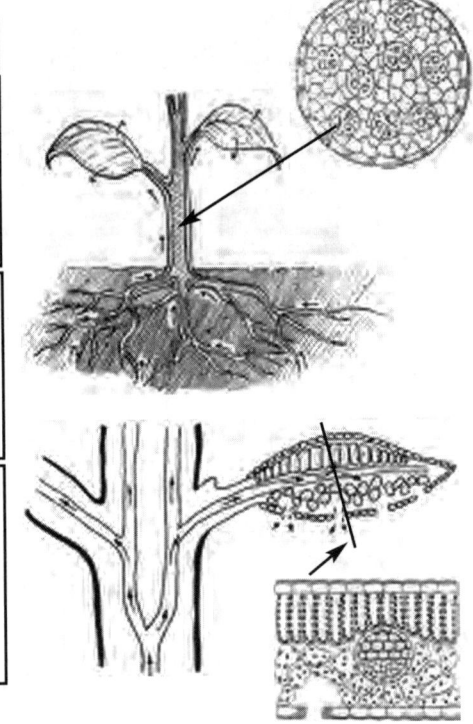

PLANT DOCTOR

You are the plant doctor.

13 List which plant structures are *not* working properly based on the following observations.

OBSERVATION	PLANT STRUCTURE
The plant is dehydrated even though the soil is wet.	a _____
The leaves are dehydrated even though the water is getting into the plant.	b _____
The root cells are starving even though the leaves are making food.	c _____
The plant is not making food or releasing oxygen	d _____

14 Compare the structure and function of the root systems of the two plants at the right.

How do the root structures differ in their function to anchor each plant? _____

_____

_____

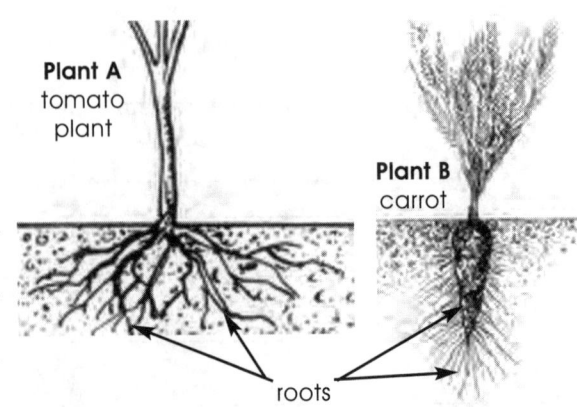

Plant A
tomato plant

Plant B
carrot

roots

15 Three human organ systems are pictured at the right. Explain how these organ systems are related.

a   A and B: _____

_____

b   B and C: _____

_____

A
Respiratory
System - Human

B
Circulatory
System - Human

C
Digestive
System - Human

ALIKE BUT DIFFERENT.
DIFFERENT BUT ALIKE.

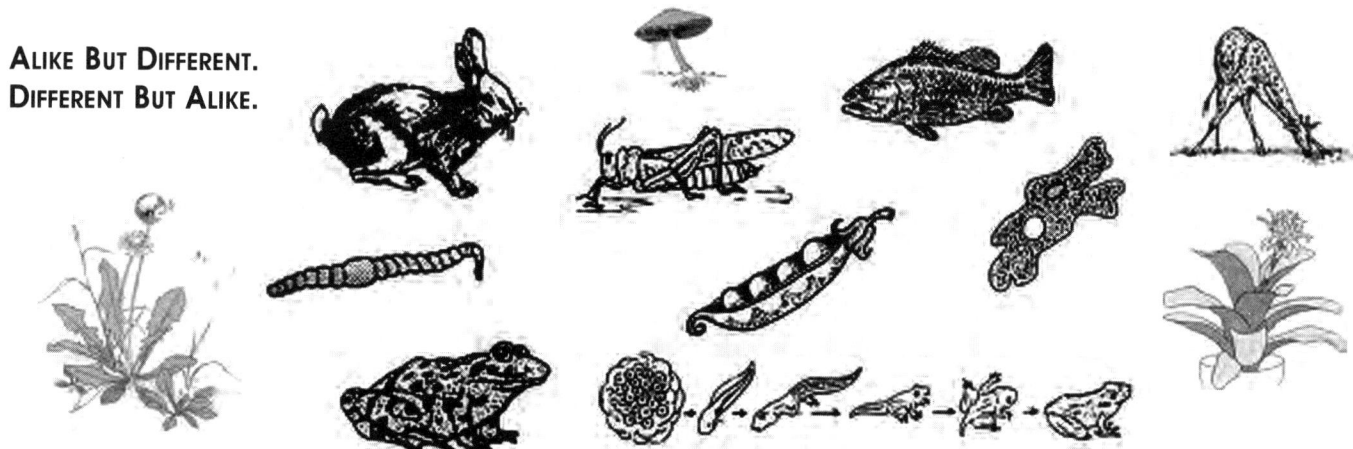

The number of different types of living things on Earth is amazing. Humans do not need to look for aliens from other planets to see some very unusual life forms. As animals and plants adapted to the different environments on Earth they developed a great variety of body plans and internal structures. Each organism adapted and became what it is to best survive in its **habitat**, its particular place in the ecosystem (interaction of living and nonliving world).

However, organisms are not all as different as they might seem. For example, multicellular animals often have similar organs and systems for carrying out their major life activities.

16 Why might organisms that live in similar environments develop similar adaptations? _____

_____

17 Why might organisms that live in different environments develop different adaptations? _____

_____

18 List FIVE structures or behaviors that fish and humans have in common. _____

_____

19 List FIVE structures or behaviors that fish and humans have that are different from one another. _____

_____

SORTING IT ALL OUT

All living things are related to each other. To better understand this, in the 1750s, Swedish biologist **Carolus Linnaeus** developed a **classification** system of all known plants and animals. He also developed a system of assigning a single scientific name to each plant and animal. This system, called binomial nomenclature, assigns a two-word Latin name to each organism. The classification system is based on whether the organisms share the same characteristics. These characteristics could be from the cellular to organism level. Internal or external characteristics of living things are used to organize them into groups.

Although science has grown more quickly in recent years, some of its ideas are from long ago. **Aristotle** (335 BC) came up with a way to sort things based on resemblances and differences that is still used today.

CLASSIFICATION GROUPS

Group	Example	Characteristics
Kingdom*	Animalia	Multicellular animals having tissues and in many cases organs
Phylum	Chordata	back supporting structure
Class	Mammalia	warm blooded, nurses young with milk
Order	Carnivora	large canine teeth
Family	Felidae	round head, claw can retract
Genus	Felis	body slender
Species**	domestica	cross between European and African wildcats (Kitty cats!)

*Largest group, most general   **Smallest group, organisms most closely related.

The scientific name of an organism is its genus and species. The scientific name of a cat would be **Felis domesticus**. The genus is always capitalized and the whole name is written in italics. (The genus is a noun and the species an adjective.)

20 If two organisms have very few characteristics in common, they are more likely to be in the same

species or kingdom. Why? _____

_____

A new organism was found in the Amazon rain forest. Scientists want to know if it is most closely related to dogs or to cats.

21 How would they go about determining the classification of the new organism? _____

_____

_____

22 What are your classification groups: a Kingdom? _____

    b Phylum? _____   c Class? _____

DISCIPLINARY CORE IDEA:

LS1.A STRUCTURE AND FUNCTION IN MULTICELLULAR ORGANISMS, THE BODY IS A SYSTEM OF MULTIPLE INTERACTING SUBSYSTEMS. THESE SUBSYSTEMS ARE GROUPS OF CELLS THAT WORK TOGETHER TO FORM TISSUES AND ORGANS THAT ARE SPECIALIZED FOR PARTICULAR BODY FUNCTIONS.

WHAT'S HAPPENING?

One of the most interesting parts of science is learning how your own body works. The body is an amazing and complex machine that allows you to survive in a complex world. The human organism has systems for digestion, respiration, reproduction, circulation, excretion, control and coordination, movement, and for protection from disease. Each system has its own major functions, but systems interact to form a whole organism. The tissues, organs, and organ systems work to keep your body supplied with its basic needs. Cell needs include nutrients, oxygen, and the removal of wastes.

© PhotoDisc

DIGESTIVE SYSTEM

System	Functions	Structures
Digestive	Mechanical (chewing and churning) and chemical (enzymes) digestion breakdown of food. This is done so that food can be absorbed into the body and transported to the cells.	teeth, esophagus stomach, small intestine, large intestine, digestive glands, pancreas, liver, gall bladder

Suppose you decide that you are going jogging later in the day, so you eat a chocolate chip cookie. Digestion of the cookie starts right away. In your **mouth**, the **teeth** cut and grind the cookie (mechanical digestion). A liquid enters your mouth. It is **saliva** which your body produces in glands. Saliva has chemicals that break down some of the starches in the cookie into sugars (chemical digestion). When you swallow, the cookie pieces are pushed down by your **esophagus** into your **stomach**. In your stomach, muscle contractions help break down the cookie (mechanical). Partial chemical digestion of the proteins in the cookie also occurs in the stomach. As the cookie enters the **small intestine**, so do many enzymes released by glands. Chemical digestion of the cookie continues. The cookie parts are now molecular in size. These may leave the small intestine and enter the blood. Food that is not digested leaves the body through the **large intestine.**

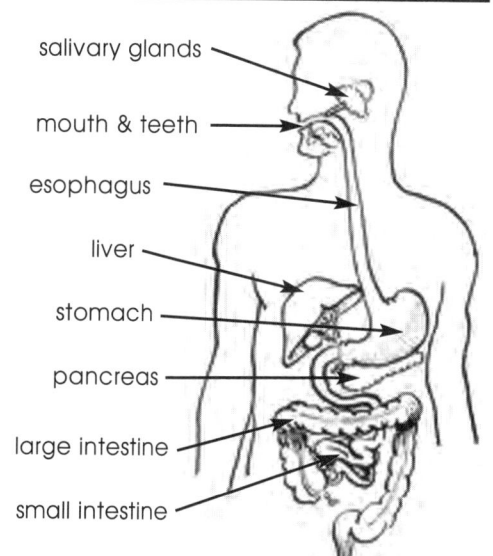

salivary glands

mouth & teeth

esophagus

liver

stomach

pancreas

large intestine

small intestine

23 What would happen if each of the following did not work properly?

   _a_   teeth: _____

   _____

   _b_   stomach: _____

   _____

   _c_   small intestine: _____

   _____

   _d_   large intestine: _____

   _____

24 What is the main purpose of digestion? (What do the cells get?) _____

   _____

   _____

25 Label the numbered structures
   and describe the digestive function of each.

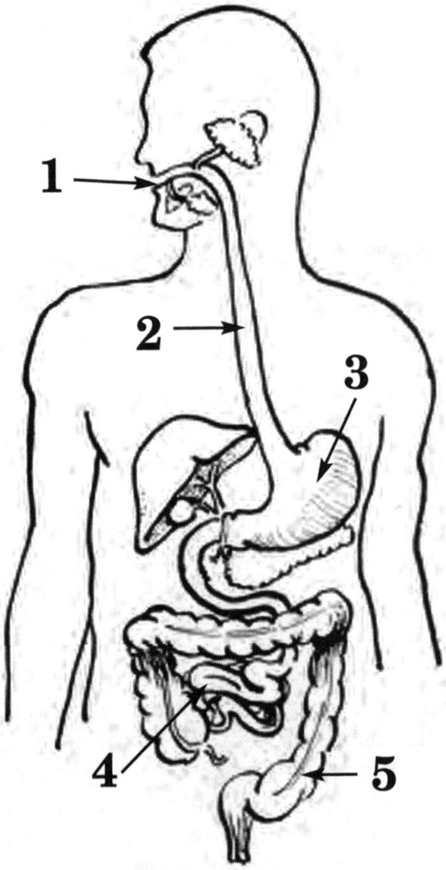

   1 _____

   _____

   _____

   2 _____

   _____

   _____

   3 _____

   _____

   _____

   4 _____

   _____

   _____

   5 _____

   _____

   _____

System	Functions	Structures
Respiratory	Carries out gas exchange. It supplies oxygen and removes carbon dioxide. This is done so that cells can use the oxygen to release energy stored in food.	nose, trachea, bronchi, alveoli capillaries, lungs, diaphragm

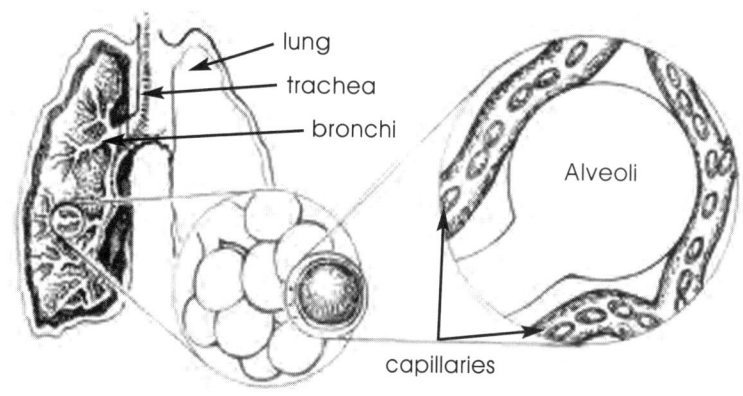

As you run, you need to get more oxygen into your body cells so you can use the energy from the food you have eaten. Your **ribs** contract, pulling upward and outward. A large muscle at the bottom of your chest, the **diaphragm**, contracts and pulls down. These actions make the chamber around your **lungs** larger. Since the area is larger, but the amount of gas in the area has remained the same, low pressure is created (Less gas per area). Air outside your body therefore becomes the higher pressure and air from the atmosphere rushes in to equalize the pressure. Air travels through your **nose**, **trachea**, **bronchi**, and into your lungs. Oxygen, from the air, leaves the lungs and enters the blood in small sacs called **alveoli** which contain **capillaries**. During the same time, carbon dioxide leaves the blood and enters the lungs. As the ribs and diaphragm relax, the chest chamber becomes smaller causing air with the waste carbon dioxide to be forced out of the lungs.

26 What is the purpose of respiration? (What do the cells get?) _____

_____

27 If you smoke, you will destroy the functioning of the air sacs in your lungs. When you breathe in, you will still draw in the same amount of oxygen, but your body will not be able to use it.

Why not? _____

_____

28 *a* When a person dies, is their last breath an inhalation or an exhalation? _____

b Why? _____

29 Label the numbered structures and describe the respiratory function of each.

1 _____

2 _____

3 _____

4 _____

The bell jar and rubber sheet model is often used to describe how the body brings gases in and out.

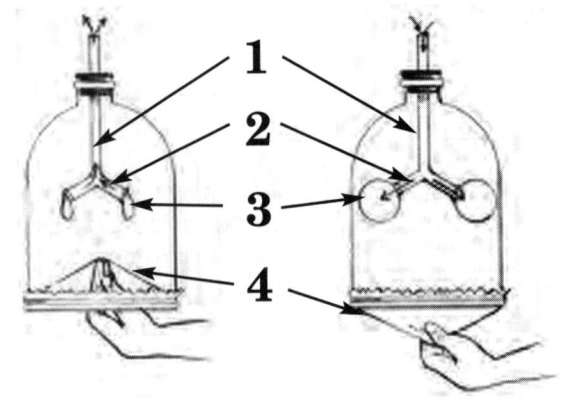

30 Describe what living structure each part of the model represents and how they function in breathing in and out.

1 _____

2 _____

3 _____

4 _____

CIRCULATORY SYSTEM

System	Functions	Structures
Circulatory	Moves substances to and from the cells. It responds to the changing demands of the cells bringing needed materials or removing cell products and wastes.	heart, lymph vessels, lymph, arteries, veins, capillaries, blood

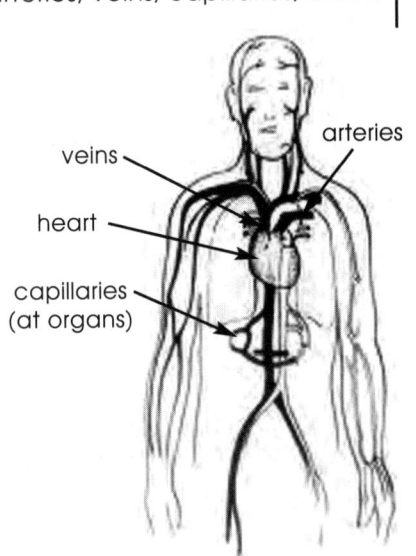

As you begin to run, your leg muscles use oxygen and food to produce energy. Soon they are low on oxygen and food. Wastes like carbon dioxide also begin to build up in the cells. The wastes pass out of the cells into the **capillaries**. Red blood cells in the capillaries are very important in transporting gases in the **blood**. The thin walled capillaries carry the wastes to larger vessels called veins. **Veins** carry the blood and the wastes back to the heart from the body's cells. The **heart** pumps the blood through thick muscular tubes called **arteries** to the lungs. In the lungs, the blood travels into capillaries where gas is again exchanged. Capillaries in the lung release carbon dioxide and pick up oxygen. The blood, now with new oxygen, travels through a vein back to the heart. The heart pumps the blood out through arteries to the body. The blood travels into capillaries and gas is exchanged as oxygen enters and carbon dioxide leaves the cells.

31 What is the main purpose of circulation? (What do the cells get?) _____

_____

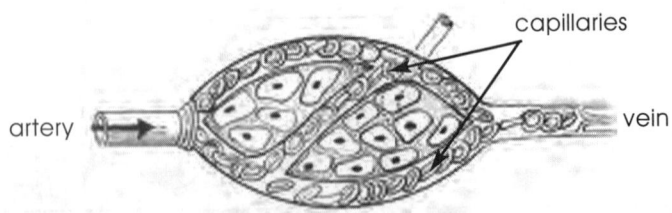

32 Which blood vessels bring blood away from the heart?_____

33 Which blood vessels bring blood toward the heart? _____

34 Which blood vessels are so thin materials can pass in and out of them? _____

HEART (OPEN)

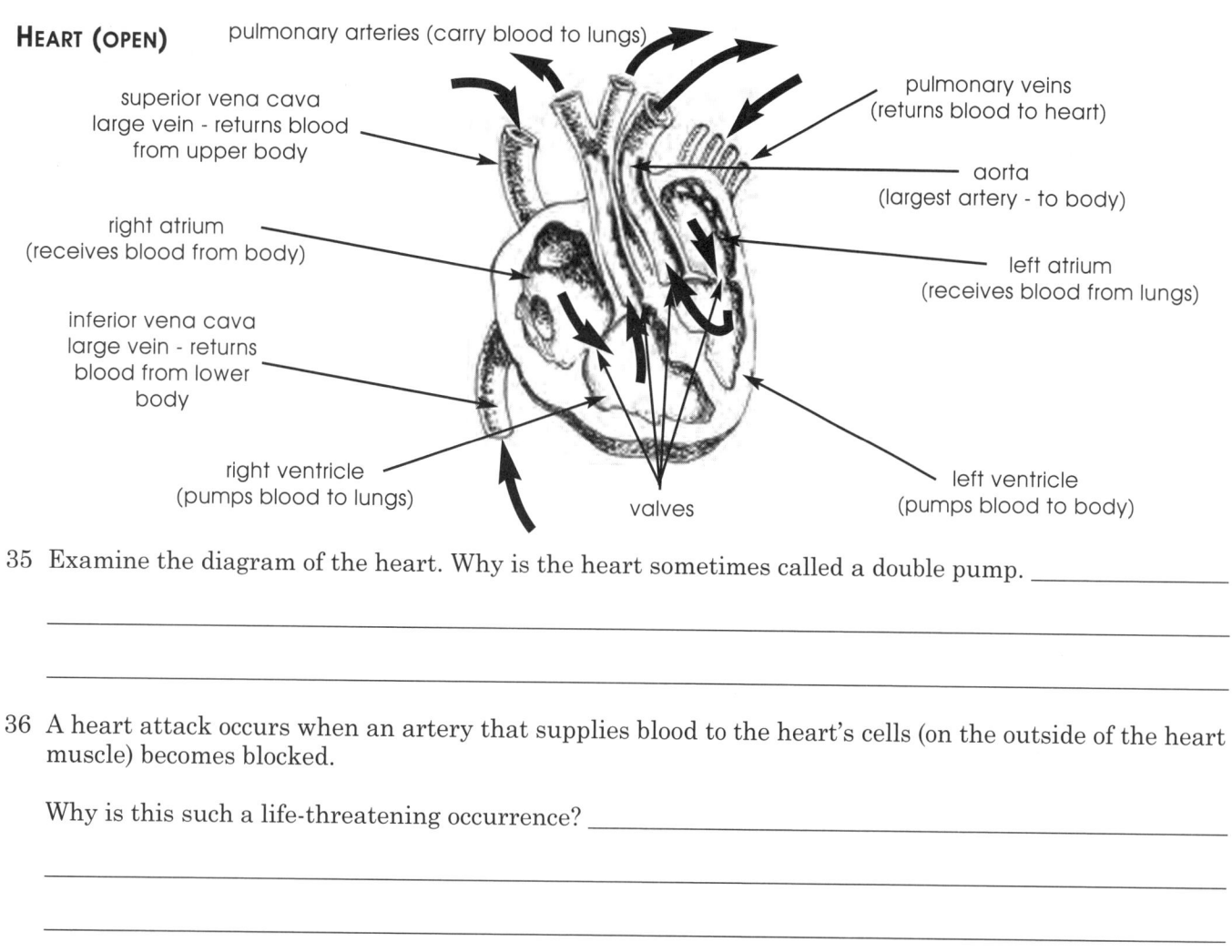

pulmonary arteries (carry blood to lungs)

superior vena cava
large vein - returns blood
from upper body

pulmonary veins
(returns blood to heart)

aorta
(largest artery - to body)

right atrium
(receives blood from body)

left atrium
(receives blood from lungs)

inferior vena cava
large vein - returns
blood from lower
body

right ventricle
(pumps blood to lungs)

valves

left ventricle
(pumps blood to body)

35 Examine the diagram of the heart. Why is the heart sometimes called a double pump. _____

_____

_____

36 A heart attack occurs when an artery that supplies blood to the heart's cells (on the outside of the heart muscle) becomes blocked.

Why is this such a life-threatening occurrence? _____

_____

_____

One of the earliest scientists to study circulation was England's **Dr. William Harvey** (1578-1657). He sought a mechanical explanation for the movement of blood in the body. A mechanical explanation would show that the body worked like a machine and that it was not "magic" as some people thought at that time. The way he did his work was important. He investigated by setting up practical experiments. He found that blood with oxygen went out one side of the heart and came back to the other side of the heart without it. The invention of the microscope helped Italian **Marcello Malpighi** (1628-1694) discover that capillaries connect the tubes taking blood to and from the heart.

IDENTIFYING PULSE POINTS & PULSE RATES

As your heart pumps blood throughout your body, you can actually feel the surges of blood as it pushes through the major arteries. One place to feel your **pulse** is at an artery near the surface of your body, called a **pulse point**.

Using the index and middle finger of your left hand, follow your ear lobe, on the left side of your head, down to a point on your neck just below your jaw bone. Press gently and move your fingers slightly right or left until you feel a gentle rhythmic pulsing on your fingers. This is the **carotid artery**. To take your pulse rate, count how many pulses you feel in one minute. Be sure to be quiet and to concentrate on your count. Take your pulse rate 3 times and calculate your average pulse rate per minute.

Pulse Rate
Minute one _____ Minute two _____ Minute three _____ Average Pulse rate _____

Using your index finger and the middle finger of one hand find the **radial artery** as it crosses the radial bone on the inside of your other wrist. Press gently and move your fingers slightly right or left until you feel a gentle rhythmic pulsing on your fingers. To take your pulse rate, count how many pulses you feel in one minute. Be sure to be quiet and to concentrate on your count. Take your pulse rate 3 times and calculate your average pulse rate per minute.

Pulse Rate
Minute one _____   Minute two _____   Minute three _____   Average Pulse rate _____

37 *a* Did the averages of your two pulse points differ? _____   Should they? _____

    b Why or why not? _____

EXCRETORY SYSTEM

System	Functions	Structures
Excretory	Gets rid of molecular sized, dissolved solid, liquid, and gaseous wastes, and removes excess heat energy (the build up of wastes can poison cells).	lungs, skin, kidneys urinary structures, large intestine, circulatory system

As you run, your body begins to heat up. Excess heat can be dangerous for cells. Sweat glands in your **skin** (below left) may go to work. They release water, salts, and other wastes. As the liquid from the sweat evaporates, it takes energy with it thereby cooling your body.

The **kidneys** filter the blood removing excess water, excess salt, and wastes from the breakdown of proteins.

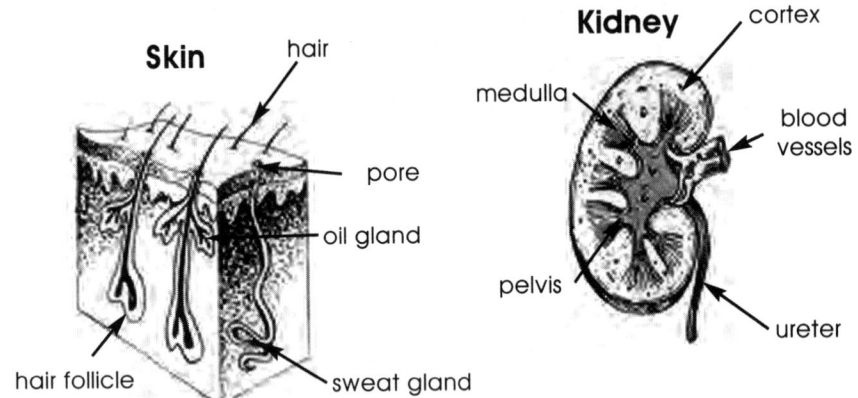

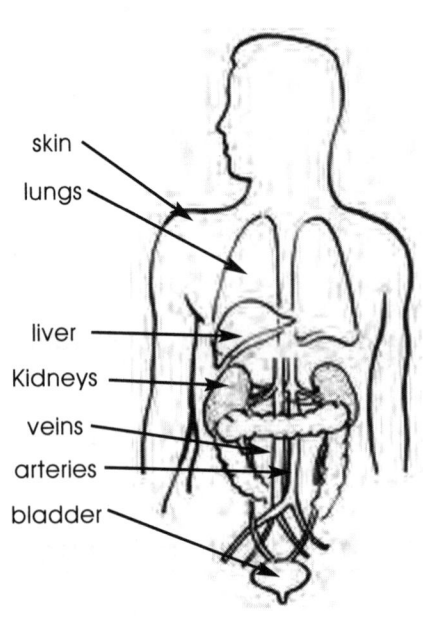

38 What is the main purpose of the excretory system? (What do the cells get rid of?) _____

_____

39 What produces the wastes that leave through the excretory system? _____

_____

_____

SKELETAL SYSTEM & LOCOMOTION

System	Functions	Structures
Skeletal	Allows movement to escape danger, obtain food and shelter, and reproduce. Provides support and protection for organs. Produces blood cells (in bone marrow).	interaction of skeletal muscles and bones coordinated by the nervous system

MUSCULAR SYSTEM & LOCOMOTION

Human Skeleton

System	Functions	Structures
Muscular	Allow for motion. Moves your skeleton, lines blood vessels and digestive tract. Heart muscle pumps blood. Some protect internal organs.	voluntary, involuntary, smooth, cardiac,

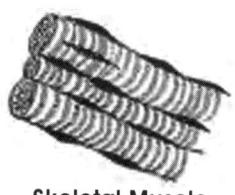

Skeletal Muscle

Smooth Muscle

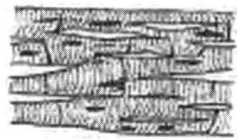

Cardiac Muscle

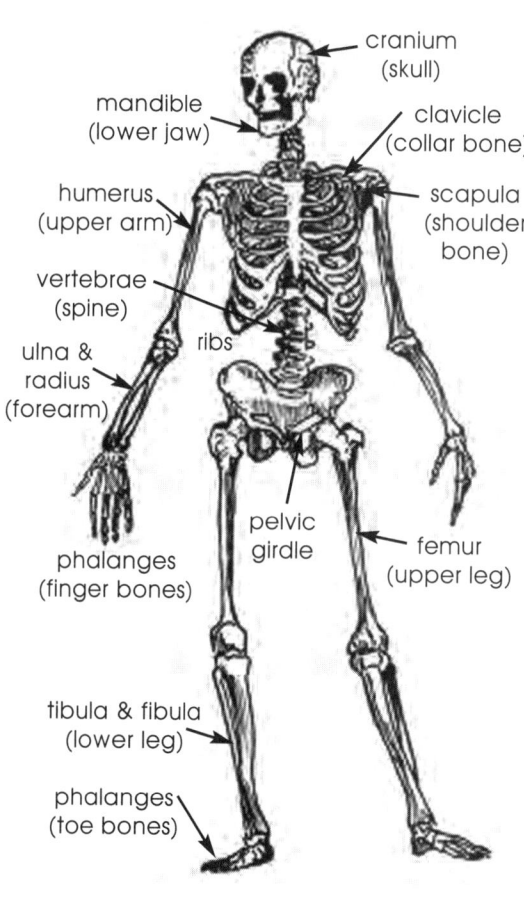

In order to run, your legs have to move. **Skeletal muscles** contract, pulling bones closer together, and your leg bends. Other skeletal muscles then contract in the opposite direction and straighten your leg. Your **nervous system** coordinates these movements so you do not trip.

Smooth muscles do not use conscious control. These muscles are found in the lining of your blood vessels and digestive tract.

The muscles of your heart are special and called **cardiac muscles**. They are very strong and pump blood to your body.

40 What is the purpose of the skeletal system? _____

_____

_____

41 What is the purpose of the muscular system? _____

_____

_____

42 Each of the three muscle types have specific functions. Choose one muscle type. What is the reason that the muscle type you chose can not do the work of the other two?

_____

_____

DISCIPLINARY CORE IDEA:

LS1.D: INFORMATION PROCESSING

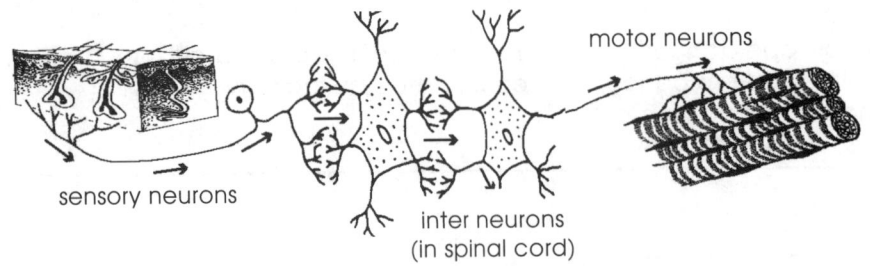

motor neurons

sensory neurons

inter neurons
(in spinal cord)

Sensory and Motor Neurons
(nerves of the muscle systems for voluntary and involuntary movement)

Sensory Neurons
(nerves of the senses: sight, touch, hearing, taste, and smelling)

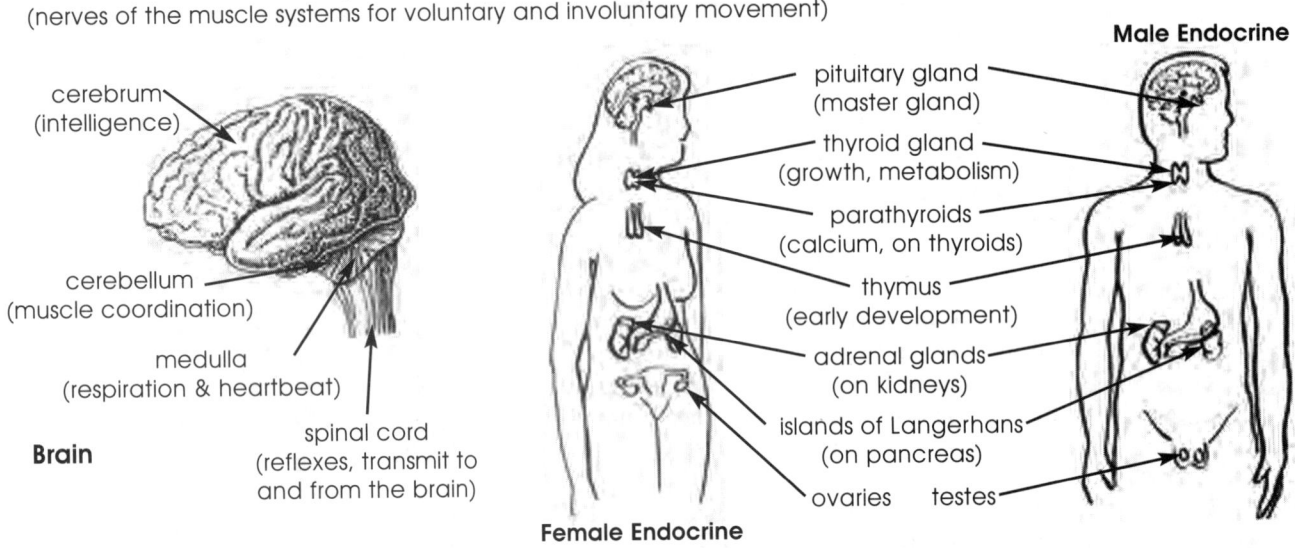

cerebrum
(intelligence)

cerebellum
(muscle coordination)

medulla
(respiration & heartbeat)

Brain

spinal cord
(reflexes, transmit to
and from the brain)

pituitary gland
(master gland)

thyroid gland
(growth, metabolism)

parathyroids
(calcium, on thyroids)

thymus
(early development)

adrenal glands
(on kidneys)

islands of Langerhans
(on pancreas)

ovaries   testes

Male Endocrine

Female Endocrine

System	Functions	Structures
Nervous and Endocrine	Interact to control and coordinate the body's responses to internal and external changes.	brain, spinal cord, nerves, glands

The **endocrine system** regulates growth, development, and reproduction. The endocrine system contains **glands** that produce **hormones** (substances that affect other parts of the body). For example, the pituitary gland at the base of your brain releases a hormone that regulates the growth of your skeleton.

Your sense organs, such as your eyes and ears, give you information about the changing world around you. As you run, light hits a fence in front of you and the light is reflected to your eyes. The light causes a reaction in your eyes that sends a signal through your **nerves** to your **brain**. Your brain interprets the signal as a low fence. The brain sends signals to many different muscles which react by sending you in a jump high over the fence.

43 What is the main purpose of the nervous and endocrine systems? _____

_____

44 What might happen to a child if their pituitary gland produced too much growth hormone? _____

_____

45 During an operation, a doctor may inject a patient with a substance that blocks the nerve impulses from the sense organs to the spinal cord or brain. The sense organs themselves are not affected. If the sense organs pick up the pain signal during the operation why does the person not feel the pain?

_____

MAIL CALL

Think of the endocrine system as postal mail. It takes a little while for you to receive the message. Think of the nervous system as e-mail. It provides a faster path for messages.

IMMUNE SYSTEM

As you jump the fence, a tree branch that you did not see hits your leg and causes a cut. **Bacteria** on the branch and in the air enter the cut. The body sends extra blood to the infected area. It brings **white blood cells** and **antibodies** which attack the bacteria, destroying it so it will not harm you.

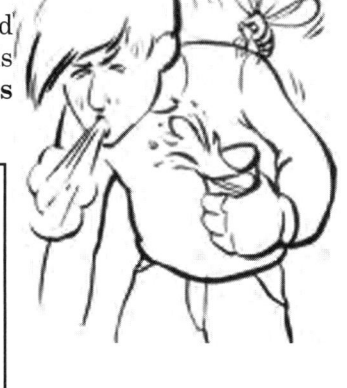

System	Functions	Structures
Immune	Protects the body from **infectious diseases** (diseases that are caused by organisms that attack the body). A **disease** is the breakdown in structure or function of an organism. It produces cells and molecules that identfy and destroy microbes that enter the body.	specialized cells, and the molecules

46 What is the main function of the immune system? _____

_____

47 AIDS causes a breakdown in the immune system. Why is this such a dangerous disease? _____

_____

48 In some cases, a body will produce too many white blood cells.

Why might this be a problem for an individual? _____

_____

49 Why is so much importance placed on cleaning a cut when you get hurt? _____

_____

_____

REPRODUCTIVE SYSTEMS

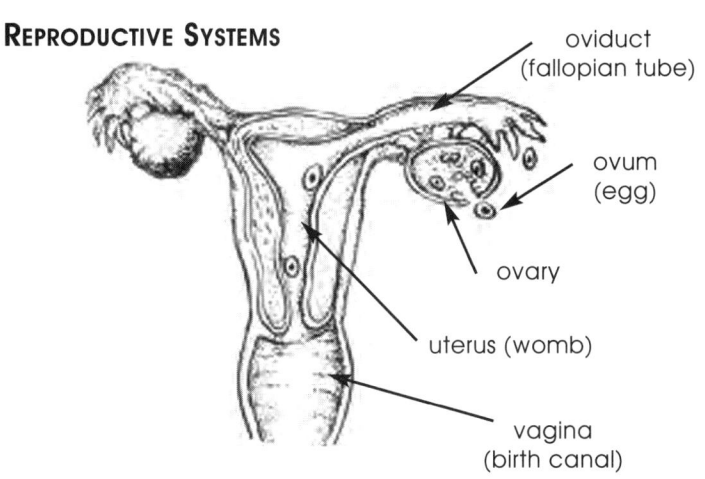

Female Reproduction

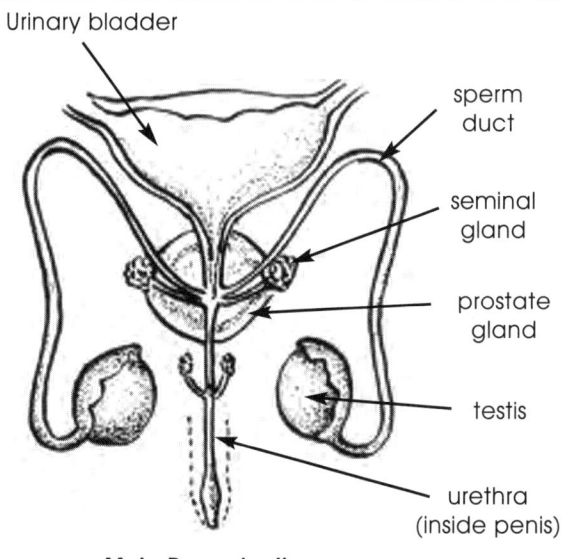

Male Reproduction

System	Functions	Structures
Reproductive	Produces sex cells necessary for the production of offspring.	Male and female (structures differ)
FEMALE	Produces female sex cells (eggs). Usually one egg is produced every 28 days.	Ovaries
	Where eggs travels from the ovaries. Organ where fertilization occurs.	Oviducts (also called Fallopian Tubes)
	Where fertilized eggs attach and develop. If the egg is not fertilized, the egg and the uterine lining pass out through the process of Menstruation	Uterus
	Where male sex cells (sperm) are deposited	Vagina (also called birth canal)
MALE	Produce male sex cells (sperm)	Testes
	Small tubes that carry the sperm to the penis	Sperm ducts
	Add secretions to sperm	Prostrate and Seminal glands
	Transfers sperm cells from the male to the female vagina	penis

The reproductive system is the only place where special cells called sex cells are made. These cells contain one half of the normal number of chromosomes. They carry one half of the genetic information for a new individual. When the **male (sperm)** and **female (egg)** sex cells unite the new cell called the **fertilized egg** or **zygote** have a full set of chromosomes. This new cell can then develop into a human being.

50 Where are the female sex cells produced? _____

51 Where are the male sex cells produced? _____

52 How are sex cells different from other body cells? _____

_____

Directions: Read each question carefully. Examine the choices and circle the best answer.

1 If a new type of living thing is discovered, of which basic unit of structure will it most likely be composed?
(1) cells   (2) organs   (3) organ systems   (4) tissues

2 Which structure is the main control center of the cell?
(1) chloroplast   (2) cytoplasm   (3) mitochondria   (4) nucleus

3 Which cell would be most efficient at transferring messages from one point to another?
(1) skeletal   (2) muscle   (3) nerve   (4) skin

4 The destruction of its chlorophyll would most directly affect a plant's ability to
(1) take in minerals   (3) produce seeds
(2) transport water   (4) undergo photosynthesis

5 The main function of the respiration system is to get oxygen to the
(1) blood   (2) cells   (3) lungs   (4) trachea

6 Which statement best describes arteries?
(1) They have thick walls and transport blood away from the heart.
(2) They have thick walls and transport blood toward the heart.
(3) They have thin walls and transport blood away from the heart.
(4) They have thin walls and transport blood toward the heart.

7 Emphysema is a disease that reduces the number and elasticity of alveoli. Its main effect is on which body system?
(1) circulatory   (2) endocrine   (3) locomotion   (4) respiration

8 Which form of life is an exception to the cell theory?
(1) bacterium   (2) plant   (3) amoeba   (4) virus

9 The African antelope is classified as Gazella thompsonii. The common name "gazelle" is derived from which part of the scientific name?
(1) kingdom   (2) phylum   (3) genus   (4) species

10 Which cell structure is most directly involved in protein synthesis?
(1) cell membrane   (2) centriole   (3) cell wall   (4) ribosome

11 Hormones in the human body
(1) are produced by every cell   (3) travel through ducts
(2) are produced only by the pituitary gland   (4) may affect areas far from where they were produced

12 In humans, most of the building blocks necessary for the building of cells come from
(1) end-products of digestion   (3) products of cellular respiration
(2) products of excretion   (4) breakdown of white blood cells

13 Which adjustment does the human body make in response to an increase in environmental temperature?
(1) secretion of insulin   (3) increased perspiration
(2) storage of fat   (4) increased urine

14 In humans, one difference between the nervous system and the endocrine system is that
   (1) nerve responses are of longer duration than endocrine responses
   (2) nerve responses are more rapid than endocrine reposes
   (3) only the nervous system plays a role in keeping the body regulated
   (4) the endocrine system is found in only one part of the body

15 Which characteristics of organisms are most commonly used to place them in a specific kingdom?
   (1) color, sex, and habitat   (3) type of nutrition and basic structure
   (2) behavior and length   (4) rate of reproduction and shape

16 The chemical regulators secreted by the endocrine system are known as
   (1) endocrines   (2) hormones   (3) enzymes   (4) neurons

Base your answer to Question 17 on the diagram at the right of the human arm and your knowledge of the living environment.

17 Movement of the lower arm bones in the direction indicated by the arrow results from contraction of which muscle(s)?
   (1) 1   (3) 1 and 2
   (2) 2   (4) neither 1 or 2

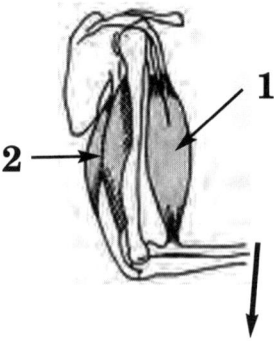

18 The graph at the right shows the effect of environmental temperature on a girl's skin temperature and on her internal body temperature.

   Which statement best describes what happens as the environmental temperature increases from 20°C to30°C?
   (1) Both the internal and skin temperatures reach 40°C
   (2) Both the internal and skin temperatures increase by about 7°C
   (3) The skin temperature decreases to about 30°C
   (4) The internal temperature increases by about 1°C

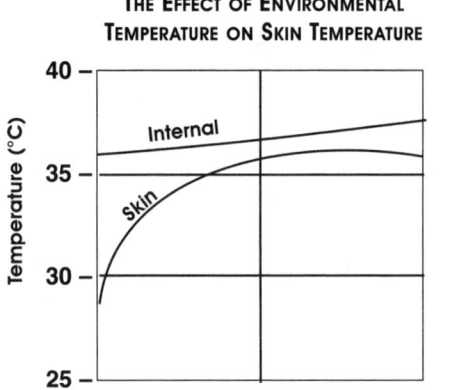

THE EFFECT OF ENVIRONMENTAL
TEMPERATURE ON SKIN TEMPERATURE

19 Which part of the human digestive system produces digestive enzymes?
   (1) esophagus   (2) appendix   (3) small intestine   (4) large intestine

20 In humans, the organ that most directly regulates the concentration of water in the blood is the
   (1) heart   (2) liver   (3) pancreas   (4) kidneyx

THE QUESTIONS THAT FOLLOW TEST YOUR UNDERSTANDING OF SCIENTIFIC PRACTICES, CORES IDEAS IN SCIENCE AND CROSS-CUTTING CONCEPTS SUCH AS PATTERNS AND CAUSE AND EFFECT RELATIONSHIPS.

SHORT ANSWER

1 The chart at the right shows the blood test of Patient X.

Based on the results of the blood test, which of Patient X's body systems is probably not working properly? (2 points)

BLOOD TEST OF PATIENT *X*		
BLOOD CONTENTS	**NORMAL**	**PATIENT *X***
Carbon Dioxide	10	20
Urea	5	7
Salt	1	2
Water	20	25

_____

Explain your answer. _____

_____

_____

2 What relationship about the effect of smoking on lung capacity does this graph demonstrate? (2 points)

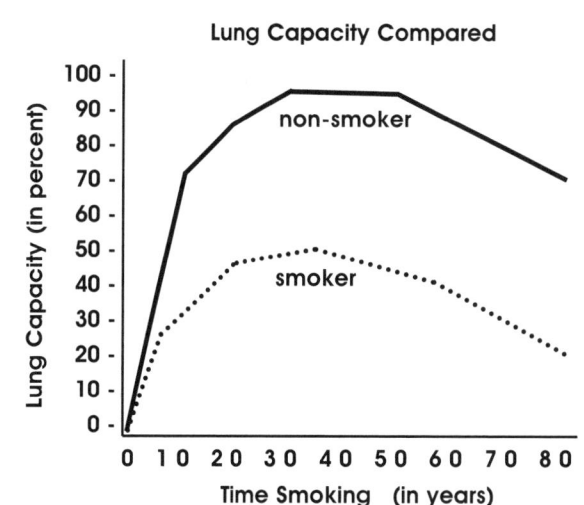

- -

- -

- -

- -

- -

- -

SHORT ESSAY

1 As you are writing the answer to this question, many systems in your body will be involved. Describe how the systems interact to allow you to answer this question. Include at least 3 systems and 5 interactions. (5 points)

_____

_____

_____

_____

HEREDITY:
INHERITANCE AND VARIATION OF TRAITS

DISCIPLINARY CORE IDEAS:

LS3.A: INHERITANCE OF TRAITS

LS3.B: VARIATION OF TRAITS

LS3.A INHERITANCE OF TRAITS

BLUEPRINTS...BLUEPRINTS...WHOSE GOT THE BLUEPRINTS?

Before a house is built, a blueprint – the plans for what the house will be like – must be drawn up. Detailed blueprints tell the builder exactly how big each piece of wood in the house will be, where the electrical outlets will go, etc. The builder could use the same set of blueprints to build many identical houses.

Every organism requires a set of instructions, like blueprints, to specify what traits the organism will develop. The 'blueprints' for an organism are found in the chromosomes of its cells. When organisms reproduce these instructions are passed from one generation to the next as chromosomes are passed to the new organism.

DECODING THE BLUEPRINTS

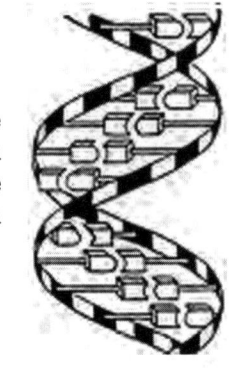

The chromosomes of cells are composed of a molecule called **DNA**. Scientists believe hereditary information is passed on based on the particular structure of the DNA molecule in the organism. The way hereditary information is stored in your body was uncovered by the work of two scientists, American **James Watson** and Briton **Francis Crick**. In 1953, Watson and Crick looked at the available evidence and came up with a model of the DNA molecule.

53 Where is the hereditary information located in the cell? _____

54 What molecule contains the genetic code? _____

A model in science is used to help scientists think about how something in nature might actually work. In order to be a good model, it must predict and/or explain what will happen in the real world. Watson and Crick proposed that the structure of the DNA molecule was like a ladder. Normally in cells, DNA would be found as a twisted ladder called a **double helix**.

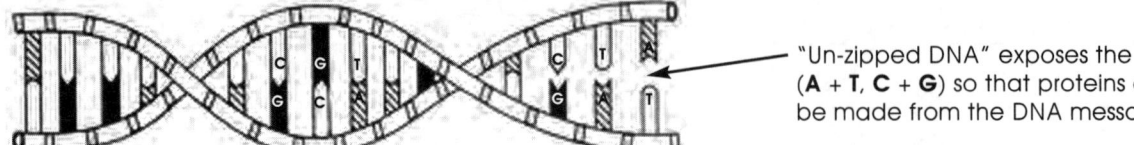

"Un-zipped DNA" exposes the bases (**A** + **T**, **C** + **G**) so that proteins can be made from the DNA message.

The rungs of the ladder are made up of pairs of four different substances called **nitrogen bases** (*A, T, C,* and *G*). The *A* base always paired with the *T*, and the *G* base always paired with the *C*. The order of the bases were thought to provide the hereditary code because their order directs the production of proteins that make up the organism.

READING YOUR AUTOBIOGRAPHY

If you were to take a book and cut each line in a strip and then tape each strip end to end, you would have a very long strip of paper that told you a story. This is what the DNA molecule is like. Along its length, the order of the bases provide the information the cell needs to produce an entire new individual. Just as a story is made up of sentences, so is the DNA molecule divided into sections. Each unit of information that contains the code for a particular trait is called a **gene**.

The gene provides the code for the organism to build a protein. The protein will help produce some characteristic in the organism. In order to produce the code for all of your traits, each of your cells has many thousands of different genes.

55 What is a gene? _____

_____

56 What is DNA? _____

CHROMOSOME PAIRS

Chromosomes come in pairs. Each pair of chromosomes has the gene for a particular trait (characteristic) in the same position along its structure. Therefore, it is a pair of genes that determine the traits of an organism. This pair of genes is called alleles. A single inherited trait may be determined by one pair of genes or by many pairs of genes.

Think of it like this: An elementary school teacher lined her students line up in pairs. (Think of each line as one chromosome.) The teacher gave out jobs based on the students' position in the line. For example, the first pair of students always cleaned the board. The second pair determined how the board was to be cleaned. The third pair of students always collected milk money. The fourth and fifth pair of students worked together to clean the room. Some traits are determined by a single gene, while others are determined by more than one gene.

57 A single inherited trait may be determined by _____ pair or _____ pairs of _____.

PASS IT ON

When organisms reproduce, genetic instructions are passed from one generation to the next. Passing of the 'blueprints' from parent to offspring means that similar structures and functions will be found in both. Dogs have puppies and cats have kittens. How do organisms insure that their offspring will be like them?

In **asexual reproduction**, all genes given to the offspring come from one parent. Since the offspring receives a copy of the parent's genes, it is genetically identical to the parent. Follow the steps of asexual reproduction in a single-celled organism, the amoeba.

"And as amoebas, you'll have no problems recruiting other sales reps ... just keep dividing and selling, dividing and selling..."

Cartoon by Larson on NYS Regents Biology Exam

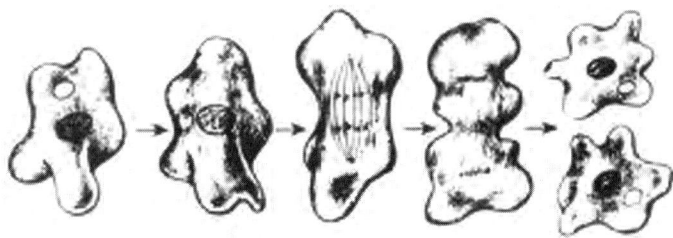

The amoeba uses fission (splitting into two) to reproduce.

ASEXUAL REPRODUCTION IN OTHER ORGANISMS

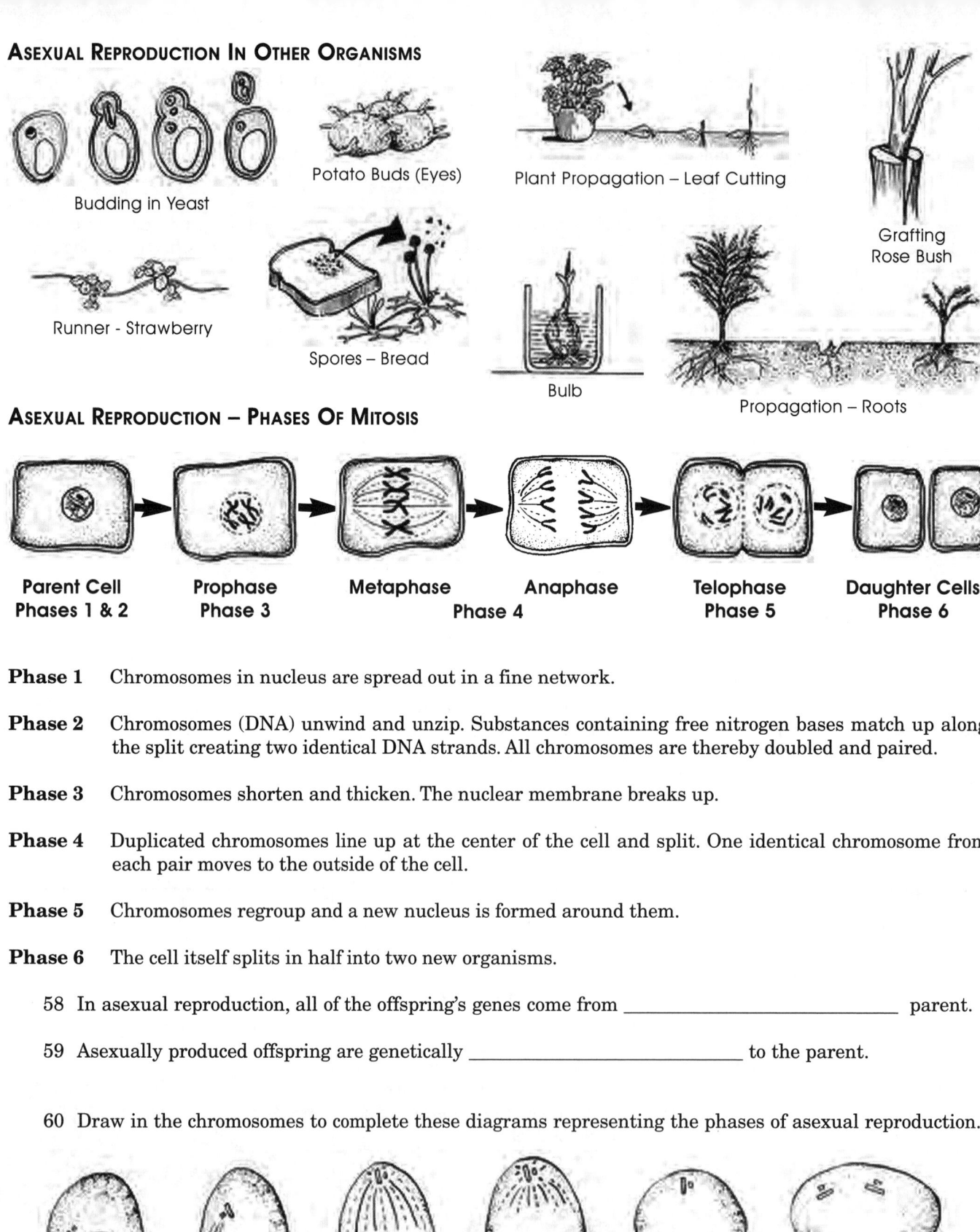

Budding in Yeast

Potato Buds (Eyes)

Plant Propagation – Leaf Cutting

Grafting
Rose Bush

Runner - Strawberry

Spores – Bread

Bulb

Propagation – Roots

ASEXUAL REPRODUCTION – PHASES OF MITOSIS

Parent Cell
Phases 1 & 2

Prophase
Phase 3

Metaphase
Phase 4

Anaphase

Telophase
Phase 5

Daughter Cells
Phase 6

Phase 1   Chromosomes in nucleus are spread out in a fine network.

Phase 2   Chromosomes (DNA) unwind and unzip. Substances containing free nitrogen bases match up along the split creating two identical DNA strands. All chromosomes are thereby doubled and paired.

Phase 3   Chromosomes shorten and thicken. The nuclear membrane breaks up.

Phase 4   Duplicated chromosomes line up at the center of the cell and split. One identical chromosome from each pair moves to the outside of the cell.

Phase 5   Chromosomes regroup and a new nucleus is formed around them.

Phase 6   The cell itself splits in half into two new organisms.

58   In asexual reproduction, all of the offspring's genes come from _____ parent.

59   Asexually produced offspring are genetically _____ to the parent.

60   Draw in the chromosomes to complete these diagrams representing the phases of asexual reproduction.

In **sexual reproduction**, half of the DNA comes from each parent. Special cells produced by each parent, called **sex cells**, contain only one half of the genetic material. Sex cells contain one chromosome from each pair of chromosomes. Only when two sex cells unite are the chromosome pairs complete. Since you get one of each chromosome pair from each parent, half of your genetic material for every trait come from each parent. What follows are the steps of sexual reproduction. The production of sperm sex cells includes:

Phase 1 Chromosomes in nucleus are spread out in a fine network.

Phase 2 Chromosomes (DNA) unwind and unzip. Substances containing free bases match up along the split, creating two identical DNA strands. All chromosomes are thereby doubled and in identical pairs.

Phase 3 Chromosomes shorten and thicken. The nuclear membrane breaks up.

Phase 4 Duplicated chromosome pairs line up at the center of the cell. One doubled chromosome from each pair moves to each pole. The cell divides so that each cell has two sets of genes from one side of the original chromosome pair.

Phase 5 Duplicated chromosomes line up at the center of the cell and split. One identical chromosome from each pair moves to the outside of the cell.

Phase 6 Chromosomes regroup and a new nucleus is formed around them.

Phase 7 The cell itself splits in half into two sex cells. (This time the DNA had not duplicated)

Normal pairs of chromosomes are in each sex cell.

Chromosome pairs double.

Original sex cell divides, restoring the original pairs of chromosomes.

These sex cells divide again.

Each sperm cell now has just one half of the original sex cell's number of chromosomes.

In the case of the human male, each original body cell contains 46 total chromosomes (23 pairs). After meiosis, each sperm produced has only 23 chromosomes (one from each of the original pairs.)

One original cell forms four sperm cells. Each has one half of the number of original chromosomes. A chromosome is mainly a long strand of DNA. In cells that are not sex cells, these strands come in pairs. Sex cells have only one strand of each pair.

In sexual reproduction, sex cells, often the sperm cell from the male and the egg cell from the female, unite during fertilization. Since half of the offspring's genes come from each parent, it is not identical to either parent.

61 In sexual reproduction, what percentage of the DNA comes from each parent. _____%

62 Are sexually produced offspring genetically identical to either parent? _____

63 What would happen if a sex cell only divided once before fertilization? _____

_____

_____

DISCIPLINARY CORE IDEAS:
LS3.B: VARIATION OF TRAITS
"MENDEL, MENDEL, HOW DOES YOUR GARDEN GROW?"

If an organism's traits are determined from the genetic code it receives from its parents, should you be able to predict what the offspring will be like if you know the genetic codes of the parents?

The study of how heredity works began long ago. In 1865, **Gregor Mendel** an Austrian monk, published his work on the inheritance of garden peas. Mendel was a high school teacher who kept a small garden plot. His most famous work was on peas he grew there. He may have chosen garden peas to study because different pea plants showed different characteristics. Some pea plants were short and bushy, others were tall and climbing. Some plants produced yellow seeds and some produced green seeds. Mendel could control the fertilization of the pea plants by controlling pollination. Pollen containing the male sex cell is simply placed on the female part of the plant which leads to the egg cell.

Gregor Mendel

Mendel proposed several hypotheses to describe the patterns of inheritance that he observed. He proposed that the various hereditary characteristics of the pea plant are controlled by factors that occur in pairs. Mendel used the term "factor" to describe what was affecting the **traits**. Today these factors are called "genes."

For the trait of height, two forms, tall and short exist in pea plants. When pea plants having different forms of the same trait were crossed, sometimes the offspring would resemble only one of the parents. Mendel called this trait that hid the other **dominant**. For example, when Mendel crossed short pea plants with tall pea plants the offspring were always tall. Mendel formulated a hypothesis, **the principle of dominance and recessiveness**. This principle states that one factor in a pair may mask or prevent the expression of the other. Mendel called the trait that masked or hid the other "dominant." He called the trait that was hidden and did not appear "**recessive**."

64 A factor that masks the expression of another factor is said to be _____ .

65 A factor that is masked by another factor is said to be _____ .

Look more closely at what happened in Mendel's experiment. Hereditary factors occur in pairs. Therefore, each trait has a pair of genes that affect it. If you use T torepresent each of the genes in a tall plant, then you can represent the tall plant as TT. Or, if you use t to represent each of the genes of the pair in a short plant, then you can represent the short plant as tt. In a non-sex cell, a tall plant would have the genes TT. In organisms that reproduce sexually, as pea plants, only one half of the chromosomes are found in the sex cells. Therefore, half the number of genes are found in the sex cell. The tall plant's sex cells would each have one T. The short plant's sex cells would each have one t.

Pure tall _____ Pure short

During fertilization, each parent contributes one of each pair of genes. Mendel called this the **principle of segregation**. A pair of factors is segregated (separated) during the formation of sex cells.

Hybrid tall _____ Hybrid tall

When both genes of a pair for a particular trait are the same, the organism is said to be pure for that trait. TT would be pure tall. tt would be pure short. TT could also be called pure dominant and tt pure recessive.

Mendel crossed TT and tt plants. Each offspring got one gene of the pair from each parent. Therefore, each offspring had the genes Tt. When an organism has different genes for a trait it is considered a **hybrid**. Tt is hybrid for tall. Since T is dominant the offspring would look just like their tall parents.

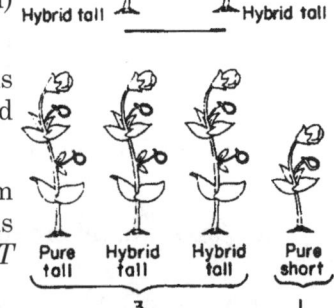

Pure tall   Hybrid tall   Hybrid tall   Pure short

3   1

Mendel did many experiments. He kept accurate records of all of his work. He worked methodically – in many cases, only one trait at a time. One reason for Mendel's success is that he worked scientifically. He used scientific methods and systematically tried to explain the nature of the inheritance of pea plants.

66 Label each of the following gene pairs as pure or hybrid.

AA _____ Aa _____ aa _____

MIXING IT UP -BLENDING

It turned out that not all genes are dominant or recessive. In 1900, **Karl Correns**, a German botanist, found that in some traits, both genes of the pair for a trait can influence the trait in an individual. This type of heredity is called **incomplete dominance** or **blending**. The effects of both members of the gene pair are 'blended' so that the offspring does not resemble either of its parents.

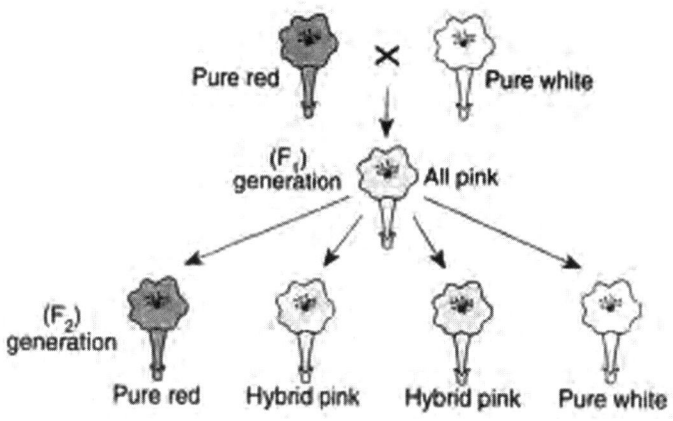

Correns worked with four-o'clock flowers. When red four-o'clock flowers are crossed with white four-o'clock flowers, the offspring are pink. To represent the genes, the red parent would be *rr* and the white parent *ww*. The offspring would get one of the gene pair from each parent and be *rw*. Notice that for incomplete dominance, genes are all represented with lower case letters and that differing gene pairs are represented by different letters.

67 Some inherited traits seem to appear as a mixture of traits from both parents.

This type of inheritance is called _____ .

LIVING ENVIRONMENT SKILL

DESIGN AND USE A PUNNETT SQUARE OR PEDIGREE CHART TO PREDICT THE PROBABILITY OF CERTAIN TRAITS.

To understand what types of offspring will occur, scientists use models of genetic inheritance. These models can help to determine the probability or chance of a trait being expressed. One model is a grid system developed by **R.C. Punnett**. It is called a **Punnett Square**. The gene pair from the female parent is put across the top of the grid. The gene pair from the male parent is written down the left side of the grid.

You fill in the grid by placing one half of each gene pair in the corresponding box. For example, in Mendel's cross of tall and short pea plants, if male short plants were crossed with female tall plants, the Punnett Squares would look like this.

_____

PUNNETT SQUARE METHOD

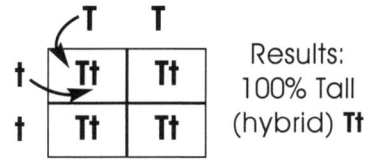

As you can see in the **first cross** on the previous page, each box represents 25% of all possible offspring. This model can therefore help you determine the probability of traits being expressed. In this case 100% (4 x 25%) of the offspring would be tall (*Tt*), but not pure tall.

In the **second cross**, two hybrids are crossed (*Tt* x *Tt*). The probability is that 75% of the offspring will be tall (*TT, Tt, Tt* – both pure and hybrid) and that 25% of the offspring will be short (*tt*).

Remember this chart just helps you predict. You cannot actually tell which sex cells will unite. Actual results may differ from predictions especially when there are only a small number of offspring.

Second Cross
Hybrid Tall (Tt) with
Hybrid Tall (Tt)

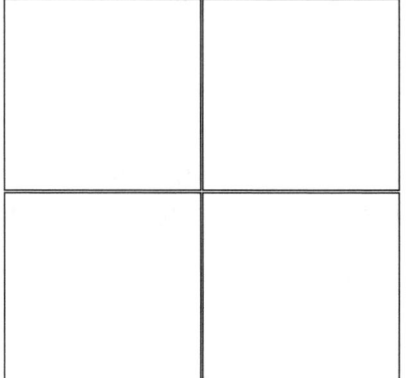

Two Hybrid Cross
Hybrid Tall (Tt) with Hybrid Tall (Tt)

	T	t
T	TT	Tt
t	Tt	tt

Results:
25% Tall (pure) **TT**
50% Tall (hybrid) **Tt**
25% Short (pure) **tt**

P generation
pure tall female pea plant x
hybrid tall male pea plant

68 Draw the Punnett Square for the cross of a pure tall female pea plant with a hybrid tall male.

69 What percent of the offspring will be tall? _____

70 What percent of the offspring will be short? _____

71 Draw the Punnett Square for the crossing of the two hybrid green peas shown in the diagram. How can yellow pea pods appear if both parents have green pea pods?

_____

_____

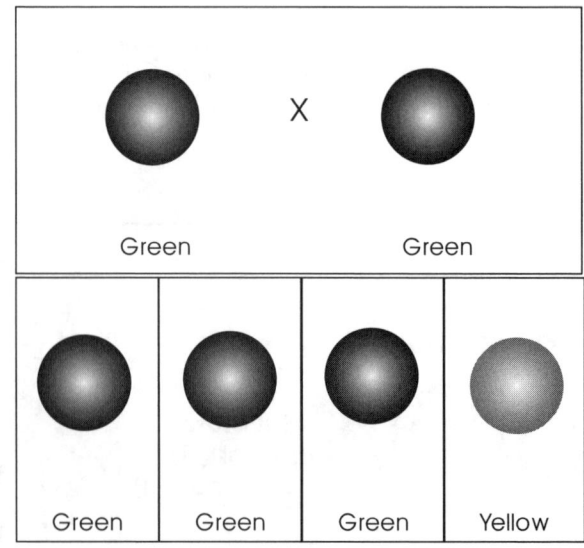

CHECKING THE FAMILY TREE

Another model for understanding genetic inheritance is a **pedigree chart**. In a pedigree chart, each level represents a new generation. Males are represented as squares and females as circles.

In this diagram, the parents (First Generation, or level one) have four offspring, two males and two females (level two). In the second generation (level two), one male and one female offspring mate. In the third

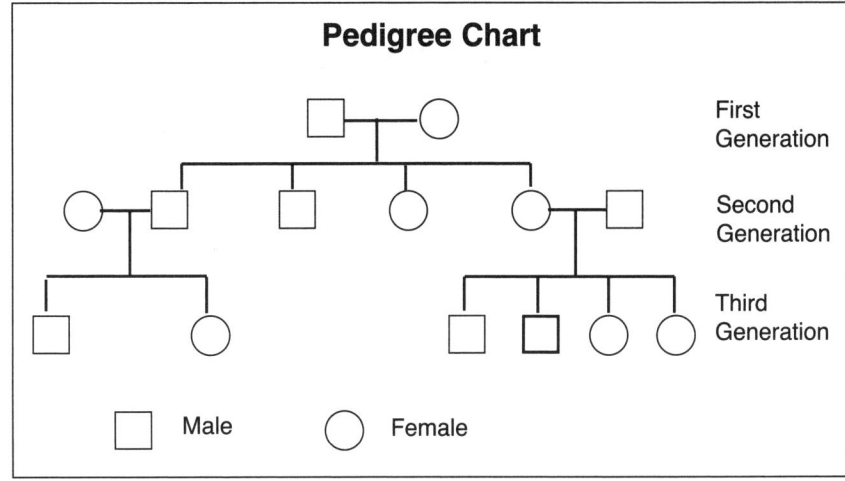

Pedigree Chart

First Generation

Second Generation

Third Generation

☐ Male   ○ Female

generation (level three), you see that the male offspring from the second generation had one male and one female offspring. The female from the second generation had two male and two female offspring.

Base your answer to the following question on the diagram of the pedigree chart at the right. Shaded represents brown fur. Unshaded represents white fur.

72 What is the probability of the offspring in the third

generation being brown? (Use a Punnett Square.)

_____

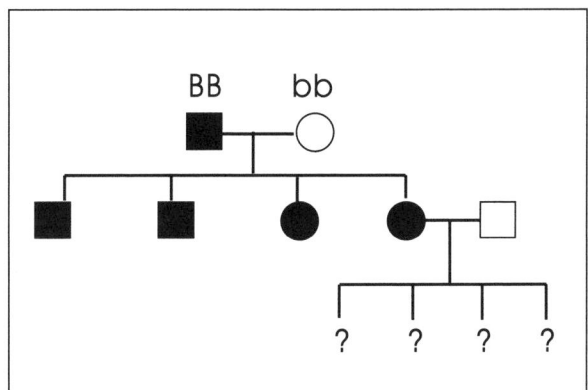

White forelock trait is seen in people as a streak of white hair. Examine the pedigree chart for the appearance of white forelock trait in this family. Use the pedigrees chart to answer the following questions.

73 Could the father be described as pure or hybrid? _____

_____

74 What percentage of the parent's offspring would you expect to show white forelock trait? (Use a Punnett Square.)

_____

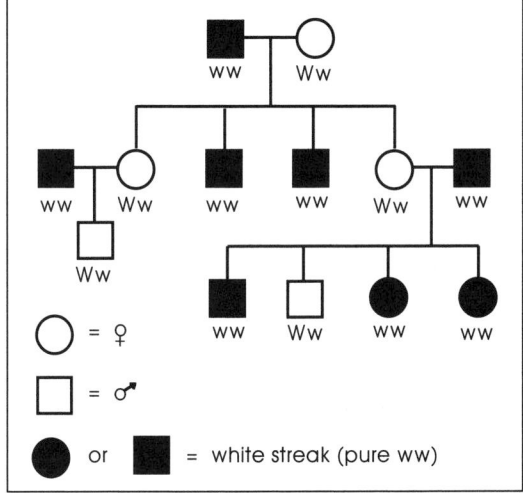

75 What percentage of the actual offspring showed white forelock trait? _____

76 Why do the children of the third generation show such a high percentage of white forelock trait? _____

_____

_____

MULTIPLE CHOICE

1   Hereditary information is contained in the
(1) cell membrane   (2) cytoplasm   (3) genes   (4) vacuoles

2   A single unit of genetic information is called a
(1) chromosome   (2) gene   (3) nucleus   (4) protein

3   A single inherited trait is determined by one or more pairs of
(1) cells   (2) chromosomes   (3) genes   (4) proteins

4   Each human cell contains the genes for how many traits?
(1) one   (2) two   (3) hundreds   (4) thousands

5   In asexual reproduction, the offspring receives genes from
(1) one parent   (2) two parents   (3) three parents   (4) many organisms

6   Genetically, how do offspring of asexual reproduction compare to their parent?
(1) identical   (2) similar   (3) very different   (4) no similarities

7   In sexual reproduction, how much genetic material does the offspring receive from each parent?
(1) none   (2) 1/4   (3) 1/2   (4) 2 sets

8   Genetically, how do offspring of sexual reproduction compare to their parents?
(1) identical   (2) no similarities   (3) a combination   (4) identical to the father

9   An organism with two identical genes for a trait is said to be
(1) dominant   (2) recessive   (3) hybrid   (4) pure

10 Crossing a red flowered plant with a white flowered plant produces a plant with pink flowers.
This is an example of
(1) blending   (2) dominance   (3) mixing   (4) recessiveness

11 Which of the following gene pairs would show the recessive trait?
(1) gg   (2) Gg   (3) GG   (4) GD

12 In screech owls, red feathers are dominant over grey feathers. If Rr and Rr owls are mated (where R is the gene for red and r is the gene for grey) what percentage of their offspring would be expected to have red feathers?
(1) 25%   (2) 50%   (3) 75%   (4) 100%

13 If the pattern of inheritance for a trait is incomplete dominance, then an organism with two different members of the gene pair for the trait would show
(1) the recessive trait, only   (3) a blend of the dominant and recessive traits
(2) the dominant trait, only   (4) an incompletely formed trait

14 When red cattle are mated with white cattle the offspring all come out white with spots of brown.
If two spotted brown cows are mated, *RW* x *RW*, what percent of red cows would be expected?
(1) 25%   (2) 50%   (3) 75%   (4) 100%

15 In a certain species of army ant, the gene for long jaws (*M*) is dominant over the gene for short jaws (m). If a biologist wanted to produce ants with long jaws only, which ants should be crossed.
(1) *Mm* x *Mm*   (2) *Mm* x *mm*   (3) *MM* x *Mm*   (4) *mm* x *mm*

16 Our current theory of inheritance states
(1) genes are made up of sugars   (3) external characteristics are not inherited
(2) hereditary factors exist on chromosomes   (4) inheritance is determined by the mother

17 A DNA molecule is composed of
   (1) one base (2) two different bases (3) three different bases (4) four different bases

18 If one side of a DNA molecule has the bases *G-T-T* then the corresponding side would have the bases
   (1) *A-T-T* (2) *G-T-T* (3) *C-A-A* (4) *T-A-A*

19 The basic principles of genetics were established in the 19th century by
   (1) Francis Crick (2) Robert Hooke (3) Jean Lamarck (4) Gregor Mendel

20 Most genetic material occurs in the cell structure known as the
   (1) nucleus (2) vacuole (3) membrane (4) lysosomex

THE QUESTIONS THAT FOLLOW TEST YOUR UNDERSTANDING OF SCIENTIFIC PRACTICES, CORES IDEAS IN SCIENCE AND CROSS-CUTTING CONCEPTS SUCH AS PATTERNS AND CAUSE AND EFFECT RELATIONSHIPS.

SHORT ANSWER

1 Humans have five fingers on each hand. One time a child was born with six fingers on each hand. When the child grew up and had children, the children also had six fingers on each hand. Explain how this sort of change in the parent can be passed on to the offspring? (2 points)

_____

_____

_____

2 A scientist wanted to know if a newly discovered one-celled organism reproduced asexually or sexually. What sort of observations would the scientist have to make to determine if the organism was asexual? Sexual? (2 points)

_____

_____

_____

SHORT ESSAY

1 In order to determine an individual's genes for a trait, scientists do a test cross. A test cross mates the organism in question with a pure recessive organism. Scientists wish to know if a pea plant is pure or hybrid for yellow color (yellow is dominant, green is recessive).

Draw two Punnett Squares. Using them, explain how the probability of yellow and green offspring differ between pure and hybrid individuals when they are mated in a test cross. (5 points)

DISCIPLINARY CORE IDEAS:

LS4.B: NATURAL SELECTION

LS4.A: EVIDENCE OF COMMON ANCESTRY AND DIVERSITY

LS4.C: ADAPTATION

© PhotoDisc

A small white moth landed on a tree covered with black soot from a nearby factory. Normally, its white color blended into the white bark of the tree and allowed it to hide from the birds that eat it. Because of the pollution, it became easy prey for a hungry bird. A black variety of the same type of moth landed on the tree. There were not many of these black moths because they had been easy prey for hungry birds when they landed on the white trees. Now, because of the pollution, the black moth variety hid more easily.

77 If both black and white varieties of this moth are born normally, how will pollution affect the number of

each type that survive? _____

_____

JURASSIC PARK - CLOSED

Gigantic dinosaurs once roamed Earth. Many other organisms that existed are no longer found today. What happened to these living things? The answer to this question comes to us in the study of evolution. **Evolution** is the change in a species over time. What causes this change?

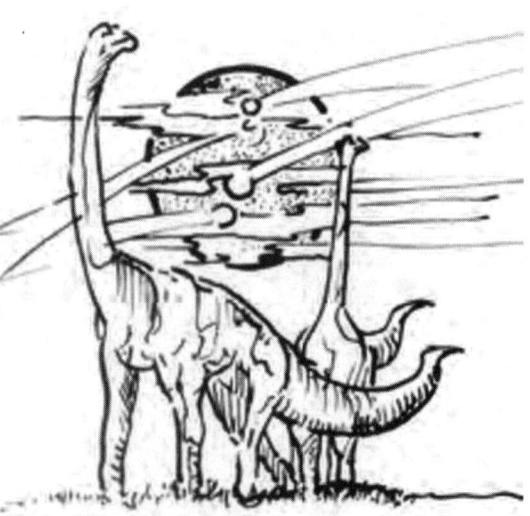

Scientists have studied the variety of living things for hundreds of years. They have tried to explain why there are so many different types of organisms.

Evolution is key to understanding life on Earth. Evidence for evolution is found in the fossil record, by comparing DNA, and in the developmental patterns of living things. For example, **Karl Ernst Von Baer** (German, 1792-1876) studied the growth of organisms from the egg stage. He noted that organisms in the same groups or phylums showed similar patterns of development.

One explanation was put forth by an 18th c. French scientist named **Jean Baptiste Lamarck**. His idea was called the *Theory of Use and Disuse*. Lamarck believed that an organism could get a new organ if it needed it. He also believed if an organ was used more it got bigger. Finally, he proposed that if a parent acquired a new trait during its lifetime, it could be passed on to the offspring.

Lamarck's ideas proved wrong. For example, just because a giraffe needs a longer neck, it will not get even longer. Just because a person does exercise to build up their muscles, doesn't mean their child will be born with bigger muscles. Lamarck should not be thought of badly because his ideas were proven wrong since the work of one scientist helps the thinking of others. Science grows as ideas are tested, evidence is gathered, and new ideas are formed.

78 The pierced ears on a person will not be passed on to their children. However, children often look like their parents. Why might this be so? _____

_____

_____

DISCIPLINARY CORE IDEAS:

LS4.C: ADAPTATION BY NATURAL SELECTION ACTING OVER GENERATIONS IS ONE IMPORTANT PROCESS BY WHICH SPECIES CHANGE OVER TIME IN RESPONSE TO CHANGES IN ENVIRONMENTAL CONDITIONS. TRAITS THAT SUPPORT SUCCESSFUL SURVIVAL AND REPRODUCTION BECOME MORE COMMON AND ARE PASSED ON TO OFFSPRING.

DARWIN

Nineteenth century English Scientist **Charles Darwin** (1809-1882) spent many years making observations about the reasons for the variety in living things upon Earth. His work illustrates the fact that sometimes it takes many years for an explanation of the natural world to be developed.

Darwin's *Theory of Natural Selection* is based on several ideas.

- Organisms produce more offspring than can survive.

- The number of organisms in a population tends to stay the same. This suggests that a struggle for survival occurs. The struggle is greatest between organisms that are similar and have the same needs.

- The individuals that survive are the best adapted – best fit – to their environment. (Adaptations that help an organism survive may be differences in structure, behaviors, or in how the organism's body works.) They are best fit or adapted because they have certain differences, or **variations**, that help them to survive better than others.

Charles Darwin

- Individuals that survive longer are more likely to reproduce. They may transfer these beneficial differences to their offspring.

- The beneficial variation may be passed on generation after generation.

79 According to Darwin's theory, what proves whether an organism is "fit?" _____

_____

_____

The diagram below represents the evolution of the modern horse.

Eohippus   Mesohippus   Merychippus   Pliohippus   Equus

80 *a*   How would Lamarck explain the changes in the horse over time? _____

_____

b   How would Darwin explain the changes in the horse over time? _____

_____

Sometimes ideas hold science back. The ancient Greek philosopher **Aristotle** (335 BC) said that humans were the most "perfect" living thing. This statement delayed the acceptance of evolution. Aristotle's idea was that "perfect" means closest to being like a human. Today, it might said that "perfect" means an organism perfectly fits its specific environment.

Let us take a closer look at how evolution occurs. In any population of organisms, there are differences or variations among the individuals. A change in the environmental conditions can affect which of these varieties best fits the new environment.

Survival of the organism may depend on a particular trait that it has. For example, in a large city, officials began using a rat poison in large quantities to kill off their large rat population. Rats compete with each other for food, so it would be an advantage to be able to eat the poisoned food and survive. Only some of the millions of rats in the city were able to withstand the effects of the poison. When these rats reproduced, their offspring that resisted the poison, also survived. After several generations, a large percent of the rat population in the city were the poison resisting variety, that could survive eating the poison. As you can see, a change in the environment can affect the survival of an individual organism with a particular trait.

81 *a*   What factor was responsible for the competition between rats? _____

b   How did the competition lead to a new type of rat? _____

_____

THE ANTIBIOTIC QUESTION

Antibiotics are a group of medicines given to help the body fight off infections. Scientists have found that bacteria are evolving varieties that are not affected by some of the more commonly used antibiotics.

82 Why is it suggested that doctors should not prescribe some types of antibiotics, even though they work

very well? _____

_____

VARIETY IS THE SPICE OF LIFE

What causes variety in a population of organisms? There are two main sources of variation within a species, sexual reproduction and **mutations**.

Whenever organisms reproduce sexually, the offspring receives genetic material from both parents. This new combination of genetic material creates a new unique individual. Therefore, no two individuals produced by sexual reproduction are completely (100%) alike, although "identical twins" are very close.

Mutations are changes in genes that are part of chromosomes. A change in a gene can change the protein being produced and therefore change the structure or function (trait) of an organism. Mutations are rare. Every once in a long while, a mutation might produce a variety that may be more fit to survive.

Small differences between parent and offspring can build up, generation after generation, until the new generations are very different from their ancestors.

83 If an environment is changing, why might organisms that reproduce sexually have an advantage

over those that reproduce asexually? _____

_____

_____

DESIGNER GENES

Human activities have affected the variations of species on Earth in many ways. **Genetic engineering** is the alteration of an organism's genetic material to eliminate undesirable traits or produce desirable traits. For thousands of years, a form of genetic engineering, **selective breeding**, was been practiced. Farmers bred selected types of corn in an attempt to get a better corn plant. For example, a farmer might breed a variety of corn with large kernels with a variety that produced a large number of kernels in order to develop a variety with both positive characteristics.

Through advances in the study of cells and genetics, genetic engineering has become a major area of research in regard to food production and medicines. Scientists can now transfer preferred genes from one organism to another to create a new variety with the preferred trait. For example, some crops have been given genetic material (a gene) to help them to produce chemicals which keep insects away.

84 What is genetic engineering? _____

_____

_____

85 How might genetic engineering be a benefit to humans? _____

_____

_____

VARIETY TO DIVERSITY TO EXTINCTION

Review why individual organisms and species change over time.

- Individuals in a population vary due to new combinations of genetic material caused by mutations or sexual reproduction.

- A change in the environment might cause an organism with a particular trait to survive better, to be better adapted, than others.

- If the **adaptation** is passed on to offspring, they also will be more likely to survive.

- A new population begins to form as more and more of the organisms with the favored trait survive, while the others decline.

- If a species (a type of organism) does not adapt enough to environmental changes, all of its kind may die off. This is called **extinction**. Extinction is common. The fossils found in sedimentary rock show many examples of organisms that became extinct.

86 What factors might lead to an organism becoming extinct? _____

_____

_____

Currently in the United States, many organisms are listed on the *Federal Endangered Species List*. This is a list of organisms that are in danger of becoming extinct.

87 What sort of things can people do to keep an organism from becoming extinct? _____

_____

_____

ADJUSTING TO THE ENVIRONMENT

Structure (form and shape) and function (life activities) are related in organisms.

Birds are a highly successful class of vertebrates. Much of their success is related to their adaptation of flight. However, the feet and beaks of birds can tell us much about how they have adapted to specific environments.

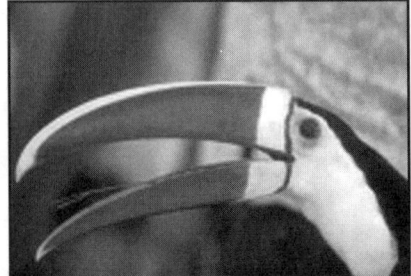

© PhotoDisc

Diagram 1 - Variations in bird beaks

Chart 1 - Variations In Beaks

Use in eating	Characteristics
seed	stout, cone shaped, thick at one end and pointed at the other
Cone-seed	tips of the beak cross to help pry out seeds
Flesh	hooked to tear meat
Fish	long spear shape or hooked
Fish	pouch or sac
Insect (in wood)	strong chisel shape
Insect (in air)	compressed, thick at base
Insect (on plant)	slender, pointed like a small pair of tweezers
Nectar	long tube-like
Straining	flat somewhat hooked
Shore organisms	long, slender

Diagram 2 - Variations in legs and feet

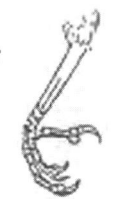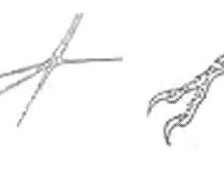

Chart 2 - Variations In Feet

Use	Characteristics
swimming or wading	toes usually lobed or webbed
holding prey	strong, widely separated, curved claws
perching	3 toes in front, 1 in back
climbing trees	2 toes in front, 2 toes in back or 2 toes in front, 1 toe in back
scratching	3 toes in front, 1 toe in back

Use your powers of observation and the information in Charts 1 and 2 and Diagrams 1 and 2 to make inferences about the lives of several birds. Try to determine the following information for the 3 birds shown.

88 Red-Winged Blackbird

Habitat _____

Function of beak _____

Function of feet _____

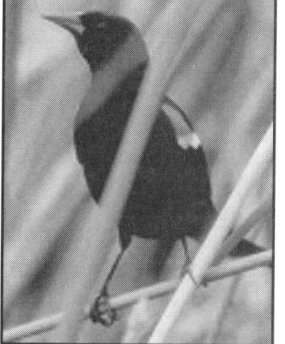

Red-Winged Blackbird

© PhotoDisc

89 Eagle

Habitat _____

Function of beak _____

Function of feet _____

Eagle

© Everything Alaska
www.everythingalaska.com

90 Heron

Habitat _____

Function of beak _____

Function of feet _____

Heron

DISCIPLINARY CORE IDEAS:

LS4.A: EVIDENCE OF COMMON ANCESTRY AND DIVERSITY

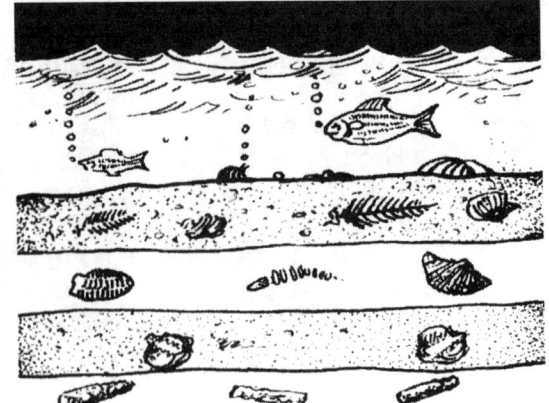

One major source of evidence supporting the theory of evolution comes from studying the layers of rock underground. In some places, sedimentary rock was formed by layers of material piling up on top of each other. As more and more layers are piled up, the tremendous pressure from the weight of the materials above cause the lower layers to be crushed together into rock. Therefore, lower layers represent the oldest layers of the rock.

Scientist may also use radioactive dating to determine the age of a fossil. Since the atoms of certain substances decay into other substances at a certain rate they can compare the amount of radiation remaining in or around a fossil to help determine its age.

Inside the many layers of sedimentary rock, there is evidence of the long history of life on Earth. As layers of materials were laid down, sometimes an organism's remains may have become part of the material. These remains of an organism that once lived, such as imprints from the organism, are called **fossils**.

As layers of sedimentary rock are examined, you can see through the fossil record that life forms have changed over time. The fossil record shows the existence, diversity, extinctions, and changes to many life forms throughout the history of life on Earth. By comparing how organisms living today and organisms in the fossil record compare in structure we can make inferences about evolutionary history and lines of descent. More recently deposited layers of sedimentary rock are likely to contain fossils resembling species that exist today.

As predicted by Darwin's Theory of Evolution, the slow changes in organisms - over a long period of time as environments change - can be seen in the fossil record. Fossils also support the fact that species may become extinct. Some fossils come from organisms that no longer exist today.

91 Based on the diagram at the right, which fossil comes from the organism that existed most recently?

_____

_____

92 Why would earthquakes and other movements of Earth's surface make it harder to understand the fossil record?

_____

_____

LIFE SCIENCE REVIEW 3

MULTIPLE CHOICE

1   Which would cause the most variation in a particular species?
    (1) asexual reproduction   (3) sexual reproduction
    (2) need   (4) budding

2   Competition among organisms is greatest when they are
    (1) larger   (2) different   (3) similar   (4) cooperative

3   Characteristics from one generation are passed on to the next generation based on the transfer of
    (1) acquired characteristics   (2) blood   (3) DNA   (4) need

4   Extinction may be caused by
    (1) adaptations to the environment   (3) lack of competition
    (2) changes in the environment   (4) mutations

Base your answer to question 5 on the diagram at the right of rock layers in the floor of the ocean.

5   Which layer would contain fossils that most resemble existing species?
    (1) A   (3) C
    (2) B   (4) D

6   If all the species originated on the primitive Earth at the same time and did not change, which pattern would be expected in the fossil record?
    (1) The lower rock layers would contain fossils of only the simpler organisms
    (2) The upper rock layer would contain fossils of only the simpler organisms
    (3) The upper rock layer would contain fossils of only the more complex organisms
    (4) Both the upper and lower rock layers would contain fossils of the more complex organisms

7   The blood of the rhesus monkey contains proteins similar to those found in the blood of humans. This observation suggests that rhesus monkeys and humans
    (1) belong to the same species   (3) have identical chromosomes
    (2) may have a common ancestor   (4) occupy the same habitat

8   Evolution refers to a change over a long period of time in a
    (1) fossil   (2) climate   (3) rock   (4) species

9   Which evolutionary concept is best illustrated by the cartoon at the right?
    (1) production of mutations   (3) survival of the fittest
    (2) use and disuse   (4) lock and key

10 The diagram at the right represents some stages in the development of the modern horse, according to evolutionary theory. The diagram is based on
(1) examination of fossilized structures of primitive horses
(2) biochemical analysis of growth hormones of primitive horses
(3) examination of structures in a modern horse before birth
(4) biochemical analysis of the DNA structure of a modern horse

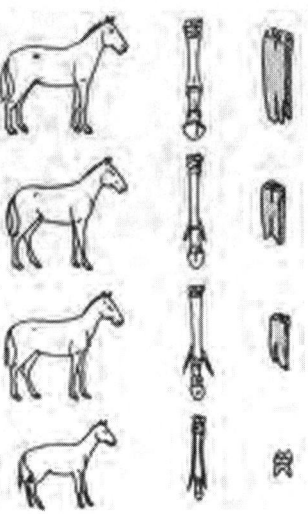

11 Darwin observed that finches belonging to different species on the Galapagos island have many similar physical characteristics. This supports the conclusion that finches
(1) have the ability to interbreed
(2) acquired traits through use and disuse
(3) occupy the same niche on the island
(4) came from a common ancestor

12 Lamarck's theory of evolution includes the concept that new organs in any species appear as a result of
(1) mutations   (2) needs   (3) environment   (4) natural variations

13 According to the modern theory of evolution, variations in a species can best be explained by
(1) mutations and sexual reproduction
(2) asexual reproduction
(3) vegetative reproduction
(4) use and disuse

14 One possible reason many plant and animal species have become extinct is that they could not
(1) reproduce sexually
(2) carry on aerobic respiration
(3) adapt to environmental changes
(4) pass on acquired traits

15 Which phrase best defines evolution?
(1) a steady state in a community
(2) a sudden replacement of one community by another
(3) a geographic or reproductive isolation of organisms
(4) a process of change in organisms over a period of time

16 Which statement would most likely be in agreement with Lamarck's theory of evolution?
(1) Black moths have evolved in an area because they were better adapted to the environment and had high survival rates.
(2) Geographic barriers may lead to isolation and production of a new species.
(3) Monkeys have long arms because their ancestors stretched their arms reaching for food.
(4) Most variations in animals and plants are due to mutations.

17 Organisms with favorable traits reproduce more successfully than organisms with less favorable traits. This statement best describes the concept of
(1) overproduction   (3) inheritance
(2) use and disuse   (4) survival of the fittest

18 How does natural selection operate to cause change in a population?
(1) The members of the population are equally able to survive any environmental change.
(2) The members of a population differ so that only some survive when the environment changes.
(3) The members of a population do not adapt to environmental changes.
(4) All the members of the population adapt to environmental changes.

19 Evolution is often represented as a tree similar to the one shown in the diagram at the right. This diagram suggests
 (1) Different groups of organisms may have similar characteristics because of common ancestry.
 (2) Because of biochemical differences, no two groups of organisms could have a common ancestor.
 (3) Evolution is a predictable event that happens every few years, adding new groups of organisms to the tree.
 (4) Only the best adapted organisms will survive from generation to generation.

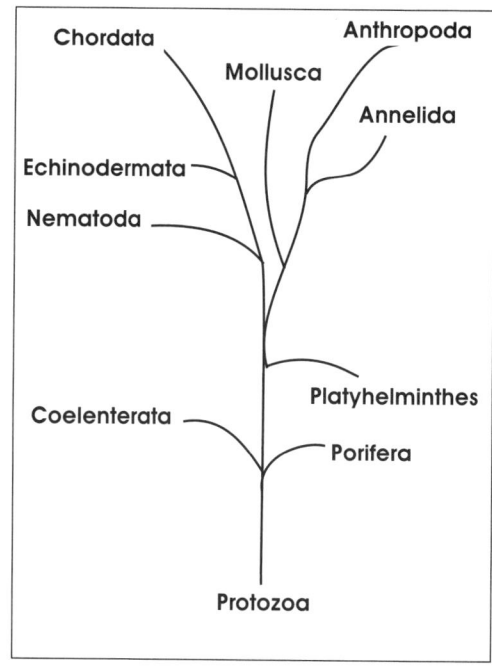

20 Breeders wishing to create a new breed of dog that is larger than its current form would
 (1) exercise the parents
 (2) over feed the parents
 (3) select parents that are large to breed
 (4) select one small and one large parent to breed

THE QUESTIONS THAT FOLLOW TEST YOUR UNDERSTANDING OF SCIENTIFIC PRACTICES, CORES IDEAS IN SCIENCE AND CROSS-CUTTING CONCEPTS SUCH AS PATTERNS AND CAUSE AND EFFECT RELATIONSHIPS.

SHORT ANSWER

1 Based on the graph at the right, how does the number of fires affect the natural selection of trees in this environment? (2 points)

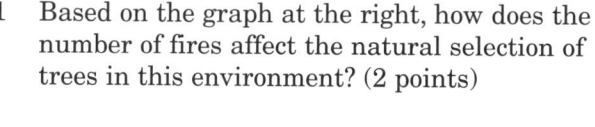

EFFECT OF NUMBER OF FIRES ON NATURAL SELECTION IN ENVIRONMENT

2 Breeders wish to create a new type of dog that can sniff out drugs better than any current breed. Describe the procedures that they might go through to create such a new breed. (2 points)

SHORT ESSAY

1 Use Darwin's Theory of Evolution to explain how a new type of human might develop, which can survive high ultraviolet radiation that will occur if the ozone layer is lost. (5 points)

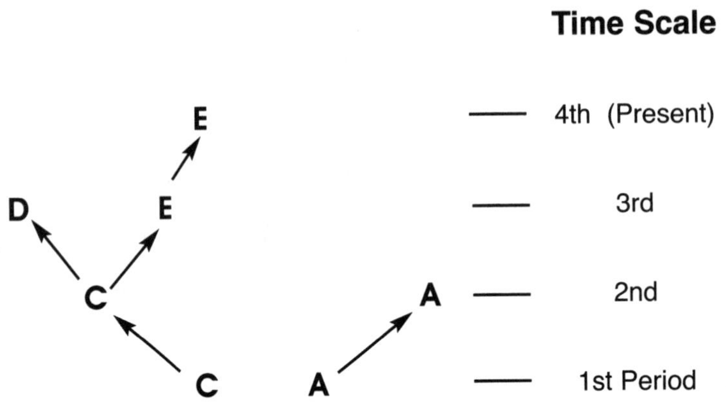

_____

_____

_____

_____

_____

Time Scale

2 Use the diagram above to interpret the evolutionary history of species *A* and *C*. Describe what happened to each species. Include your explanation for one occurrence. (5 points)

_____

_____

_____

_____

_____

_____

_____

GROWTH AND DEVELOPMENT

DISCIPLINARY CORE IDEAS:

LS1.B: GROWTH AND DEVELOPMENT OF ORGANISMS

PERFORMANCE INDICATOR 4.1
OBSERVE AND DESCRIBE THE VARIATIONS IN REPRODUCTIVE PATTERNS OF ORGANISMS, INCLUDING ASEXUAL AND SEXUAL REPRODUCTION.

Sarina was going through some old family photos and was shocked to find a picture of a baby that looked just like she did as a baby. She showed the photo to her mom who said it was a picture of her grandmother, Sarina's great grandmother. Mom said that maybe Sarina would grow up to look like her great grandmother. Sarina went back to find more pictures.

How do individual organisms create future generations? Without reproduction of new offspring, a species cannot survive and continue. All living things go through a life cycle involving the production of new individuals. All living things follow an orderly sequence of events as they develop from birth into the adult form.

PERFORMANCE INDICATOR 4.2
EXPLAIN THE ROLE OF SPERM AND EGG CELLS IN SEXUAL REPRODUCTION.

IT TAKES TWO - SOMETIMES
Reproduction is the process by which living things produce other living things like themselves. There are two major types of reproduction, sexual and asexual. Some organisms produce asexually, some sexually and some can reproduce both ways.

ASEXUAL REPRODUCTION
Asexual reproduction is the creation of a new individual without the joining of two cells. Types of asexual reproduction include:

- **Fission**: the splitting of a one-celled organism into two equal halves, with equal genetic and cell material, each becoming a new individual (example: fission of amoeba).

- **Budding**: the splitting of a one-celled organism into two halves, with equal genetic and unequal cell material, each becoming a new individual (example: budding in yeast).

- **Vegetative Propagation**: In some multicellular plants, a part of a plant can be separated and grown into another individual.

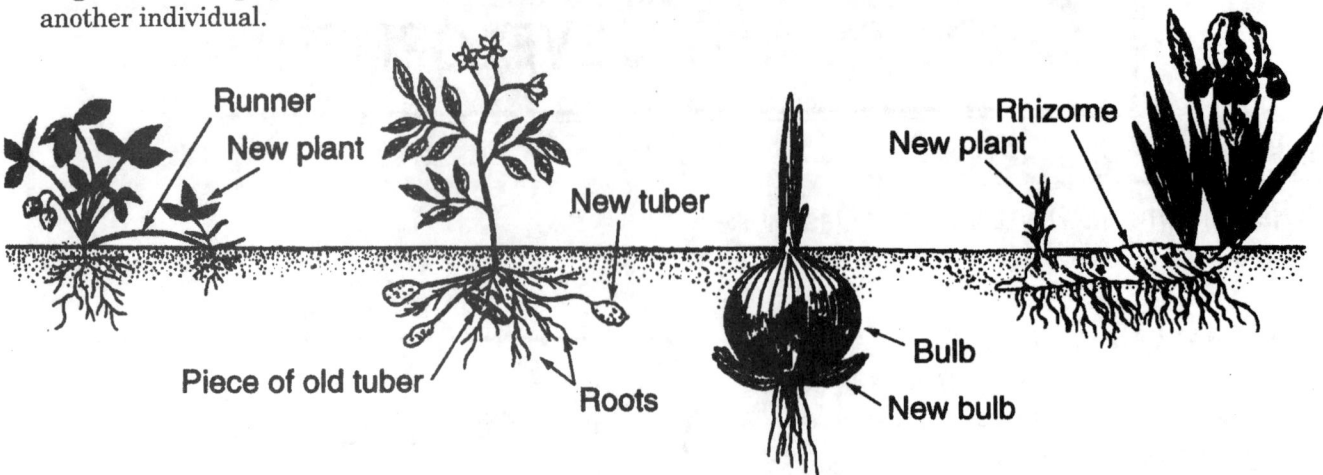

- **Cloning**: Cloning is a general term for asexual reproduction. Recent interest in this area comes from scientists using cloning to produce important gene products. Some genes or groups of genes can be taken from an organism and placed in another species. When the cells multiply, the new genes also are multiplied. For example, the gene for producing insulin has been placed in bacterial cells. These cells manufacture insulin for use by diabetics.

SEXUAL REPRODUCTION

Sexual reproduction involves the joining of two cells to begin the development of a new individual. In many species that reproduce by sexual reproduction, the sex cells are different. One is the sperm (from the male) and one is the egg (from the female).

The sex cells are very special cells. They are created through a special process of division so that they each end up with one-half of the genetic information needed for a new organism. For a new organism to be complete, the male cell – the **sperm**, and the female cell – the **egg**, must unite. The joining of the sperm and egg is called **fertilization**. A fertilized egg (**zygote**) contains a full set of genetic information, one half of the information from each parent. Fertilization can occur inside or outside of an organism's body, depending on the type of organism. In most fish, fertilization occurs externally, in the water around the fish. In reptiles, fertilization occurs internally, inside the female's body.

93 On the lines given, identify the type of reproduction shown in each of the following diagrams (sexual or asexual reproduction).

Reproduction: a _____

chromosome

fertilized egg

egg

Reproduction: b _____

sperm

Reproduction: c _____

Runner
New plant

Reproduction: d _____

Reproduction: e _____

DISCIPLINARY CORE IDEAS:

LS1.B: GROWTH AND DEVELOPMENT OF ORGANISMS

MULTIPLYING BY DIVIDING

Why does a fertilized frog egg cell grow into a frog? When formed by either asexual or sexual reproduction, how is a new organism given the directions, for what it is to do and become? The answers come from looking at how cells divide.

In **mitotic cell division**, each new daughter cell normally receives a complete and identical set of chromosomes. Chromosomes contain the genetic information that directs an organism's growth and development. During mitotic cell division, the chromosomes are duplicated by the cell. The duplicated pair separate. One identical set of chromosomes goes to each of the resulting daughter cells. Therefore, each new cell gets identical hereditary information.

In multicellular organisms these new cells allow for growth, maintenance, and repair of body cells. In some one-celled organisms, this type of division is a form of asexual reproduction.

In **meiotic cell division**, each new daughter cell normally receives one-half of a complete set of chromosomes. Cells with one-half of a set of chromosomes are called **sex cells.** A male sex cell is called a sperm. A female sex cell is called an egg. Both sperm and eggs contain one-half of the hereditary information needed by a new organism. Only when they unite, when the egg is fertilized, is the whole set of chromosomes united and a new organism formed.

94 Why is mitotic cell division so important for the growth and repair of a multicellular organism? _____

_____

_____

The diagram at the right represents the life cycle of a green alga.

> **Note:** the letter "n" in the diagram represents one (1) set of chromosomes. A complete set is represented by "2n."

95 Which letter represents the phase where cells are produced by meiotic division? _____

96 Which letter represents fertilization? _____

R.G. Harrison (1907) and Fell (1928) found that even after a type of cell is removed from a body the cells will continue to grow and divide retaining their traits. For example, a skin cell will produce more skin cells. Therefore, cells seem to have an internal regulating system. **Cancer** is a general name for diseases caused by unregulated cell growth. It is caused by a change in the genes that regulate cell growth.

97 Why would production of abnormal cells cause problems for a living thing? _____

_____

_____

FROM ONE TO MANY - CHANGE WILL DO YOU GOOD

If every cell were identical to every other cell, then a multicellular organism would just be a big pile of similar cells. A major benefit of being multicellular is that cells are different. The cells can specialize and become more advanced. In many multicellular organisms, there is a series of complex changes as the organism develops into the specific living thing that it is to become.

BUILDING A BETTER SCIENCE STUDENT

In humans, the fertilized egg, now called an **embryo**, undergoes many cell divisions until a small cluster of cells is formed. The cluster of cells forms a single layer of cells that becomes indented so that there are now two layers of cells. A third layer of cells develops between the first two layers. These three layers of cells begin to produce cells that are different from each of the other layers. They will become different tissues, organs, and organ systems in a human body. For example, the cells in the outer layer may become skin cells. The cells in the middle may form internal organs. The cells in the inner layer may become cells in the digestive system.

In humans, organs begin to form during the first three months of development. All organs and body features develop by six months. During the last three months of pregnancy, organs and features develop so that they can function properly after birth. In just nine months, the organism grows from one cell and an entire new science student is born.

Well, maybe it is not a *completely* developed science student. As humans grow from infant to young adult, their bodies change. Children are not just small adults. Thinking and behavior also change. Your body will change as you mature into an adult. Your reproductive organs will become functional. Boys' voices will deepen and facial hair will begin to appear. The shape of a girl's body will change. All this is just the natural process of developing into an adult.

Use the diagram which represents stages in vertebrate development to answer the following questions.

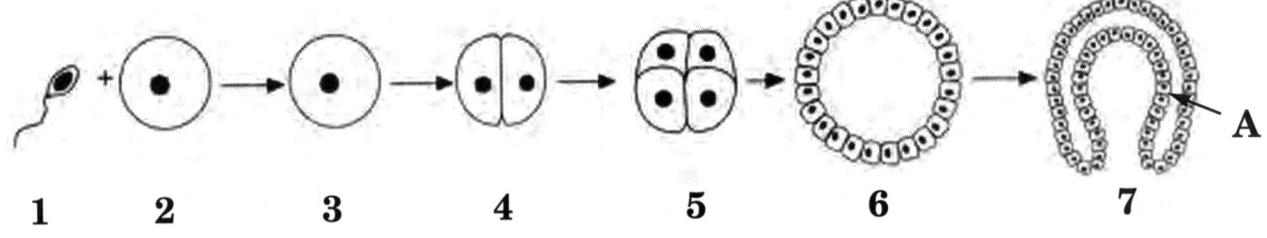

98 Which stage represents the first cell with a complete set of chromosomes? _____

99 Which stage represents a fertilized egg? _____

100 What reproductive process occurs between stage 1 and 2? _____

101 How many sets of genetic information does structure 1 carry? _____

102 What might be some uses for these cells developing at *A*? _____

_____

_____

_____

UNUSUAL BABY PICTURES PART ONE - THE ANIMALS

Not all animals develop in the same way. In some species, the young resemble the adult. In others they do not. For example, some insects have a three stage life history called **incomplete metamorphosis**. The three stages are egg, nymph, and adult. The nymph that hatches from the egg looks very much like the adult form. As it gets older, it will develop wings and reproductive structures and become an adult.

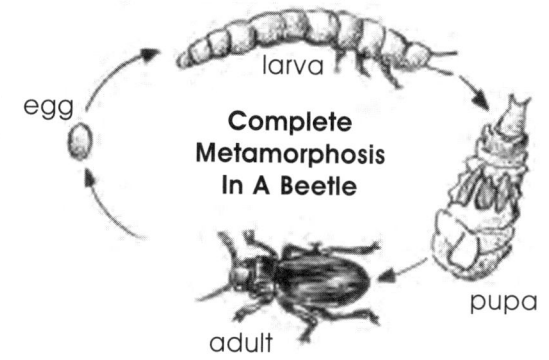

Other insects have a four stage life history called **complete metamorphosis**. The stages of this cycle are egg, larva, pupa, and adult. The larva that hatches from the egg does not look like the adult form. It is worm-like or a caterpillar-like depending on the type of insect. This stage is a time of eating and growing. The larva often does not eat the same type food as the adult. The larva forms a shell or case around itself and changes into the pupa stage. In the pupa, the insect changes into the adult from.

103 As compared to complete metamorphosis, what stages are missing from incomplete metamorphosis?

_____

104 As compared to incomplete metamorphosis, how might complete metamorphosis improve an organism's

chances for survival? _____

_____

105 What type of metamorphosis does the diagram at the right

represent? _____

106 What evidence did you have for your choice? _____

_____

_____

_____

METAMORPHOSIS

Some amphibians also undergo complete metamorphosis as they mature. The leopard frog lays eggs in the spring. The male frog spreads sperm over the eggs as they pass out of the females body. Therefore, the fertilization of the eggs takes place outside of the female's body. The eggs hatch into tiny tadpoles. Tadpoles seem more like fish than frogs. They are a fish-like stage with fins and gills.

Hind legs develop, then the front legs and the tadpole changes into its typical frog-like shape. This metamorphosis, change in form, takes place in about one month.

These two examples, insects and amphibians, demonstrate how different animals have different behaviors which help them increase their odds of reproducing.

Development of a Frog

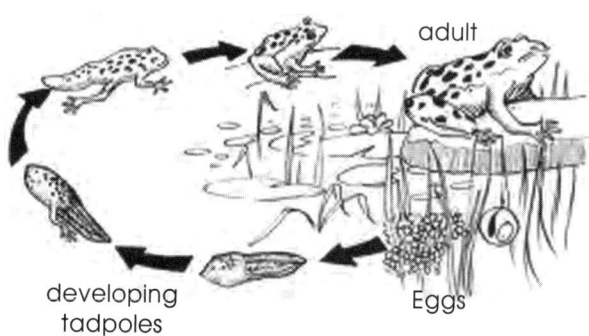

UNUSUAL BABY PICTURES PART TWO - THE PLANTS

Not all plants develop in the same way. Some plants, like ferns and mosses, develop after fertilization (**gametophyte**) into the part of the plant that produces spores (sporophyte). The spores develop a "plantlike" structure that does not "look like" a fern. From that structure come eggs and sperm that after fertilization, grow into the leafy plant, known as the fern.

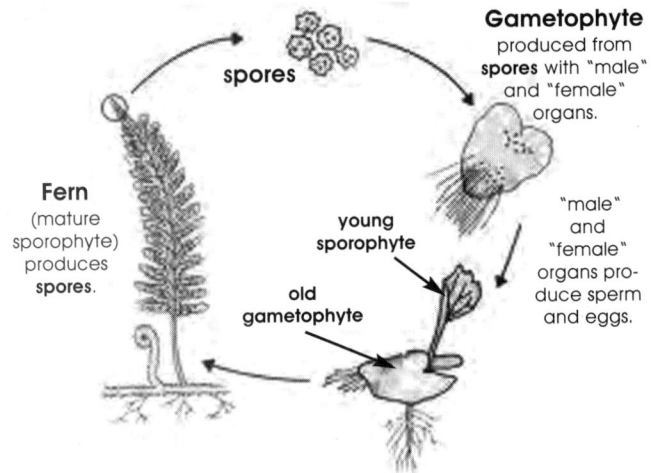

spores

Gametophyte
produced from
spores with "male"
and "female"
organs.

Fern
(mature
sporophyte)
produces
spores.

young
sporophyte

"male"
and
"female"
organs pro-
duce sperm
and eggs.

old
gametophyte

A new sporophyte grows from the fertilized
egg (zygote) from the old gametophyte.

© PhotoDisc

Some plants, called seed plants, develop after fertilization into a seed. A **seed** is a young plant surrounded by stored food and covered with a protective coat. Seed plants vary greatly in how they develop from the seed. Some seed plants like pine trees have seeds that are exposed. Other seed plants like apple trees have fruit around the young plant.

Some plants and animals have become very dependent upon each other for survival. Many plants need insects like bees and butterflies to carry their pollen to other plants to fertilize them so that seeds may form. The bees and butterflies depend on the plants for food like pollen and nectar.

107 What advantages do seeds have over spores?_____

_____

_____

It is important to remember that it is not only genetics that will affect the growth and development of an organism. Environmental conditions play a role as well. For example, a plant may have the genetics to grow well but without the correct amounts of water, sunlight, minerals, it may never reach its potential.

MULTIPLE CHOICE

1 The joining of cells that begins the development of a new organism is a type of reproduction called
   (1) budding   (2) fission   (3) sexual   (4) vegetative

2 The merging of a sperm cell and an egg cell is called
   (1) asexual reproduction   (2) budding   (3) development   (4) fertilization

3 How much of a new individual's genetic information is carried in a sex cell?
   (1) 1/4   (2) 1/2   (3) all   (4) twice what is needed

4 Cells identical to the parent cell are created by the process of
   (1) fertilization   (2) meiosis   (3) mitosis   (4) mutation

5 Which represents the correct sequence of development for complete metamorphosis?
   (1) egg, larva, pupa, adult   (3) egg, pupa, larva, adult
   (2) egg, nymph, adult   (4) adult, egg, pupa, larva

6 A colony of organisms is found, all containing identical genes. This group of organisms most likely reproduces
   (1) asexually   (2) sexually   (3) through fertilization   (4) slowly

7 The type of reproduction that happens when an organism reproduces by splitting in two equal halves is called
   (1) budding   (2) grafting   (3) fission   (4) sexual

8 Fertilization of various organisms may occur
   (1) only internally   (3) internally or externally
   (2) only externally   (4) with only one parent cell

9 The human female sex cell is called a (an)
   (1) egg   (2) ovary   (3) sperm   (4) embryo

10 The genetic information found in any human body cell will
   (1) be identical   (3) differ according to cell function
   (2) match its parents   (4) differ according to cell location

11 In multicellular organisms, growth and maintenance are accomplished through
   (1) fertilization   (2) meiosis   (3) mitosis   (4) the production of sex cells

12 Seed bearing plants have an advantage over other types of plants because seeds contain
   (1) extra water   (2) many plants   (3) stored food   (4) a new plant

13 The diagram at the right represents a cell process. If the cell in diagram 1 contains 4 chromosomes, what is the number of chromosomes in each cell in diagram 3?
   (1) 8   (3) 16
   (2) 2   (4) 4

Diagram 1   Diagram 2   Diagram 3

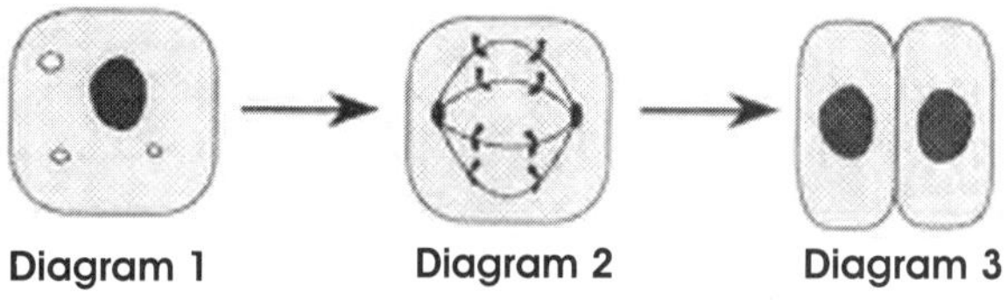

14 In many aquatic vertebrates, reproduction involves external fertilization. What is a characteristic of this type of fertilization?
(1) Sex cells join outside of the body of the female.
(2) Sex cells join in the reproductive tract of the female
(3) Offspring produced have twice as many chromosomes as each of the parents
(4) Offspring produced have only half the number of chromosomes as each of the parents.

15 Which statement is true regarding plants produced by vegetative propagation?
(1) They normally exhibit only recessive characteristics.
(2) They normally have only one half the number of chromosomes.
(3) They normally obtain most of their nourishment from the seed.
(4) They are normally genetically identical to the parent.

16 Which event would most probably result in twins that did not have the same genetic information?
(1) One egg is fertilized by two sperm cells.   (3) Two eggs are each fertilized by separate sperm cells.
(2) Two eggs are fertilized by one sperm cell.   (4) Two eggs developed without fertilization.

17 The accompanying diagrams represent two different cells undergoing mitotic cell division. Which statement about these divisions are true?

Diagram A   Diagram B

(1) Both divisions could occur in a human.
(2) Division A could occur in a grasshopper and Division B could occur in a turtle.
(3) Division A could occur in a bean plant and Division B could occur in a maple tree.
(4) Division A could occur in a grasshopper and Division B could occur in a maple tree.

18 Cancer is a disease characterized by the
(1) uncontrolled division of abnormal cells
(2) unlimited production of abnormal sex cells
(3) uncontrolled replication and lining up of chromosomes
(4) limited production of normal sex cells

19 The process of budding is characterized by
(1) the formation of a cell wall   (3) a decrease in the chromosome number
(2) an unequal division of cytoplasm   (4) a large number of nuclei

20 Each body cell of a chimpanzee contains 48 chromosomes. How many chromosomes would normally be present in a sex cell produced by this chimpanzee?
(1) 24   (2) 36   (3) 48   (4) 96

THE QUESTIONS THAT FOLLOW TEST YOUR UNDERSTANDING OF SCIENTIFIC PRACTICES, CORES IDEAS IN SCIENCE AND CROSS-CUTTING CONCEPTS SUCH AS PATTERNS AND CAUSE AND EFFECT RELATIONSHIPS.

SHORT ANSWER

1 Looking at a little baby, a person remarked, "It has its mother's nose and its father's ears." Using your knowledge of reproduction and development explain how this could be so. (2 points)

_____

_____

_____

_____

2   Which diagram at the right represents correct meiotic division?

_____ Explain why you chose that letter. (2 points)

_____

_____

Diagram A   **Diagram B**

SHORT ESSAY

1   Organism *A* is one-celled and has 2 pairs (4 total) chromosomes.

a   Describe and diagram the process of cell division if Organism *A* were to
    reproduce asexually. (2 points)

Organism A

Asexual Reproduction Diagrams:

b   Describe and diagram the process of cell division if Organism *A* were to reproduce sexually. (3 points)

Sexual Reproduction Diagrams:

ECOSYSTEM: INTERACTIONS, ENERGY AND DYNAMICS

DISCIPLINARY CORE IDEAS:

LS3.D: ENERGY IN CHEMICAL PROCESSES AND EVERYDAY LIFE

COMPARE THE WAY A VARIETY OF LIVING SPECIMENS CARRY OUT BASIC LIFE FUNCTIONS AND MAINTAIN DYNAMIC EQUILIBRIUM.

A BALANCING ACT

It was mid-afternoon and school was almost over. Caity felt a grumbling in her stomach. She leaned over towards Austin and whispered, "I'm starving! I can't wait to go home and get something to eat." Austin nodded. He had not been hungry himself, but now that Caity had mentioned food, his stomach began grumbling too.

Organisms respond to things both inside and outside of their body. The world presents living things with a constantly changing environment. To survive, they must be able to respond and adjust to it.

© PhotoDisc

YOU ARE HUNGRY AGAIN

One need that is constantly changing is the need for energy. Depending on an organism's activities, its need for energy might vary greatly. Think about how much energy you would need if you had to run a mile. Even when you sit perfectly still or when you are asleep, your cells need energy constantly just to remain alive. Cells of living things are constantly making or converting energy for their own use. Energy may come from combining oxygen with food, in the form of glucose, in the cell's mitochondria. Water and carbon dioxide are waste products created by this process. The process can be written as a formula:

$$\text{glucose } + \text{ oxygen } \longrightarrow \text{ water } + \text{ carbon dioxide } + \text{ energy}$$

Certain organisms and certain animal cells can make energy without oxygen. In this case, food (glucose) is broken down into lactic acid. This way of getting energy does not produce as much energy as when oxygen is used.

108 Think back on what you have learned thus far. How could you determine that you were using energy

even when you sleep? _____

_____

109 Why is oxygen so important to most living things? _____

_____

SHOPPING AT THE "SUN SUPERMARKET"

How do organisms obtain the nutrients they need for energy? Organisms have adapted many different ways to get their energy. Scientists have divided organisms into groups based on how they get their energy.

PRODUCERS

Producers make their own food. Green plants are an example of producers. Plants, algae (including phytoplankton) and many microorganisms are examples of producers. They use **chlorophyll** to trap the Sun's energy and convert into chemical energy in the form of sugar. The sugar may be used immediately or stored and used for energy or growth later on. This process is called photosynthesis. The formula for photosynthesis can be written as:

word formula:

$$\text{Carbon dioxide} + \text{Water} + \text{light (used for energy)} \xrightarrow{\text{chlorophyll}} \text{Glucose} + \text{Oxygen} + \text{water}$$

chemical formula:

$$CO_2 + H_2O + \text{energy} \xrightarrow{\text{chlorophyll}} C_6H_{12}O_6 + O_2 + H_2O$$

CONSUMERS

Consumers are organisms that cannot make their own food. All animals are consumers. Animals get their food by eating food that has already been made by other living things. Consumers can be divided into several groups based on where they get their food.

Herbivores are animals that get energy by using the energy stored in the plants that they eat. An example of a herbivore is a deer.

Carnivores are animals that get their energy by using the energy stored in the animals that they eat. A lion is both a carnivore (meat eater) and predator.

Omnivores are animals that get their energy by using the energy stored in the plants and animals that they eat. An example of an omnivore is a human.

Decomposers are organisms that get their energy by using the energy stored in the wastes or dead organisms that they eat. An example of a decomposer is a mushroom.

110 On the line next to the "method" of obtaining energy, put the letter (*a - e*) from the "consuming" column which best matches the method.

Method (Example)	Energy from consuming
___ carnivore (tiger)	*a* dead organisms
___ decomposer (mushroom)	*b* animals
___ herbivore (deer)	*c* makes own food
___ omnivore (bear)	*d* plants
___ producer (corn)	*e* plants and animals

111 Why are producers considered the starting point for all energy used by living things? _____

_____

_____

112 Make a list of 12 animals that live in your area. Classify them as herbivores, carnivores, or ominivores.

Herbivores _____ _____ _____ _____

Carnivores _____ _____ _____ _____

Omnivores _____ _____ _____ _____

Look at your four herbivores. Create smaller, more specific categories. Develop three additional categories and identify a common characteristic. Group your herbivores into the new categories.

Herbivores
characteristic _____ _____ _____

animal _____ _____ _____

animal _____ _____ _____

WAIT ONE HOUR BEFORE YOU SWIM

Parents say, "Don't go swimming for one hour after you eat or you will get cramps." Well, it is sort of true. After you eat, your body sends lots of blood and energy to the digestive system so it can digest your food. It does not have a lot of blood and energy to send to the muscles of your arms and legs if you start vigorous swimming. If you were swimming a long distance after eating, a signal that you did not have enough energy for both would be cramps in your muscles.

Your body regulates itself by sensing what is happening inside your body. It then changes how your body is working so that it can continue to work properly. The regulation of your body is controlled by both your nerves and hormones. For example, your body produces a hormone called insulin. Insulin regulates how much sugar is in the blood. Normally, insulin will help by storing extra sugar for you or releasing extra sugar if you need more energy. However, if your body produces too much insulin, blood sugar that should go to your cells goes to your liver for storage. If this happens, you will feel tired and lack energy. If your body produces too little insulin, the extra sugar in your blood stays there and cannot be used by cells for energy.

I CAME, I SAW, I RESPONDED

Organisms respond to stimuli in their environment. You respond to little messages, such as hearing a twig creak behind you and to big messages, such as turning and seeing a car coming toward you. The ability to sense and respond to messages is important to the survival of an organism.

113 Describe all the things your body needs to do in order to sense and respond to crossing a busy

intersection. _____

_____

DISCIPLINARY CORE IDEAS:

LS1.C: ORGANIZATION FOR MATTER AND ENERGY FLOW IN ORGANISMS

DESCRIBE THE IMPORTANCE OF MAJOR NUTRIENTS, VITAMINS, AND MINERALS IN MAINTAINING HEALTH AND PROMOTING GROWTH, AND EXPLAIN THE NEED FOR A CONSTANT INPUT OF ENERGY FOR LIVING ORGANISMS.

YOU ARE WHAT YOU EAT

Why do you eat? A young child may say that it eats because it is hungry. Why is it hungry? Its body needs a great deal of different materials to help it to grow, repair, and just continue working. These materials are called

food. Food provides the molecules that serve as the fuel and building materials for all organisms. For example, if you eat a hamburger, some of that hamburger may be used to provide you with energy to run around the next day, and some of the hamburger may actually be changed into the cells that make up your muscle tissue.

A NUTRITIOUS MENU

Each second of every day, your body carries out a tremendous number of chemical reactions that are the basis for you being alive. The sum of all the chemical reactions in your body is called **metabolism**. For you to provide your body with all the raw materials you need for metabolism, you must eat. Food may contain many different substances, each of which may have a very important use for a living thing. The following chart is a list of some important groups of substances needed by living thing.

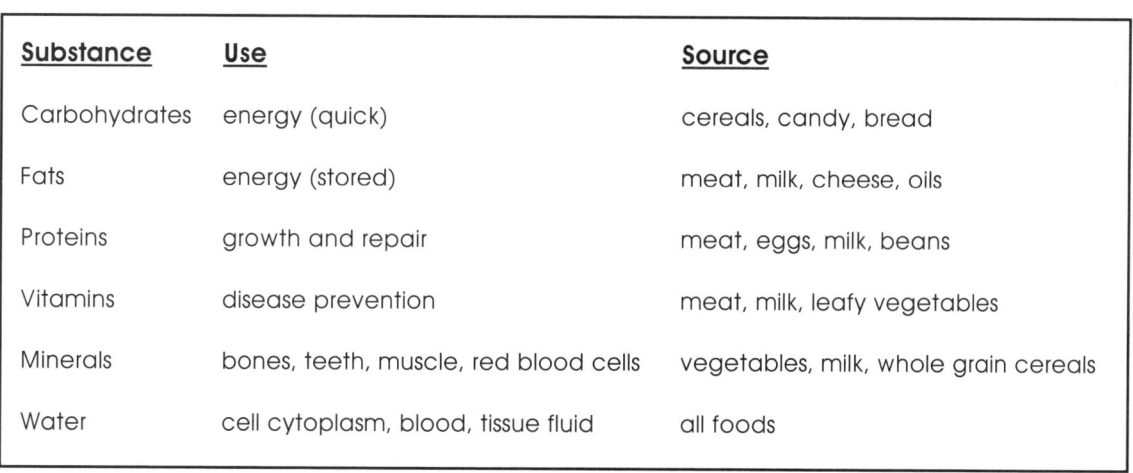

Substance	Use	Source
Carbohydrates	energy (quick)	cereals, candy, bread
Fats	energy (stored)	meat, milk, cheese, oils
Proteins	growth and repair	meat, eggs, milk, beans
Vitamins	disease prevention	meat, milk, leafy vegetables
Minerals	bones, teeth, muscle, red blood cells	vegetables, milk, whole grain cereals
Water	cell cytoplasm, blood, tissue fluid	all foods

To maintain a healthy state, all living things must have a minimal amount of each nutrient. The amount of each nutrient an organism needs will vary based on many factors such as type of organism, size, age, sex, activity level and how well the organism's body is working. If an organism does not get the correct amount of any one of the nutrients, it may gain weight it does not need, lose weight it needs, or even become diseased.

Many people are concerned that young people are eating too much "junk" food. Junk foods often contain large amounts of carbohydrates and fats.

114 Why would a diet containing large amounts of carbohydrates and fats not be beneficial to a young person?

_____

_____

115 Grapefruit diet? Grapefruits are good for you, but why would a diet of grapefruit alone be dangerous?

_____

_____

BRITISH LIMEYS

In the 1600s and 1700s, sailors often got a disease called **scurvy**. Scurvy caused the body to heal poorly. Capillary walls were weak so sailors bruised easily. Sores appeared in their mouth and on their gums and a sailor might lose a lot of blood. Portuguese navigator Vasco De Gama once lost almost 100 of 170 men to scurvy.

In 1753, **James Lind**, a Scottish doctor, showed that eating lemons and oranges would prevent the disease. Citrus fruits contain vitamin C which helps prevent scurvy. In 1771, **James Cook**, a British commander, became the first to prevent an outbreak of scurvy by giving his sailors fruit. In 1795, the British Navy followed Dr. Lind's advice and began issuing a daily ration of fruit or juice to all its men. The British sailors got the nickname, "Limeys" from this practice.

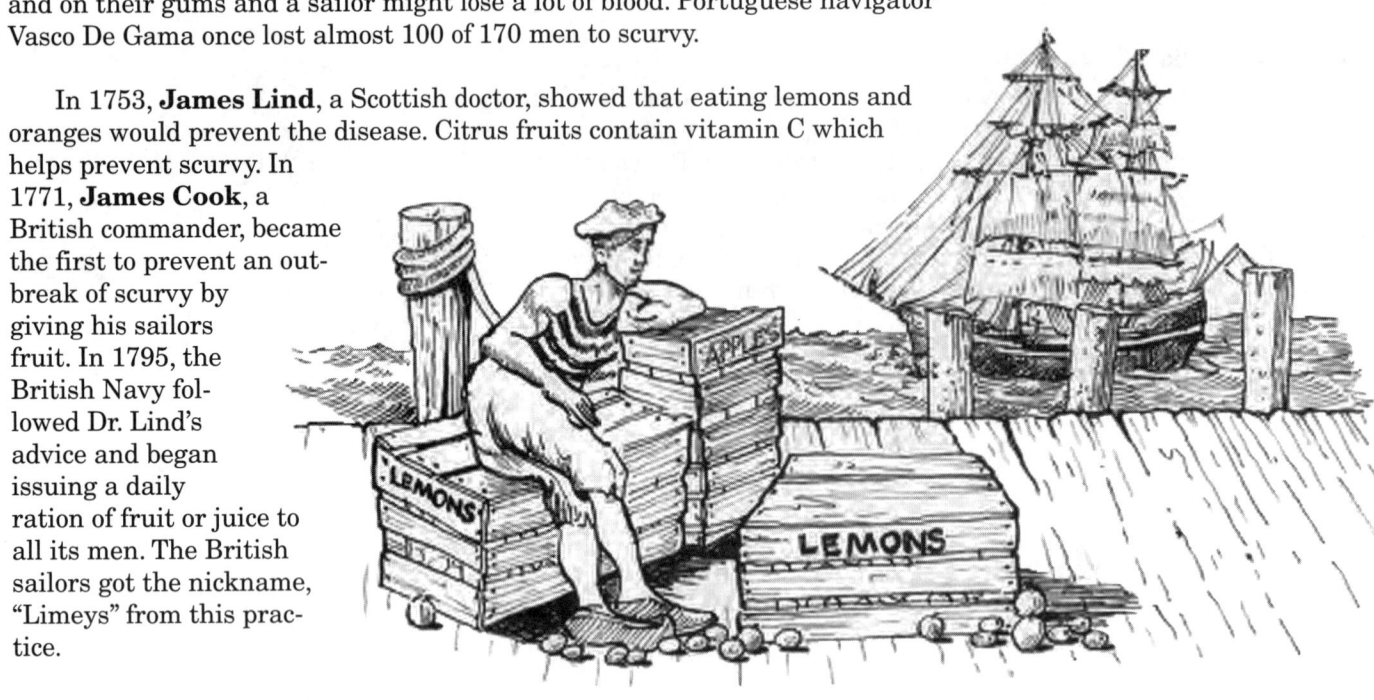

FOOD AS FUEL

Metabolism is influenced by several things. Your body's hormones regulate the rate of metabolism. Metabolism is also influenced by how much you exercise and what you eat.

Each type of food you eat provides different things for your body. For example, each type of food provides a different amount of energy for you to use. The amount of energy in foods is measured in **Calories***. A Calorie is a measurement of energy. It is the amount of energy needed to raise the temperature of 1000 grams of water one degree Celsius. A daily requirement of about 2,500 to 3,500 Calories is about average for an active teenager. The number of Calories an individual actually needs depends on many factors such as body weight, age, sex, activity level, and how efficiently the food is used.

Carbohydrates and **proteins** contain 4 Calories per gram. **Fats** contain 9 Calories per gram. Following is a list of some foods and the Calories they contain.

Food	Calories
Apple, large	.125
Hamburger sandwich	.245
Ice Cream, one half cup	.135
Apple Pie, large slice	.405
Steak, sirloin 3 oz.	.240
Strawberry Shortcake	.417

From Nutritive Value of Foods, US Dept. of Agriculture, 1985.

*There is some confusion over the term calorie. Calorie written with a small c, **calorie**, represents the amount of energy to raise 1 gram of water 1 degree Celsius. In most cases, when speaking of food, the little c is used, but it means 1000 c. It is distinguished by calling it a "food" calorie. Written with a capital C it means the energy to raise 1000 grams of water 1 degree Celsius. In this text a calorie is a calorie (1 gram, raised 1 degree Celsius) and a Calorie is a Calorie (1000 grams, raised 1 degree Celsius).

116 Carbohydrates account for almost one half of peoples' diet, yet your body has very little carbohydrate in it.

What happens to all the carbohydrates you consume? _____

_____

_____

117 Why would reducing fats in one's diet quickly reduce Calorie intake? _____

_____

_____

118 Why are desserts a problem for those trying to reduce Calorie intake? _____

_____

_____

KEEPING THE BALANCE

Many substances enter your body, many are used in various ways, and many leave your body. Your body works to maintain a balance. It keeps and uses the materials it needs and removes wastes and materials it does not use. Too much or too little of a substance may cause problems for your body. For example, even though vitamins are important for good health, they are only needed in very low quantities. Too much of certain types of vitamins can cause your body harm.

There are several factors that might upset the balance in your body. Some infectious diseases can affect the balance. What you eat is important too. You should try to develop good dietary habits that provide your body with a good balance of the raw materials it needs. Use of some substances is harmful to your body. Tobacco, alcohol, and drugs can seriously harm your body. Sometimes the effects of these substances shows up quickly, while in other cases the substance slowly destroys your body and the effect is not seen for many years. For example, smoking slowly destroys the tissues in your lungs. Smoking causes cancer, but even if a smoker does not get cancer, the smoking causes abnormal cells to form and damages the trachea, bronchi, and air sacs. Beginning with the very first cigarette you have, you upset the balance in your body.

119 List some ways you can keep your body in balance.

_____

_____

STARTING A NEW LIFE OFF RIGHT

The use of drugs, tobacco, and alcohol are dangerous during pregnancy. They are dangerous because of the damage they may cause the unborn fetus. The new human being formed starts from just one cell. All its future cells will be built from this first cell. As it grows, a small number of cells will be responsible for the development of the whole organism. Damage to these early cells may have large effects on the health of the forming infant. This is why expectant mothers must be extremely careful about what they put into their bodies during pregnancy. A healthy diet will provide the building blocks for a healthy child.

120 How can expectant mothers help their children to get a good start on life, even before birth? _____

_____

_____

LIFE SCIENCE REVIEW 5

MULTIPLE CHOICE

1 An organism that can convert simple substances into energy rich substances is a
   (1) consumer   (2) decomposer   (3) predator   (4) producer

2 Scientists checked the droppings (feces) of a new organism found in Africa and found it contained only plant material. This organism would be classified as a (an)
   (1) carnivore   (2) decomposer   (3) herbivore   (4) omnivore

3 Which nutrient would a person need the most when their body is growing most?
   (1) carbohydrate   (2) fat   (3) protein   (4) water

4 To get quick energy, a person might eat a candy bar because it contains
   (1) carbohydrate   (2) fat   (3) protein   (4) water

5 Energy can be produced when foods react with
   (1) carbon dioxide   (2) nitrogen   (3) oxygen 4 water

6 An organism that cannot make its own food is classified as a (an)
   (1) autotroph   (2) consumer   (3) green plant   (4) producer

7 An animal that eats mostly meat is called a (an)
   (1) carnivore   (2) herbivore   (3) omnivore   (4) producer

8 The tissues of a dead organism were broken down and most of the materials returned to the soil. The process was most likely the result of the work of a (an)
   (1) carnivore   (2) decomposer   (3) herbivore   (4) producer

9 The two systems that most directly control the regulation of the human body are the
   (1) digestive and respiratory   (3) muscular and skeletal
   (2) circulatory and reproductive   (4) nervous and endocrine

10 Many animals have their sense organs located on their heads because
   (1) they are useless on other body parts   (3) they need them there
   (2) it allows them to sense the new areas they enter   (4) it confuses predators

11 Fuel and building materials for animals come from
   (1) air   (2) food   (3) soil   (4) water

12 Energy in food is measured in
   (1) Calories   (2) grams   (3) joules   (4) watts

13 The number of Calories a person needs
   (1) is constant   (3) is not affected by activity
   (2) is less than 100 per day   (4) varies with age

14 Consumers are characterized by their
   (1) ability to convert light energy into chemical energy
   (2) dependence on the chloroplasts in their cells
   (3) ability to convert simple substances into complex ones
   (4) dependence on obtaining food from their environment

15 The effects of infectious disease
   (1) always appear immediately   (3) appear immediately or after a delay
   (2) always appear after a long delay (4) never show up

16 The process by which simple substances are converted into energy rich food is called
   (1) hydrolysis   (2) digestion   (3) photosynthesis   (4) respiration

17 Hormones and nerves cause responses to the environment that may differ since nervous responses are usually
   (1) for a longer duration   (2) quicker   (3) slower   (4) harmful

18 The sum of all the chemical reactions in your body is called
   (1) photosynthesis   (2) metabolism   (3) respiration   (4) synthesis

19 A nutrient needed in small quantities that plays a major role in disease prevention is
   (1) carbohydrates   (2) fats   (3) proteins   (4) vitamins

20 If a person were to take in many more Calories than they need they would
   (1) lose weight   (2) gain weight   (3) remain healthy   (4) have more energy

THE QUESTIONS THAT FOLLOW TEST YOUR UNDERSTANDING OF SCIENTIFIC PRACTICES, CORES IDEAS IN SCIENCE AND CROSS-CUTTING CONCEPTS SUCH AS PATTERNS AND CAUSE AND EFFECT RELATIONSHIPS.

SHORT ANSWER

1 Humans, fish, and plants all need oxygen to survive. Each carries on respiration with different organs and in different ways.

   a Describe a way in which respiration in humans, fish, and plants is alike. (1 point)

   _____

   _____

   b Describe a way in which respiration in humans, fish, and plants is different. (1 point)

   _____

   _____

SHORT ESSAY

1 In general, a 20-foot tall apple tree needs more water than a 5-foot tall apple tree. Beginning with your hypothesis, set up an experiment using the "size" variable to illustrate a plant's nutritional needs. Hint: In addition to water, you might take into consideration sunlight, carbon dioxide, soil type, and/or fertilizer. Remember, use only one variable. (5 points)

   _____

   _____

   _____

   _____

   _____

   _____

LIVING THINGS DEPEND ON ONE ANOTHER AND THEIR PHYSICAL ENVIRONMENT

DISCIPLINARY CORE IDEAS:

LS2.A: INTERDEPENDENT RELATIONSHIPS IN ECOSYSTEMS

You wake up slowly in the morning, stretching and yawning. You feel the warmth of your bed and do not want to get up. Your stomach growls, so you sit up and peek out the window. It looks cold, so you find a heavy shirt. Before you leave for school, you grab a glass of milk or juice and a donut.

Within the first 30 minutes of your morning, you have shown that you are totally dependent on other plants, animals, and your physical environment to survive!

ENERGY - THE ABILITY TO DO THINGS

Living things require energy to carry out the activities of life (such as breathing and walking). You are dependent on those things that provide your energy. Energy for your body comes from your food. The things you eat are used as fuel and to build and repair your body.

FOOD FOR A DAY !

List all the food you ate yesterday.

Breakfast	Lunch	Dinner	Snacks
_____	_____	_____	_____
_____	_____	_____	_____
_____	_____	_____	_____
_____	_____	_____	_____
_____	_____	_____	_____
_____	_____	_____	_____
_____	_____	_____	_____

The foods made from grain and sugars, such as cereal and bread, gave you energy. The meats and beans gave you protein to build and repair your body. Fatty foods, like oils, gave you energy to store.

If you think about it, most energy used by living things originates from the Sun. Energy from the Sun helps plants grow and humans eat plants or animals that eat plants for energy.

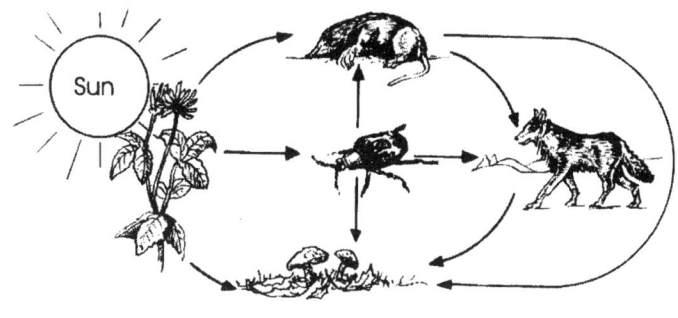

121 How does the mole depend on energy from

the Sun if it eats plants? _____

_____

Disciplinary Core Ideas:

LS1.C: Organization for Matter and Energy Flow in Organisms

Provide evidence that green plants make food and explain the significance of this process to other organisms.

Photosynthesis = Manufacturing With Light

The energy from the Sun is converted into chemical energy by chlorophyll-containing cells during the process of **photosynthesis**. This chemical energy starts in the form of sugars which the organism can use for energy or to grow. Humans obtain this chemical energy by eating the organism.

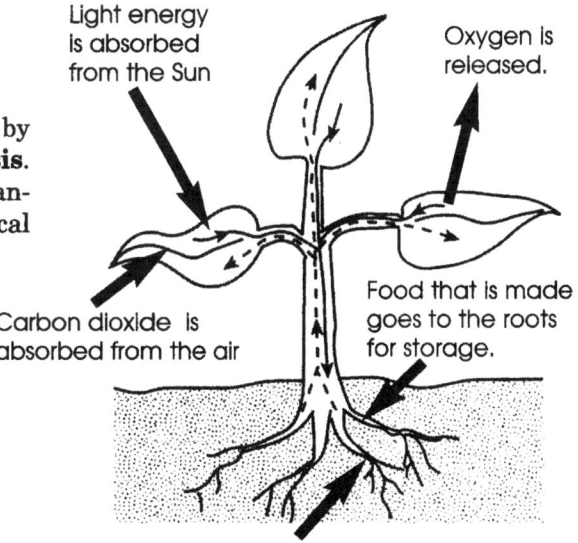

Light energy is absorbed from the Sun

Oxygen is released.

Carbon dioxide is absorbed from the air

Food that is made goes to the roots for storage.

Water is absorbed from the soil.

Using the energy from sunlight, plants make food from carbon dioxide and water. Oxygen is released as a by-product (waste).

$$\text{Carbon dioxide} + \text{water} + \text{energy from sunlight} \xrightarrow{\text{Chlorophyll}} \text{simple sugars} + \text{oxygen}$$
$$\text{(glucose)}$$

Photosynthesis
- **is the major source of atmospheric oxygen**
- **removes carbon dioxide from the atmosphere**
- **is a major source of chemical energy for use by organisms**

If you look carefully at the photosynthesis equation above, you realize how important it is to your life. Plants take two very simple and plentiful chemicals, water and carbon dioxide, and produce food for you to eat and oxygen for you to breathe. **Algae** are organisms that also contain chlorophyll. The Algae in the oceans produce most of the oxygen for Earth. Without the food for fuel and the oxygen to help burn the fuel, you would not have the energy to be reading this page.

122 Besides water and carbon dioxide, what do green plants need for photosynthesis? _____

_____

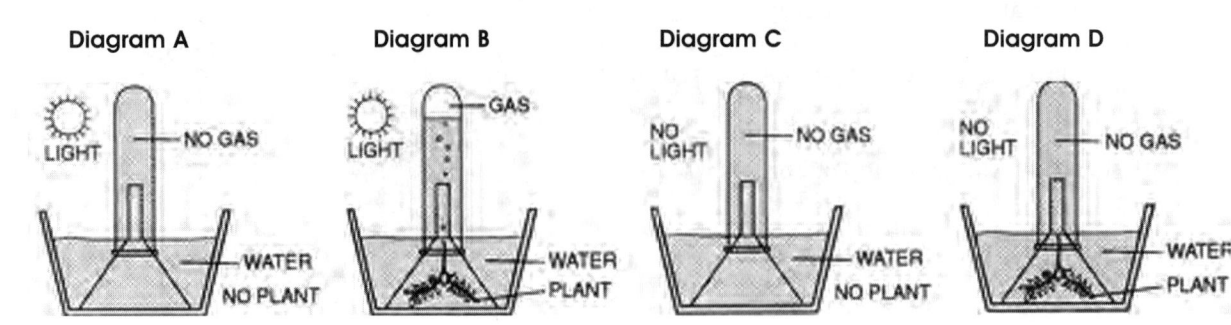

| Diagram A | Diagram B | Diagram C | Diagram D |

123 In the diagram above, what gas is being produced in *B*? _____

RECYCLING – MORE THAN JUST PAPER!

CARBON DIOXIDE / OXYGEN CYCLE
One reason water and carbon dioxide are so plentiful is that they are recycled. Carbon dioxide is used by plants, and they produce oxygen. Animals use the oxygen and produce carbon dioxide. This is known as the **carbon dioxide / oxygen cycle**.

WATER CYCLE
Water from bodies of water evaporates into the sky. Water is also released into the air by plants. Water in the air condenses and returns to Earth as rain, sleet, or snow. This water can then be used by plants again.

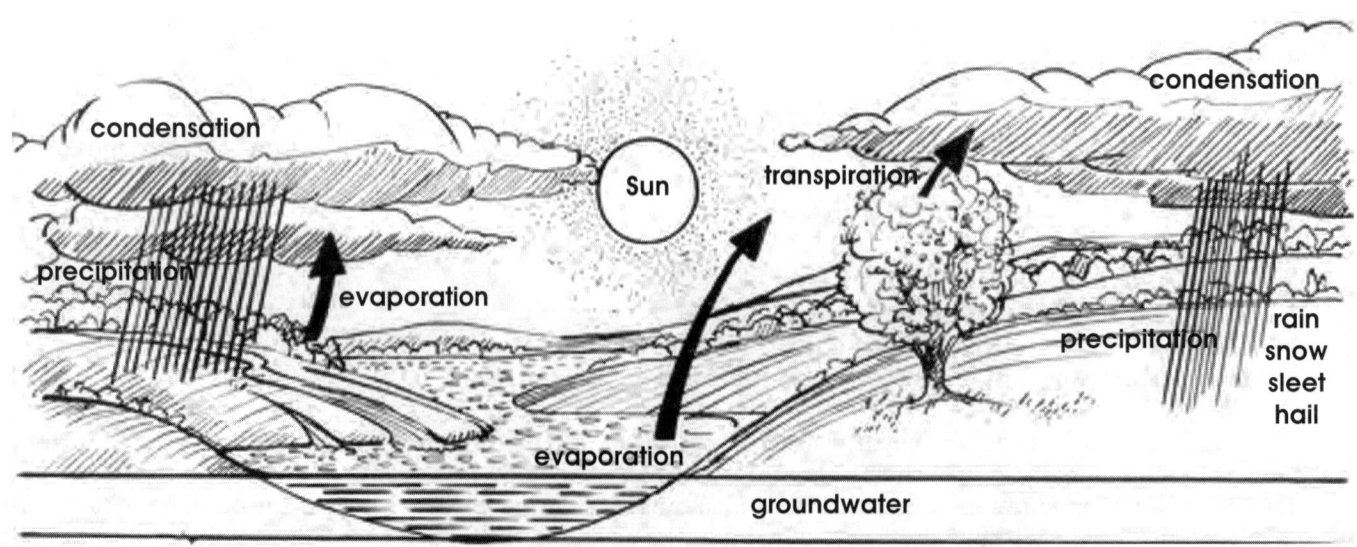

124 Label the water cycle at the right.

A _____

B _____

C _____

D _____

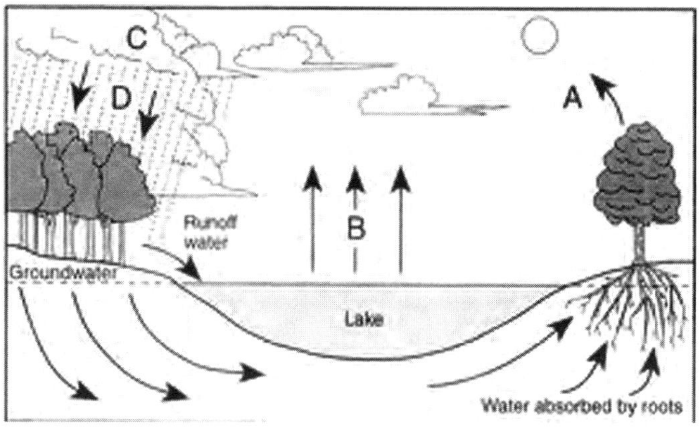

125 List two ways water returns to the air.

_____

_____

126 Why would you expect more moisture in the air near oceans or large forests? _____

_____

THE NITROGEN CYCLE

Another important substance that is recycled is nitrogen. Nitrogen is an essential element for living things and is needed to build proteins. Although most of the air you breathe is nitrogen you cannot use it in this form. The **nitrogen cycle** shown at the right, helps make nitrogen useable for living things.

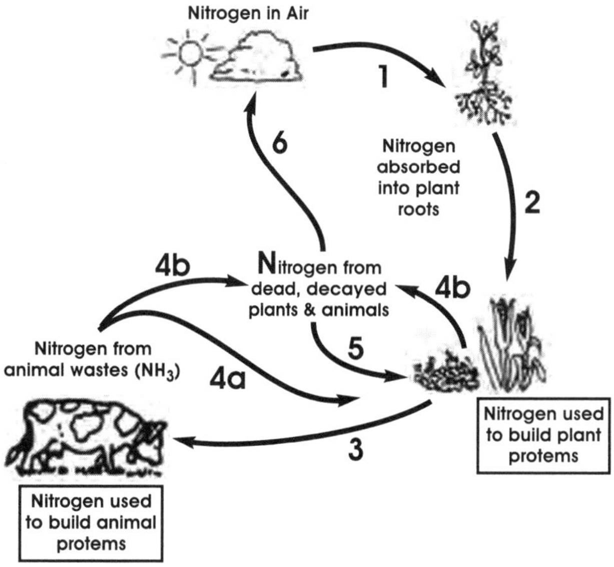

NITROGEN CYCLE ILLUSTRATION EXPLANATION

1 Bacteria in the soil and in the roots of certain plants (legumes) can take nitrogen out of the air and convert it to a useful form.

2 Plants can then absorb the nitrogen compound into their roots. This nitrogen may be used to build proteins.

3 Animals get nitrogen by eating plants and other animals. Some nitrogen is used to build proteins.

4 *a* Animals use some of the protein they take in for energy. When this happens, nitrogen is released as a waste. Bacteria convert this waste into a form useable by plants.

  b When animals and plants die, other bacteria break them down and convert the nitrogen into a useable form.

5 This form of nitrogen can be absorbed by plants and the cycle begins again.

6 Other bacteria break nitrogen-bearing substances down so that the nitrogen is released back into the atmosphere and the cycle begins again.

Follow The Flow !

INTERPRET AND/OR ILLUSTRATE THE ENERGY FLOW IN A FOOD CHAIN, ENERGY PYRAMID, OR FOOD WEB

Disciplinary Core Ideas:

LS2.A: Interdependent Relationships in Ecosystems

LS2.B: Cycle of Matter and Energy Transfer in Ecosystems

THE FLOW OF ENERGY IN AN ECOSYSTEM CAN BE VISUALIZED USING FOOD CHAINS, FOOD WEBS, AND ENERGY PYRAMIDS.

Which Way Did It Go? (The Energy That Is!)

A **food chain** shows the path of energy transfer from green plants through a series of organisms each of which eats the preceding organisms.

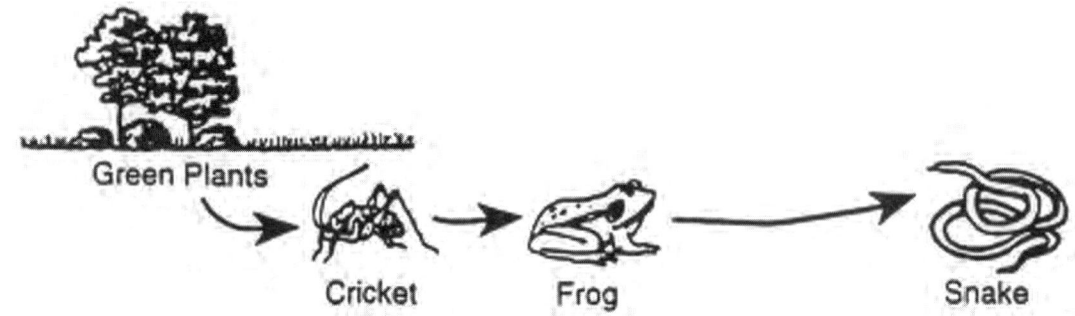

127 Where do each of the following get their energy from?

Green plants _____, cricket _____, snake _____

What A Tangled Web We Weave!

A **food web** better illustrates what really happens in an ecosystem. Since organisms eat more than one other organism, a food web showing the branching connections between an organism and its energy sources gives a better picture of how energy is transferred.

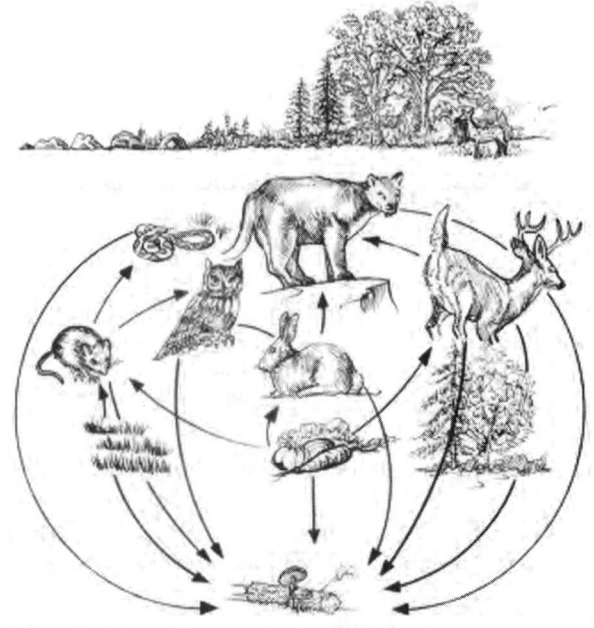

Organisms in an ecosystem can interact with each other and the physical environment in many ways. Predatory interactions may reduce the number of organisms or eliminate whole populations of organisms. For example, a rabbit thrives when the fox population is down while a fox thrives when the rabbit population is up. Mutually beneficial relationships may become so interdependent that each organism needs the other to survive. Although the types of organisms that interact in these different relations may vary in different ecosystems, the patterns of interactions are shared.

128 From what do mice get their energy? _____

129 What organisms get their energy from eating mice? _____

130 What would happen to the deer if all the plants died? _____

_____

ENERGY – USE IT & LOSE IT

An **energy pyramid** best illustrates what happens to the amount of energy left in an ecosystem as energy is transferred from one organism to another along the food chain or web. The size of each section represents how much energy it contains. At each transfer energy is lost to the system. This is because the organism eating the organism before it on the chain or web cannot use all of the energy from that organism.

131 Why must there be more grass than crickets? _____

_____

132 How do the number of owls and frogs compare? _____

_____

133 Why is there a difference in the numbers? _____

_____

_____

134 Where does the grass get its energy from? _____

An organism uses energy in order to live. Much energy is lost as heat. The energy pyramid is a clear reminder that plants are the base on which life on Earth is supported. What it does not show is significant. The energy pyramid does not show that the Sun is the source of energy for all living things. Only the relationships between the living organisms are shown.

Base your answers to the following questions on the chart.

Organism	Food consumed in one week
Cricket	. . . . .10 blades of grass
Frog	. . . . .20 crickets
Owl	. . . . .5 frogs

135 How many crickets are needed to support 5 frogs for one week?   5 x ____ = _____

136 How much grass is needed to support 100 crickets for one week?   100 x ____ = _____

Naming the types of energy users makes it easier to understand them.

- **Producers**: produce chemical energy from nonliving sources, such as light or chemical energy. Green plants are producers.

- **Consumers**: must take in (consume) other living or dead material to gain energy.

- **Decomposers**: gain energy by breaking down the complex chemicals in waste from organisms and dead organisms.

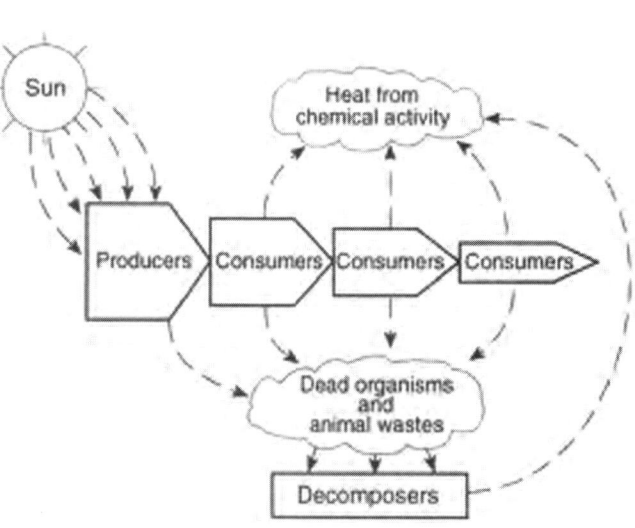

The following story will help you understand the flow of energy. The story is set in an imaginary ecosystem.

artist: Eugene Fairbanks, ©N&N 2000

It was a beautiful day! The **Jub Jub** birds flew by and landed lightly on the **Tum Tum** trees – their favorite place to munch and just hang out. Off in the distance, they could see the **Zwoks** approaching. The Jub Jubs were not afraid because the Zwoks were their friends. The Zwoks were coming to eat the fruit of the trees.

As the Zwoks sat in the shade of the Tum Tum trees – stuffing their furry faces – a **Sticky Bear** crawled out of a nearby log. The Zwoks took careful notice. Although they would share the fruit with the Sticky Bear, it was known that the Sticky Bears, besides liking Tum Tums, enjoy baby Zwoks for lunch!

Suddenly, out of nowhere a mighty **Wackyjobber** leaped! From its nostrils came a bad smelling vapor. Its giant jaws opened wide, and its ferocious claws were gleaming. The Wackyjobber grabbed the Sticky Bear, and in a flash its gooey head was gone. Without a moment's hesitation, the Zwoks reacted. Grabbing up stones, they called to the Jub Jub birds, "Lift us up, above the Wackyjobber." On strong wings, the Jub Jubs lifted the Zwoks, who then dropped the stones, killing the feared Wackyjobber.

As the day ended, the **Toolies** came by and ate most of the Wackyjobber. In a month, the **Coll** would decay the rest. Nothing would be left under the Tum Tum tree except the grave marker of the Sticky Bear that the Zwoks had made. It read, "Although nothing remains of the Sticky Bear, its energy *lives* on."

137 What was the producer in the story? _____

138 What was the decomposer in the story? _____

139 List three consumers in the story. _____

140 List a food chain described in the story. _____

_____

141 If there were too many Wackyjobbers in this land, what would happen to each of the other living

things? _____

_____

_____

Base your answers to the following questions on the diagram at the right.

142 List the producers.

_____

_____

_____

143 List the consumers. _____

_____

_____

The diagram shows levels C, B, A with organisms:
C — Mountain lion, Snake, Hawk
B — Elk, Rabbit, Field mouse
A — Trees, Shrubs, Grasses

144 Which organisms can transform light energy into chemical energy? (A, B, or C) _____

145 At which level (A, B, or C) will the smallest number of organisms be found? _____

PLANTS & ANIMALS DEPEND ON EACH OTHER & THEIR PHYSICAL ENVIRONMENT

the study of the living **environment** is a study of interactions and relationships. It is similar to players on a soccer team. each player has a different role, but together the players make a team. If one player does not do their job, the whole team was affected. When studying the living environment, do not think of each factor separately, but think about how all the parts work together.

146 Write a sentence or two describing how members of a team are interdependent on each other and on their environment (playing field).

_____

_____

_____

_____

_____

© PhotoDisc

MULTIPLE CHOICE

1 Which represents the correct flow of energy through an ecosystem?
   (1) consumer, decomposer, producer, Sun   (3) Sun, decomposer, consumer, producer
   (2) producer, consumer, decomposer, Sun   (4) Sun, producer, consumer, decomposer

2 Plants use sunlight to make
   (1) carbon dioxide   (2) fats   (3) nitrogen   (4) sugar

3 The energy used by most organisms in an ecosystem
   begins in
   (1) animals   (3) plants
   (2) bacteria   (4) soil

Use the diagram of a food web to answer questions 4 and 5.

Food Web

4 Which of the following organisms in the food web
   are consumers?
   (1) deer, rabbit, owl
   (2) grass, snake, mouse
   (3) vegetables, rabbit, owl
   (4) tree, vegetables, grass

5 Which level of organisms represents the most
   stored energy in the ecosystem?
   (1) deer, rabbit, mouse
   (2) mountain lion, snake, owl
   (3) tree, deer, mountain lion
   (4) tree, vegetables, grass

6 Energy enters most ecosystems as
   (1) animals   (2) plants   (3) soil   (4) sunlight

7 Most energy that enters an ecosystem is
   (1) eventually converted to a different form   (3) present in plant tissue but not in animal tissue
   (2) present in animal tissue and not plant   (4) stored, mostly as heat

8 A food chain represents
   (1) a list of what one organism eats   (3) the flow of energy from one organism to another
   (2) links of what animals live together   (4) the way that food is produced in an ecosystem

9 An animal is found to lack energy and be unable to grow properly. This is most likely caused by a lack of
   (1) carbon dioxide   (2) food   (3) shelter   (4) water

Base your answers to questions 10 to 14 on the diagram at the right of a food pyramid.

10 Which direction is the energy flowing?
    (1) frog - insect - weed
    (2) insect - frog - weed
    (3) snake - frog - insect
    (4) weed - insect - frog

11 The initial energy to start this pyramid comes from
    (1) decayed matter   (3) soil
    (2) snakes   (4) the Sun

12 In this pyramid, the insects are considered
    (1) decomposers   (2) carnivores   (3) consumers   (4) producers

13 The reason the energy pyramid is smaller at the top than at the base is
    (1) energy is gained at each level   (3) the snake is a producer
    (2) energy is lost at each level   (4) animals hide in the weeds

14 The reason that producers are at the base of almost all energy pyramids is
    (1) most organisms build their homes on or near producers
    (2) most organisms use food, directly or indirectly, made by producers
    (3) plants are the most abundant organisms on Earth
    (4) producers are strong and form a good base for the pyramid

15 The process by which the Sun's energy is converted into and stored as chemical energy is called
    (1) decomposition   (2) metabolism   (3) photosynthesis   (4) respiration

16 The Sun's energy is converted by plants into chemical energy in the form of
    (1) fats   (2) protein   (3) sugar   (4) vitamins

17 A group of organisms that carry out photosynthesis are
    (1) fireflies   (2) trees   (3) humans   (4) viruses

18 The Sun's rays can only penetrate the upper level of the ocean. Therefore, in most oceans photosynthesis occurs in the
    (1) top layer   (2) middle layer   (3) bottom layer   (4) air above the ocean

19 If producers were no longer able to carry out photosynthesis, which gas in the atmosphere might be reduced?
    (1) argon   (2) carbon dioxide   (3) oxygen   (4) nitrogen

20 If producers were no longer able to carry out photosynthesis, which gas in the atmosphere might be increased?
    (1) argon   (2) carbon dioxide   (3) oxygen   (4) nitrogen

THE QUESTIONS THAT FOLLOW TEST YOUR UNDERSTANDING OF SCIENTIFIC PRACTICES, CORES IDEAS IN SCIENCE AND CROSS-CUTTING CONCEPTS SUCH AS PATTERNS AND CAUSE AND EFFECT RELATIONSHIPS.

SHORT ANSWER

1 Base your answers to *a* and *b* on the graph at the right. (2 points)

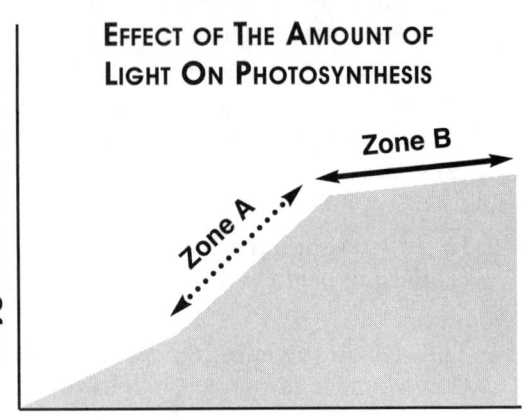

EFFECT OF THE AMOUNT OF LIGHT ON PHOTOSYNTHESIS

a As the quantity of light increases, what happens to the amount of oxygen released in Zone *A*?_____

_____

b As quantity of light increases, what happens to the amount of oxygen released in Zone *B*?

_____

Organism	Weekly Diet
Owl	20 mice
Mouse	40 grasshoppers
Grasshopper	10 blades of grass

2 Based on the chart above, how many blades of grass are needed to support the food chain so that one owl can survive for one week. SHOW ALL CALCULATIONS. (2 points)

_____

_____

SHORT ESSAY

1 Photosynthesis:

Carbon dioxide + water + energy from sunlight $\xrightarrow{\text{Chlorophyll}}$ simple sugars + oxygen

Respiration:

Food (simple sugars) + oxygen $\xrightarrow{\text{enzymes}}$ Carbon dioxide + water + energy release

Statement: Matter is recycled in ecosystems but energy is not.

Use the formulas above to support this statement with an example from an actual ecosystem. (5 points)

_____

_____

_____

_____

_____

_____

_____

ECOSYSTEM:
INTERACTIONS, ENERGY AND DYNAMICS

DISCIPLINARY CORE IDEAS:

ESS3.A: NATURAL RESOURCES

ESS3.C: HUMAN IMPACTS ON EARTH SYSTEMS

LS2.C: ECOSYSTEM DYNAMICS, FUNCTIONING, AND RESILIENCE

DESCRIBE HOW LIVING THINGS, INCLUDING HUMANS, DEPEND UPON THE LIVING AND NONLIVING ENVIRONMENT FOR THEIR SURVIVAL.

The heat becomes overwhelming. It is dry and there is little moisture in the air. Hot winds gust and swirl around you. Other than some humans and a few birds, few other animals are seen during the day.

147 Where are you? _____

© PhotoDisc

If you thought you were in a desert, you were close. You are in a city. Human development and decisions can make a tremendous impact on the conditions of the world around us. People have a responsibility to consider the impact of their actions on the environment. In order to make responsible decisions, you need to know about how things affect each other. You need to know about ecology.

INTERDEPENDENCE DAY

Ecology is the study of how all living things interact with one another and how they interact with the physical environment.

VISUALIZATION - ALONE IN A STORM

Take a moment and think about how might feel to be a wild animal caught in a thunderstorm.

Choose an animal. What would this animal feel and do as a rainstorm approached, followed by thunder, lightning, and heavy rain? Close your eyes. Take a moment to think about it.

148 Write down what the animal might feel and do. _____

_____

Sometimes, how each living thing depends on many other living things and the physical environment is forgotten. Understanding these interactions is very important to understanding the role of humans in the environment.

LET'S GET ORGANIZED!

Ecology is easier to understand if you divide it up into smaller parts.

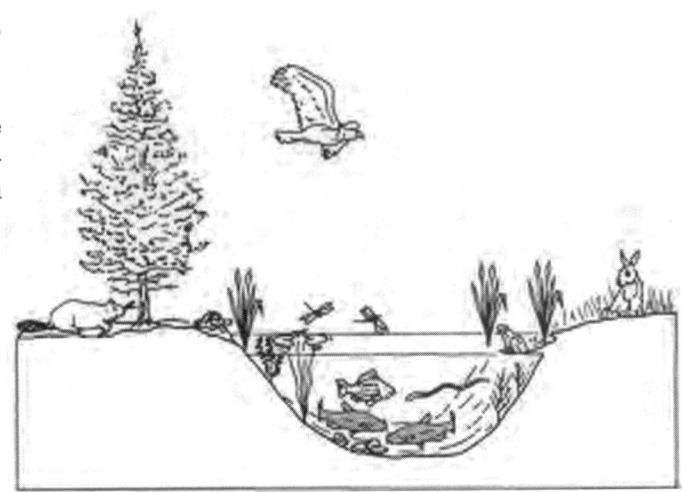

A **population** is all the members of the same type of living things in a given area. Examples: the population (number) of humans in your class or the population of trout in a lake.

149 What is the population of catfish in this pond?

_____

150 What is the population of cattails in this pond?

_____

A **community** is all the plant and animal populations interacting in a given area. A community consists only of living things. Example: A pond is a community that consists of different fish, turtles, algae, plants and insects.

151 Look back at the diagram of the pond again. List the members of the pond community. _____

_____

Physical factors are all the nonliving variables affect or are affected by the living factors. Physical factors greatly influence the type of plant life in an area, which, in turn, influences the animals which depend on them. **Climate** is the long term average weather. The amount of light and the temperature are important climatic factors. The type of **soil** effects plant growth, erosion, and available minerals. **Water** is critical to all life. It is a major component of living things. The amount of oxygen and the quality of the **air** will affect living things. **Light** is needed for photosynthesis. It also is important for sight.

152 How would the physical factors influence life in the

desert? _____

_____

153 How would the physical factors influence life in the

Arctic? _____

_____

154 How would the physical factors influence life in the

ocean? _____

_____

INTERMEDIATE SCIENCE- 3D LEARNING & ASSESSMENTS

An **ecosystem** is the community and the nonliving factors with which they interact. Think of an ecosystem as the total functioning unit, all the parts that interact and affect each other. Ecosystems are dynamic in nature. This means they change over time. Disruptions to the physical or biotic parts of an ecosystem can cause changes in all of its populations. Example: The pond ecosystem includes the community of living things and the water, rocks, sunlight, temperature and other physical factors.

155 How does the tree affect the deer?

_____

_____

156 How does the deer affect the tree?

_____

_____

157 How does the stream affect the pond community? _____

_____

158 How does the Sun affect the plants? _____

_____

POLLUTION

Pollutants are substances that are harmful to living things. Pollutants may contaminate the air, water, and soil. Waste products from industry, cars and homes often add to pollution. Careful monitoring of pollution and actions to control the causes of emissions are essential to maintaining a healthy environment.

159 What would happen if the factory produced too much carbon dioxide?

_____

_____

_____

_____

_____

_____

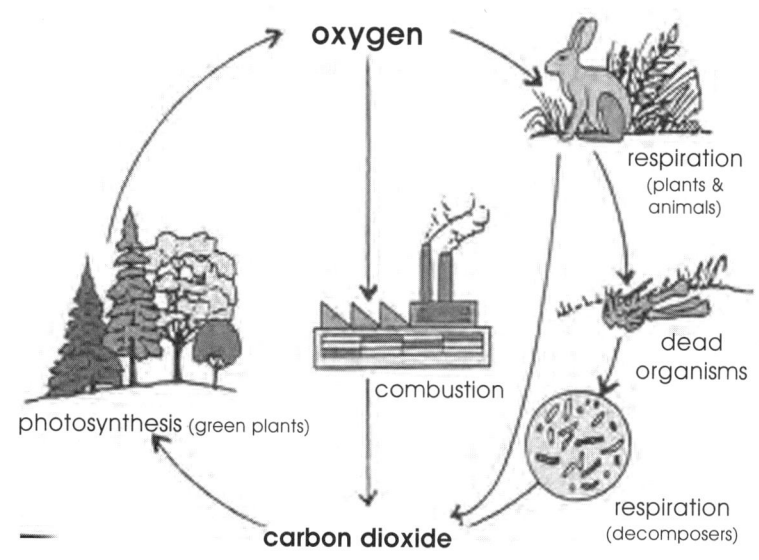

RELATIONSHIP ISSUES

Look closely at what happens among members of a community as they interact.

Rating	Interaction
+ +	Both organisms benefit from the interaction. *Example 1* - A little bird eats bugs off of a rhino's back. It gets food, and the rhino gets rid of the bugs. *Example 2* - A lichen is a living organism composed of two other living things, an alga and a fungus. The fungus helps collect water and the alga uses chlorophyll to make food. *Example 3* - A microorganism in the stomach of a termite helps it to digest its food. The microorganism is protected in the termite's stomach.
+ 0	One organism benefits and the other is not affected by the interaction. *Example* - A remora follows a shark and eats left over scraps of food. The remora gets food and the shark is unaffected.
+ −	One organism benefits and the other organism is harmed by the interaction. This is called **parasitism**. *Example* - Athlete's Foot is a fungus that lives on human skin. It gets food from your skin and causes a bad itch and a rash.
+ −	**Predation** - One organism (the predator) eats another (the prey). *Example* - An owl (predator) benefits from eating a mouse (prey).

160 Rate the following interactions:

a   A cat eats a bird. _____   *b*   A mosquito bites a dog. _____

DISCIPLINARY CORE IDEAS:

LS2.A: INTERDEPENDENT RELATIONSHIPS IN ECOSYSTEMS

ESS3.C: HUMAN IMPACTS ON EARTH SYSTEMS

LS4.D: BIODIVERSITY AND HUMANS

DESCRIBE THE EFFECTS OF ENVIRONMENTAL CHANGES ON HUMANS AND OTHER POPULATIONS.

TEN LITTLE SQUIRRELS SITTING IN A TREE

Have you ever noticed that you do not usually see ten squirrels in the same tree? Most organisms are capable of increasing their numbers at a rapid rate. But you do not see hundreds of *all types* of organisms running around the place. In most cases, there are **limiting factors** that slow down the growth of a population. There are many limiting factors such as light, food, shelter, water, climate, and predation (capturing prey).

Interactions among the community members and the physical factors around them lead to a balance. In a balanced community, factors interact to maintain a certain number of each type of living thing.

The pyramid on the right represents the number of each type of organism in an ecosystem.

161 What problem would occur, if more large fish moved into this

ecosystem? _____

_____

_____

A Balancing Act - the Climax Community

When populations exist in balance with one another, the environment is stable (stays relatively the same). This stable state is called the **climax community**. If some major change occurs, such as fire, climate change, earthquake, or invasion by other species, the balance may be upset. When a community is forced to start over, it goes through a series of changes as it regains its balance and becomes stable. The stages of change are called ecological succession. Ecological succession occurs as each new community changes the environment making it better for the next community.

What would happen if a fire swept through a northeastern forest and burned everything to the ground? The first to return would be quick growing grasses and weeds from seeds that might have been blown in by the winds. Animals that live in the grasses and weeds would also return. Birds might drop other seeds in the area and small bushes and trees would appear. Quick growing pines would begin to shade the ground from the Sun. This shade provides a good environment for young hardwoods to develop. The hardwoods would eventually outgrow the pines and shade them out. The northeastern forest would therefore be reestablished and be stable until another major change occurred.

What's up Doc?

There were no naturally occurring rabbits in Australia. People thought that due to Australia's climate and the grasses that grow there, it would be a good place to raise rabbits. However, Australia also lacks many natural predators.

162 What do you think happened to the number of rabbits in

Australia? _____

_____

163 Why could this cause a problem? _____

_____

©PhotoDisc

They Paved Paradise and put up a Parking Lot
The Pine Bush: A Case Study

The Pine Bush is a globally rare ecosystem found in upstate New York. It is unique because, as ancient glaciers left the area, a river deposited large amounts of sand in one place. This deposit of sand became the Pine Bush. It is a pine barrens ecosystem similar to those found on Long Island, New Jersey, and Florida. The sand drains very well and makes the Pine Bush very dry. Because it is so dry, natural fires often occur. Plants and animals that live in the Pine Bush have adapted to a dry, sandy, often burned area.

Since the 1940s, people have moved into the Pine Bush in great numbers. This urban growth has decreased the size of the Pine Bush. Because humans now live in the Pine Bush, the natural fires are no longer allowed and are quickly put out.

Without the natural fires, many species have begun to invade the Pine Bush. Aspens, for example, have begun to

Upland Pines

courtesy of:
NATL Photo Gallery, UFL

take over the Pine Bush because there are no natural fires to control them. Aspens grow and produce shade which prevents the light from reaching the native plants that need it. Between the land use decisions and the lack of natural fire, the Pine Bush is almost gone.

164 What could be done to save the Pine Bush? _____

_____

_____

TOO MUCH OF A BAD THING – POLLUTION

The human population has increased rapidly in the last 100 years. In 1960, there were 3 billion people. Forty years later, the population doubled to 6 billion! Today the population is over 8 billion. Supporting such a growing population has placed large demands on the environment. Since the Industrial Revolution – when people started burning large quantities of fossil fuels, such as coal and oil – a large amount of pollution has been created.

Although it might seem that Earth is so big that just one case of pollution will not affect it, when these cases start adding up, Earth's environment can be greatly changed.

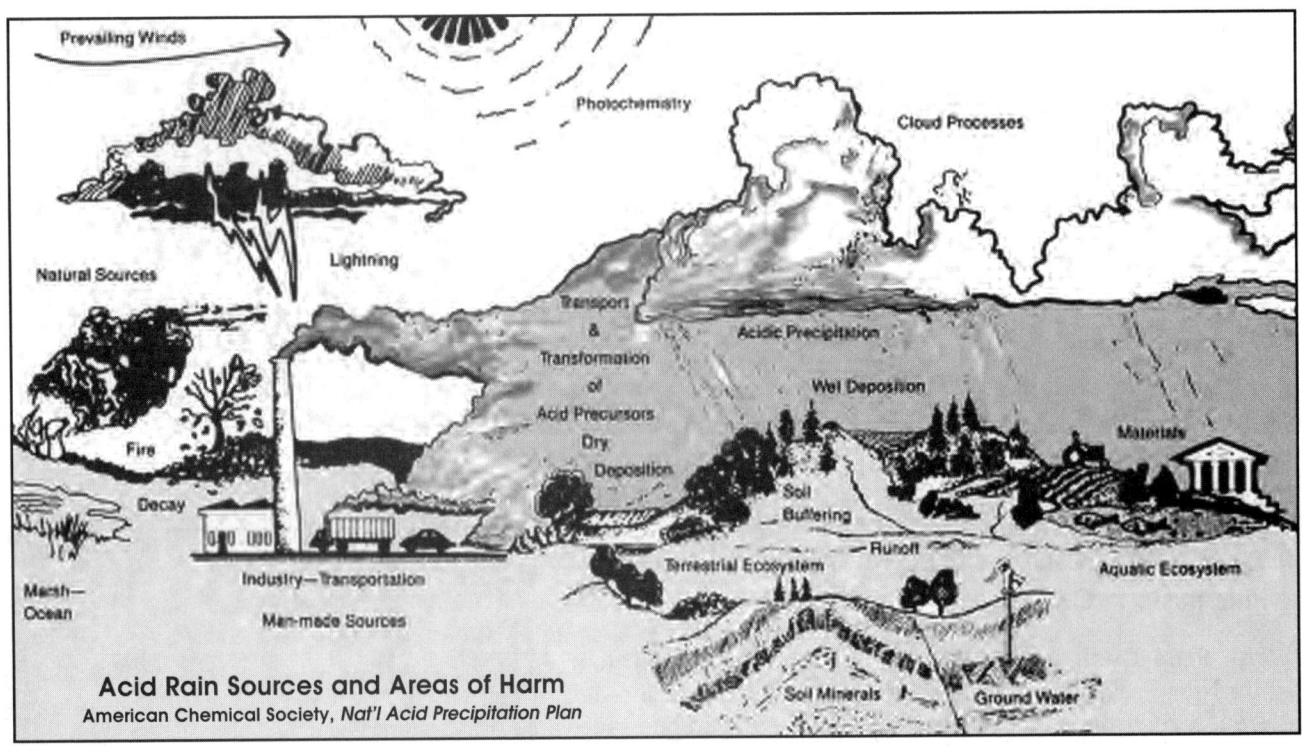

Acid Rain Sources and Areas of Harm
American Chemical Society, *Nat'l Acid Precipitation Plan*

ACID RAIN

Scientists have shown that industrial emissions combine with atmospheric moisture to form sulfuric acid. These acids may be carried long distances in the atmosphere and be deposited in precipitation such as rain or snow (**Acid Rain**). Acid Rain may affect soil, fresh water lakes, crops and forests. Over time and the changing of physical factors in an environment, acid rain may affect the organisms that live there. Over 200 lakes in the Adirondacks no longer support life because of acid rain, and sections of Canada have been ruined for years to come.

Laws have been passed to help reduce the amount of acid rain. The *Clean Air Acts* called for a reduction of sulfur dioxide released from power plants from 20 million tons in 1990 to 10 million tons in the year 2000.

GLOBAL CLIMATE CHANGE

The use of fossil fuels, such as coal, petroleum, and natural gas, produces gases such as carbon dioxide and nitrogen oxides. If these gases occur in great enough amounts, they can stop infrared radiation from escaping Earth's atmosphere. This could lead to an increase in Earth's temperature. This is called the **Greenhouse Effect**. Scientists are worried that an increase in temperature could cause coastal flooding and climate changes. Climate changes would cause major changes in Earth's present ecosystems.

U.S. News ©1992, Robert Crawford - reprinted with permission

Time Inc. ©1992 - reprinted with permission

OZONE DEPLETION

Ozone is a form of oxygen in the atmosphere that forms a layer that protects us from ultraviolet radiation. An increase in **ultraviolet radiation** may cause an increase in skin cancer, damage certain crops, and affect plankton, which is a basic component of food webs in the oceans and other water environments.

Many years ago, scientists discovered that chemicals called chlorofluorocarbons (CFCs) used in refrigerants and aerosol spray propellants reacted with ozone in the upper atmosphere and destroyed it. Not until recent years has the question of ozone depletion become an international concern.

165 Match each form of pollution with the correct letter describing its effect:

a _____	Acid Rain	*1*	allows more ultraviolet radiation to hit Earth
b _____	Global warming	*2*	allows less infrared radiation to escape Earth
c _____	Ozone depletion	*3*	changes the nature of soil and freshwater

BACK IN YOUR HANDS!

The survival of living things on Earth depends on the **conservation** (wise use) and protection of Earth's resources. It is your decisions, as a participating member of the community of your ecosystem, that will affect the world in which you live.

Scientists use the term **biodiversity** to describe the variety of species found in an area or ecosystem. The amount of diversity and the numbers of each type of organism can be used to measure the health of an ecosystem. Changes in biodiversity may affect humans. It may affect food, energy and medicines available to us. For example, if we begin to rely on only one specific species of grain for food and that species declines due to a drought, for example, the lack of diversity in our food sources would become a problem.

You are interdependent. You depend on other living things and they depend on you. Interdependence also means you have a responsibility to maintain your part of a healthy world.

166 Describe what you could do to help protect or preserve the environment where you live? _____

_____

_____

_____

LIFE SCIENCE REVIEW 7

MULTIPLE CHOICE

1 All of the groups of species that occur together at a given time and place are a (an)
(1) biosphere   (2) community   (3) ecosystem   (4) population

2 All the same type of fish that live in a pond would be part of the same
(1) biosphere   (2) community   (3) ecosystem   (4) population

3 Without disease, predators, or lack of resources, most populations would
(1) increase   (2) decrease   (3) remain the same   (4) both increase and decrease

4 Human activities can bring about positive environmental changes through
(1) conservation   (2) resource acquisition   (3) urban growth   (4) waste disposal

5 Pollution can be best reduced through the use of
(1) atomic energy   (2) burning coal   (3) larger cars   (4) solar energy

6 The community and the physical factors with which it interacts make a (an)
(1) ecosystem   (2) neighborhood   (3) niche   (4) population

7 The water, animals, and plants in and near a pond would be considered a (an)
(1) ecosystem   (2) community   (3) climate   (4) population

8 Growth of populations in an ecosystem can be limited by an increase in
(1) predation   (2) resources   (3) shelter   (4) water

9 If two organisms struggle to obtain the same type of food their relationship can be said to be
(1) competitive   (2) beneficial   (3) mutualistic   (4) parasitic

10 If one organism lives on or in another and gets its energy by harming the other, their relationship can be said to be
(1) competitive   (2) beneficial   (3) positive   (4) parasitic

11 In order to insure a healthy environment individuals should
(1) add pollutants to undisturbed areas   (3) monitor pollution in soil, water and air
(2) leave pollutants as they are   (4) place pollutants in the ocean

12 In order to insure the survival of our planet, people have a responsibility to
(1) avoid recycling materials   (3) consider the impact of their actions
(2) eliminate all plant life on this planet   (4) use as much energy as possible

13 Overpopulation of any species harms the environment due to
(1) increased use of resources   (3) production of extra resources
(2) decreased use of resources   (4) resources being recycled

14 Human activities can hurt the environment through
(1) increased removal of resources   (3) reduction of urban growth
(2) proper waste disposal   (4) responsible land use

15 A type of pollution that has damaged many lakes is
(1) acid rain   (2) fog   (3) global warming   (4) ozone depletion

16 Global Warming can be reduced if humans limit their use of
(1) atomic energy   (2) fossil fuels   (3) ozone   (4) solar energy

17 Ozone in our upper atmosphere protects us from
(1) acid rain   (2) global warming   (3) thunderstorms   (4) ultraviolet radiation

Base your answers to questions 18 through 20 on the diagrams below and your knowledge of the living environment. The diagrams show various stages of plant development on a volcanic island over a 300-year period after it was formed.

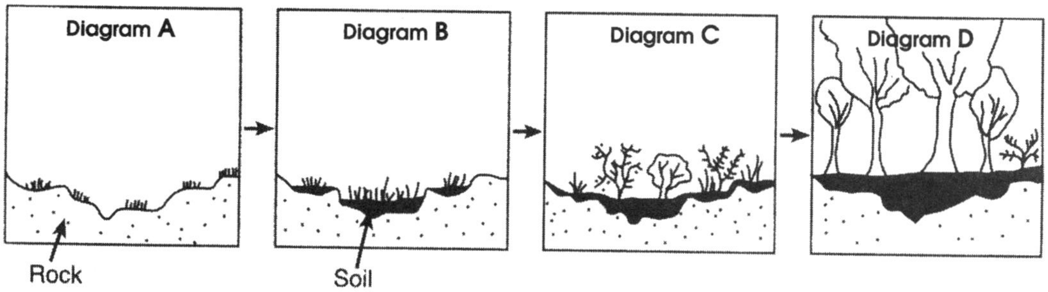

18 The first organisms to move in after the volcano erupted are seen in diagram
(1) A   (2) B   (3) C   (4) D

19 If Diagram D represents the climax plants of the region, the dominant animals would most likely include
(1) worms and lizards   (2) snakes and turtles   (3) squirrels and skunks   (4) wolves and hawks

20 Which diagram would best represent the kind of environment which would most likely have the richest soil and support the greatest variety of organisms?
(1) A   (2) B   (3) C   (4) D

THE QUESTIONS THAT FOLLOW TEST YOUR UNDERSTANDING OF SCIENTIFIC PRACTICES, CORES IDEAS IN SCIENCE AND CROSS-CUTTING CONCEPTS SUCH AS PATTERNS AND CAUSE AND EFFECT RELATIONSHIPS.

SHORT ANSWER

1 The diagram at the right shows the change in a pond over time. Why did the type of organisms that lived in the pond area change at each stage? Note: *Stages A-D* have been placed in successive order. (2 points)

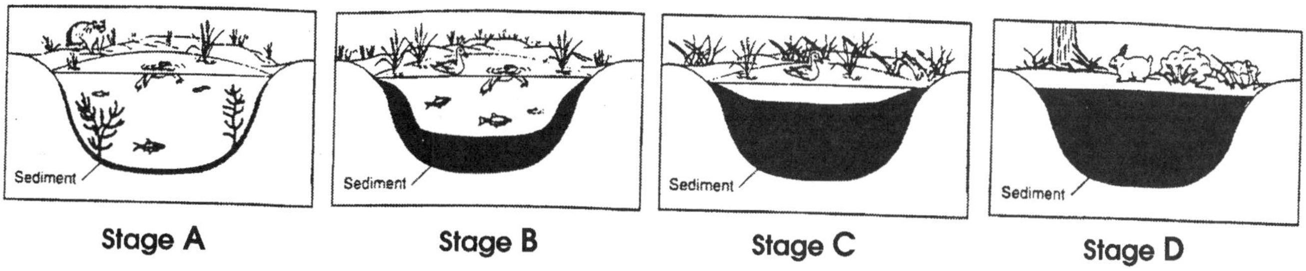

Stage *A* _____

_____

Stage *B* _____

_____

Stage *C* to *D* _____

_____

2   When burned, natural gas and crude oil products produce less pollution than coal. Describe three actions that must be taken to protect our environment before we become dependent on our coal reserves. (2 points)

Fossil Fuel		*Estimated U.S. Reserves
Crude Oil		35 Years
Natural Gas		65 Years
Coal		1,800 Years

*N.Y.S. Energy Educational Project 1997

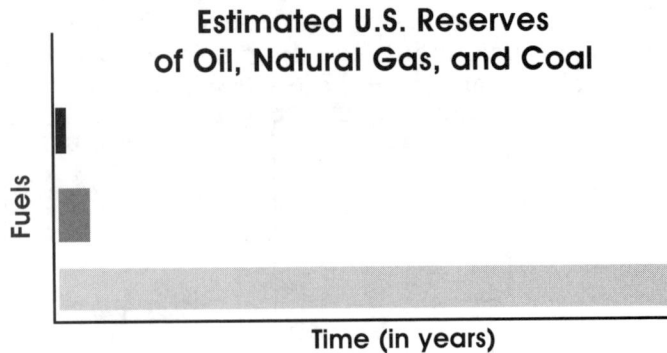

Estimated U.S. Reserves of Oil, Natural Gas, and Coal

Fuels

Time (in years)

_____

_____

_____

SHORT ESSAY

Use the **Energy Resources Chart** below and your knowledge of the **Living Environment** to answer the following questions.

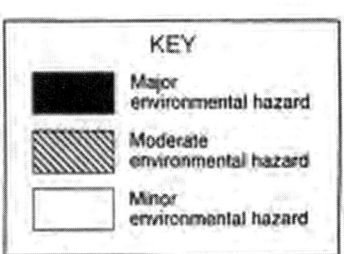

Energy Resource	Disadvantage					
	Particles	Thermal pollution	Sulfur dioxide	Carbon dioxide	Damage to land and water	Potential dangers
Coal						Fire
Oil						Fire
Natural gas						Explosion
Geothermal						Earthquake
Nuclear						Radiation
Hydroelectric						Flooding
Wind						
Solar						
Ocean						

KEY

◼ Major environmental hazard

▨ Moderate environmental hazard

☐ Minor environmental hazard

1   Give two reasons why natural gas is a good source of energy for our homes. (2 points)

_____

_____

2   Give two reasons why natural gas should not be used as a source of energy for our homes. Include in your statements its relationship with global warming. (3 points)

_____

_____

_____

Section Three

EARTH AND SPACE SCIENCE

EARTH'S PLACE IN THE UNIVERSE

DEVELOPING AND USING MODELS: PATTERNS AND APPARENT MOTION OF THE SUN, THE MOON, AND STARS.
THE SKY CAN BE OBSERVED, DESCRIBED, PREDICTED, AND EXPLAINED WITH MODELS.

DISCIPLINARY CORE IDEAS:

ESS1.A: THE UNIVERSE AND ITS STARS

ESS1.B: EARTH AND THE SOLAR SYSTEM

MODELS OF THE UNIVERSE

IMPORTANCE OF USING MODELS

Scientists make observations about the natural world. They formulate hypotheses and then test them. Sometimes their hypotheses are proven to be correct. Other times, new hypotheses are needed. Often, scientists will create a model to explain their observations.

MODELS OF THE UNIVERSE

To understand the early models of the universe, it must be understood how motion is detected. All motion is relative to a frame of reference (point or object). For example, if you are standing still on a platform at a train station and a train passes by, your measurement of its speed would be relative to your position. Suppose you measured the speed to be 50 km/hr when you are standing still. After doing this, you decide to remeasure the speed of the train while you were walking at a rate of 5 km/hr in the same direction as the train. Now, to your frame of reference, the train would seem to be going only 45 km/hr. The speed of the train appeared to change because there was also a change in your frame of reference.

1   How is motion detected? _____

_____

GEOCENTRIC MODEL

There was a time in history when Earth was considered to be the center of the universe. This was known as the **geocentric model**. It was proposed in 150 A.D. by the Greek astronomer, **Claudius Ptolemy**. Earth was believed to be the center of the universe with the Moon, Sun, planets, and constellations orbiting Earth. This model lasted for hundreds of years because it could satisfactorily explain most celestial motion.

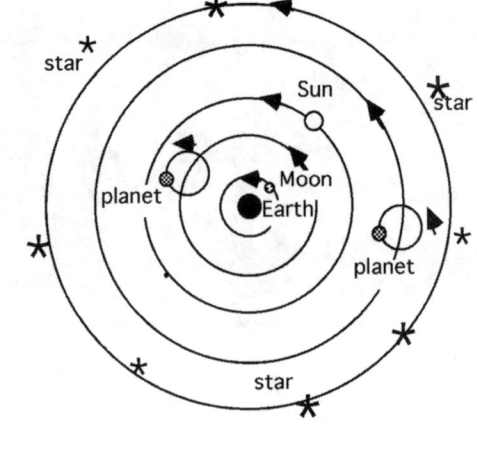

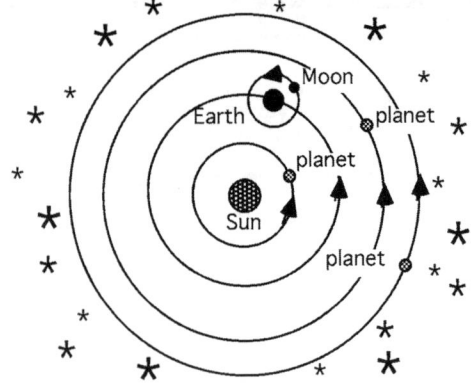

HELIOCENTRIC MODEL

One of the weaknesses of the geocentric model was that it could not easily explain the apparent backward motion of some of the planets. However, it was not until 1543 that **Nicholas Copernicus** proposed the **heliocentric model**. It placed the Sun in the center of the universe. This model was superior to the geocentric model because it more accurately explained the apparent backward motion of the planets.

2   What is the difference between the heliocentric and geocentric models of the universe? _____

_____

_____

DISCIPLINARY CORE IDEAS:

ESS1.A: THE UNIVERSE AND ITS STARS

ESS1.B: EARTH AND THE SOLAR SYSTEM

MODERN DAY MODEL OF THE UNIVERSE

Today's model of the universe reflects the idea that all celestial bodies are in motion relative to Earth and one another. In the age when Copernicus lived, there was no method for measuring star distances. The immense size of the universe that we know exists today was unimaginable. Both the heliocentric and geocentric models showed the solar system to make up the entire universe with the constellations placed at its farthest boundary. Today, we know that the Sun is one of millions of stars that reside off one arm of the **Milky Way Galaxy** (home for the solar system). The Sun is an average size star, though over a million times greater in volume than Earth.

The Milky Way Galaxy is one of billions of galaxies which are all in motion relative to one another. There is even recent evidence that supports the hypothesis that new solar systems are being formed in some of these galaxies.

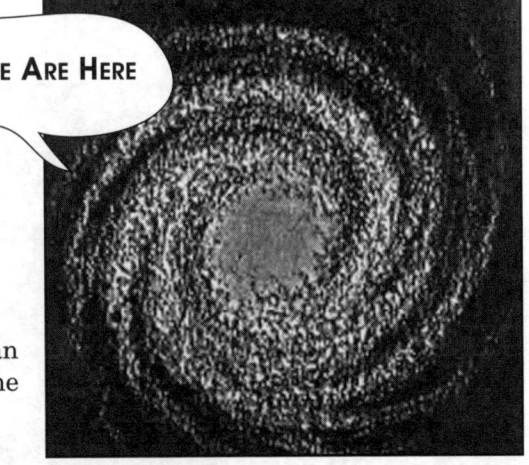

OUR PRESENT DAY UNDERSTANDING OF THE UNIVERSE SHOWS HOW SCIENCE WORKS.

When observations about the natural world no longer support an accepted model, the theory is changed or modified by scientists. The new theory reflects the new observations.

3   Why do scientists use models? _____

_____

MOTIONS OF EARTH

What causes night and day? Why does the shape of the Moon appear to change periodically in the evening sky? Why are there seasons?

Earth both rotates on its axis and revolves around the Sun. The motions of Earth are used to measure time in days and years. One complete rotation of Earth equals a day, while one revolution around the Sun accounts for one year.

EARTH'S ROTATION (DAY & NIGHT)
EVIDENCE THAT EARTH ROTATES

Evidence that Earth rotates was determined in 1851 by the French scientist, **Jean Foucault**. He suspended a large metal ball by a 60 meter steel cable from the dome ceiling of the Pantheon in Paris. Once a pendulum is set in motion, it will only swing in that direction. After one hour, the pendulum's plane of motion appeared to change. Foucault was able to attribute this change in motion to the rotation of Earth.

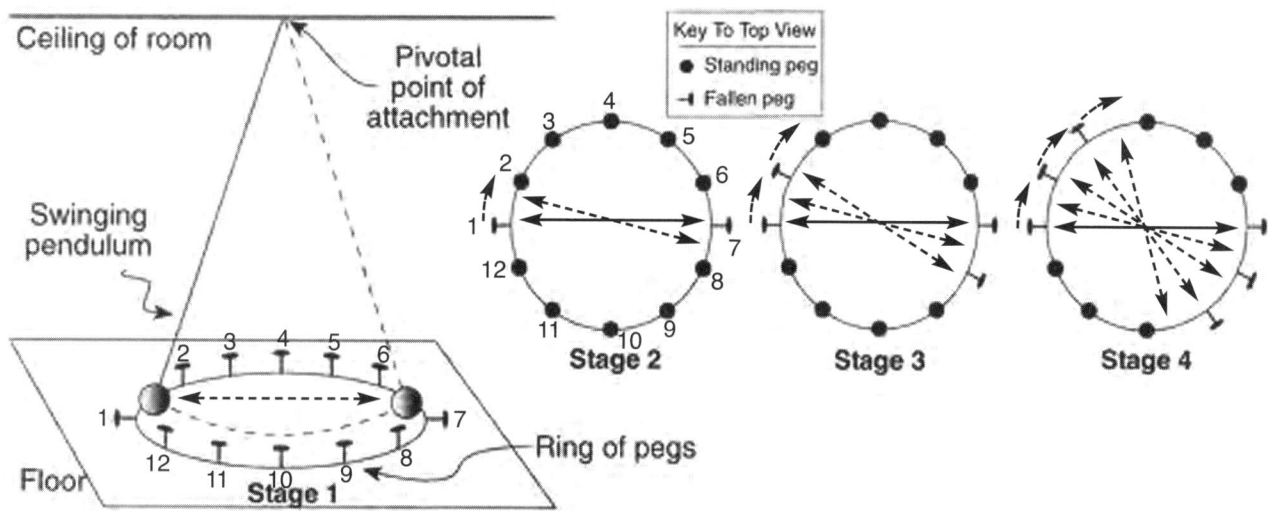

A modern piece of evidence that supports the idea that Earth rotates is satellite pictures. For example, a satellite placed in a polar orbit can take pictures of places at different meridians (degrees of longitude). If a camera was placed on Earth to take continual pictures of the night sky, Earth's motion would create **star trails** (arcs of light). The size of the arc was directly related to the amount of time the camera was photographing the evening sky.

CFHT Star trails – 5 September 2000
Canada-France-Hawaii Telescope

4   Describe two pieces of evidence that support the idea that Earth is rotating. _____

_____

_____

RISING & SETTING OF THE SUN (A DAY)

Since ancient times, humans have woven together many stories to explain the rising and setting of the Sun. Foucault was able to show that the Earth rotates. Due to his work and that of others, it is now known that the rising and setting of the Sun has to do with the rotation of Earth on its axis. As in all motion, there is a need for a frame of reference.

When the Sun is used as a frame of reference, it is called a **solar day**. A solar day is the amount of time needed for Earth to rotate from one given meridian (lines of longitude) back to the same meridian. This takes about 24 hours. The reason has to do with the size of the circumference of Earth at the equator and the rate of its rotation. The circumference is a little over 38,400 km (24,000 miles) at the equator and Earth's rotational speed at this point is about 1600 km/hr (1000 mi/hour). Therefore, Earth rotates about 15 degrees every hour. It should take about 24 hours to complete one rotation (a complete circle).

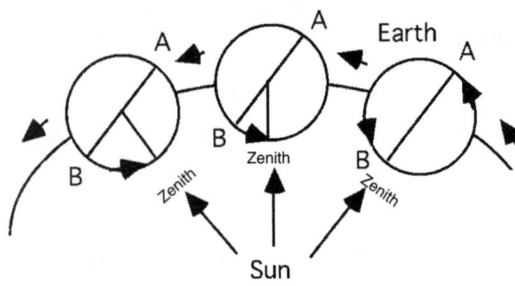

Earth's rotation and revolution are counter-clockwise. As Earth revolves, Earth must rotate more than 360° for the Sun's zenith to return to the same Earth longitude.

The U.S. Naval Observatory, which keeps time for the United States, uses **sidereal day**, because it is more accurate. In this case, a star is used as the frame of reference for when Earth has completed one rotation on its **axis** as compared to that star's position (23 hours, 56 minutes, 4.09 seconds in units of mean solar time).

5   What is the difference between a solar and sidereal day? _____

_____

_____

EFFECTS OF EARTH'S ROTATION

Earth's rotation affects the apparent motion of the winds and ocean currents. Remember, motion is detected by a frame of reference. An observer on Earth will view objects moving over the surface of Earth as being deflected from a straight path.

In the Northern Hemisphere, the winds and ocean currents appear to be deflected to the right because Earth is viewed as rotating counterclockwise. Viewed from the Southern Hemisphere, the winds and ocean currents are deflected to the left because Earth appears to be moving clockwise. This effect of the motion of winds and the ocean currents is called the **Coriolis Effect** – named after the French mathematician, Gaspard Coriolis who first described this effect in the 19th century.

Coriolis Effect
A ball being thrown between two persons riding on a rotating platform illustrates the Coriolis effect.
A – On a stationary platform, the pitcher throws the baseball to the catcher in an apparent straight line.

B – As the platform rotates counter-clockwise, the pitcher throws the baseball in the direction of the catcher, but the turning platform has carried the catcher counter-clockwise away from the incoming ball.

C – From above the platform, the baseball appears to have curved to the right as the platform moved counterclockwise to the left, thus demonstrating the Coriolis Effect on the surface of the rotating Earth.

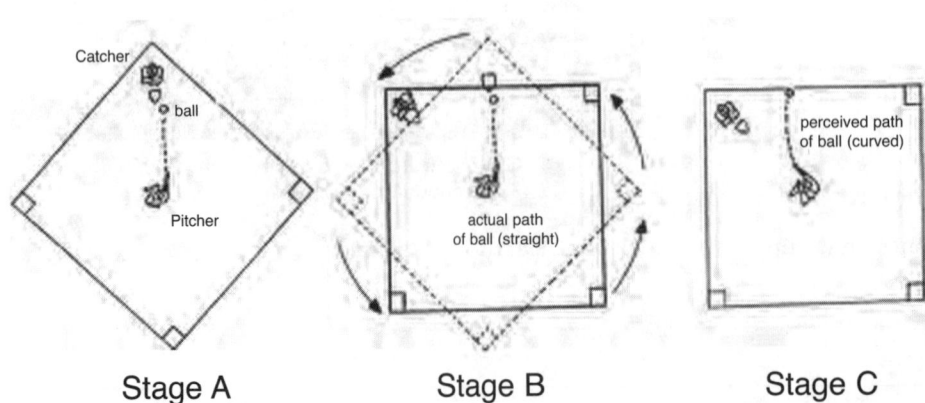

Stage A   Stage B   Stage C

6 What effect does Earth's rotation have on ocean currents and wind?

_____

_____

_____

_____

_____

CORIOLIS EFFECT

In the Northern Hemisphere what a sight
A moving object drifts to the right.

And generally, storms swirl counterclockwise
While seawater chooses to circle clockwise.

If all of this seems a bit much to rehearse
Remember, the Southern Hemisphere does it all in reverse.

EARTH'S REVOLUTION (A YEAR AND ITS SEASONS)
EVIDENCE THAT EARTH REVOLVES AROUND THE SUN

The earliest proof that Earth is revolving in space around the Sun can be traced to the first use of telescopes. Since stars are so far away, their relative position in the night sky to one another appears fixed. However, over the course of a year, telescopes showed the position between a relatively close star and one much farther away appeared to change in relationship to one another. This could best be explained if Earth was moving around the Sun.

Another observation that supported the model of an Earth revolving around the Sun was that the **constellations** (patterns of stars) appeared to move. During a year, the constellations visible to the eye in the evening sky change cyclically (repeating pattern).

Observation of Earth's closest star, the Sun, showed it to appear larger during the winter months and smaller as summer approached. This cyclic change could best be understood if Earth was revolving around the Sun. As the Earth revolves around the Sun, the altitude of the east–west path of the Sun across the Earth changes. During winter months in the Northern Hemisphere, the altitude is lower and so closer to the horizon. During summer, the altitude becomes higher. The change in altitude of the Sun makes it appear to be larger or smaller.

The strongest evidence that Earth is revolving around the Sun is the changing **seasons**. Earth rotates on an axis. This is an imaginary line from the **north** to **south pole**. Earth's axis is tilted about 23 $\frac{1}{2}$° away from its vertical. As

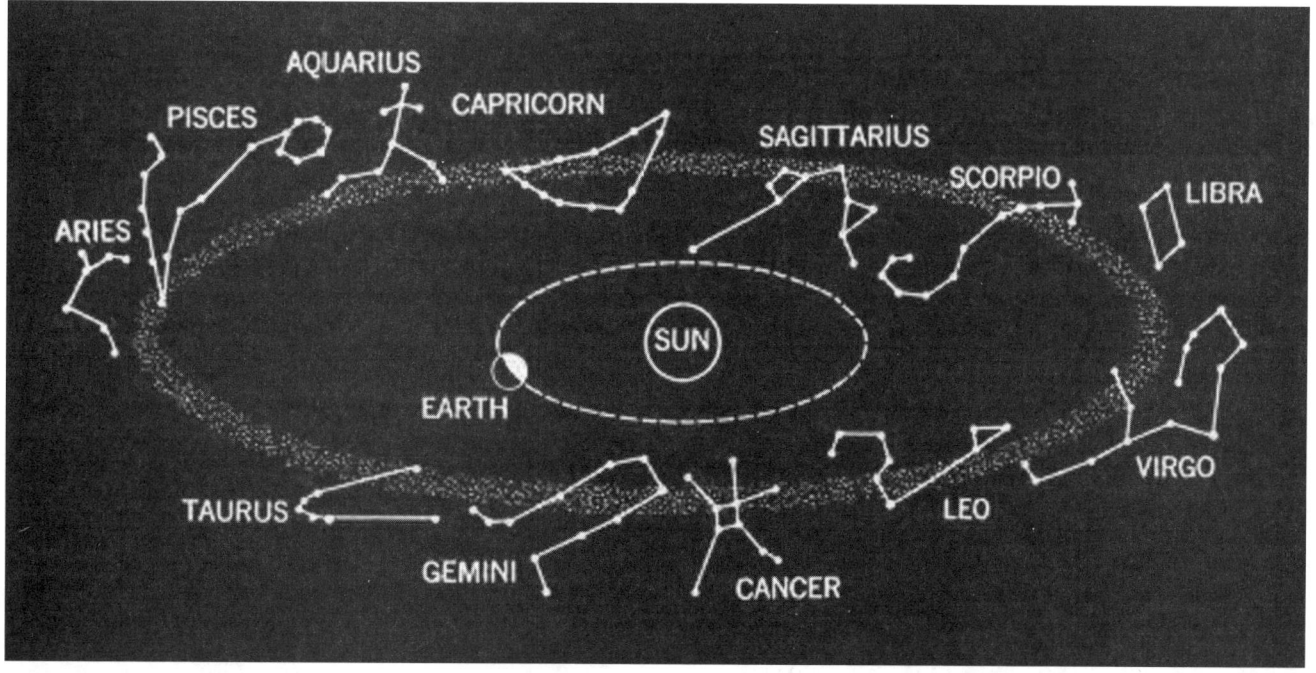

Earth moves around the Sun (about 93,000,000 miles away), the axis remains parallel to its previous position. This has the effect of causing Earth to receive varying amounts of **solar radiation** (sunlight).

When the tilt of Earth allows the Northern Hemisphere to receive its *maximum* solar radiation (most direct), the first day of **summer** is beginning in the Northern Hemisphere (**summer solstice**, June 21st). The Northern Hemisphere also experiences its longest day of sunlight on this date.

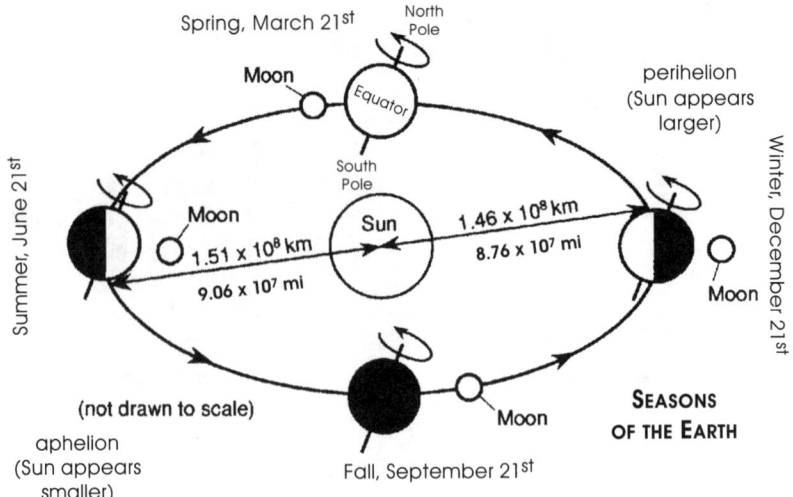

When the tilt of Earth allows the Northern Hemisphere to receive its *least* solar radiation, the first day of **winter** is beginning in the Northern Hemisphere (**Winter Solstice**, December 21st).

The first day of **fall** (**Autumnal Equinox**, September 22nd) has about an equal amount of darkness and daylight.

The first day of **spring** (**Vernal Equinox**, March 21st) has an about equal amount of darkness and daylight.

The seasons are in reverse in the Southern Hemisphere. When the Northern Hemisphere is tilted most toward the Sun, the Southern Hemisphere must be tilted farthest away.

Therefore, the tilt of Earth is the reason why Earth experiences seasons.

7 Describe two evidences that support the model that Earth revolves around the Sun. _____

_____

_____

8 What causes the changing seasons? _____

_____

DISCIPLINARY CORE IDEAS:

ESS1.B: EARTH AND THE SOLAR SYSTEM

THE IMPORTANCE OF GRAVITY AS IT RELATES TO THE MOTION OF EARTH

In 1666, the English scientist Sir Isaac Newton proposed the **Universal Law of Gravitation**. Basically, this law states that every object in the universe exerts a pull on every other object.

This law can be used to show that Earth must be round. An object weighs about the same no matter where it is weighed on Earth. This means the distance to the center of Earth (where the pull of gravity is the greatest) is the same in all places. This can only be true if Earth was round. Actually, Earth is not a perfect sphere. Earth bulges at its equator causing the diameter from east to west of Earth to be greater than the diameter from north to south (**oblate sphere**). However, the difference is only 0.3%. For all practical purposes, it is still considered a sphere. Pictures of Earth from space cannot detect this difference in size.

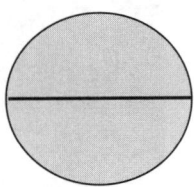

In addition to verifying the shape of Earth, gravity is what holds together the solar system, galaxy, and all of the universe. All celestial bodies of the universe are affected by the gravitational pull of other bodies.

ORBITAL MOTION

Orbital motion is caused when there is a balance between the forward motion of an object and the pull of gravity. When the solar system was formed, all objects were in motion. According to Newton's First Law of Motion, called the **Principle of Inertia**, an object in motion tends to stay in motion unless acted upon by an outside force. Since the Sun has the most amount of mass in the solar system, it exerts the strongest gravitational pull. A planet would move in a straight line, but the Sun's gravitational pull causes it to constantly change its direction. This results in a type of curved motion called **orbiting motion**.

Using Newton's **Law of Gravity**, the 17th century German astronomer **Johannes Kepler** calculated all orbits to be **ellipses** (oval shaped), though the orbit of Earth is almost a circle.

THE MOON & ITS MOTIONS.

The Moon shines because of reflected light from the Sun, not because it can give off any light of its own. It is located, an average, 240,000 miles from Earth. Although the Moon's gravitational pull is 1/6th that of Earth, its presence is felt on Earth. The gravitational pull of the Moon and Sun cause the rise and fall of waters on Earth (the tides). However, since the Sun is so far away, its effect is not as strong as the Moon's. It also pulls on the solid Earth, but that is not as apparent.

MEASURING A MONTH

The Moon is a **natural satellite** (a solid object that orbits another) of Earth with Earth's gravitational pull placing it in orbit. The motion of the Moon around Earth is used to calculate a month.

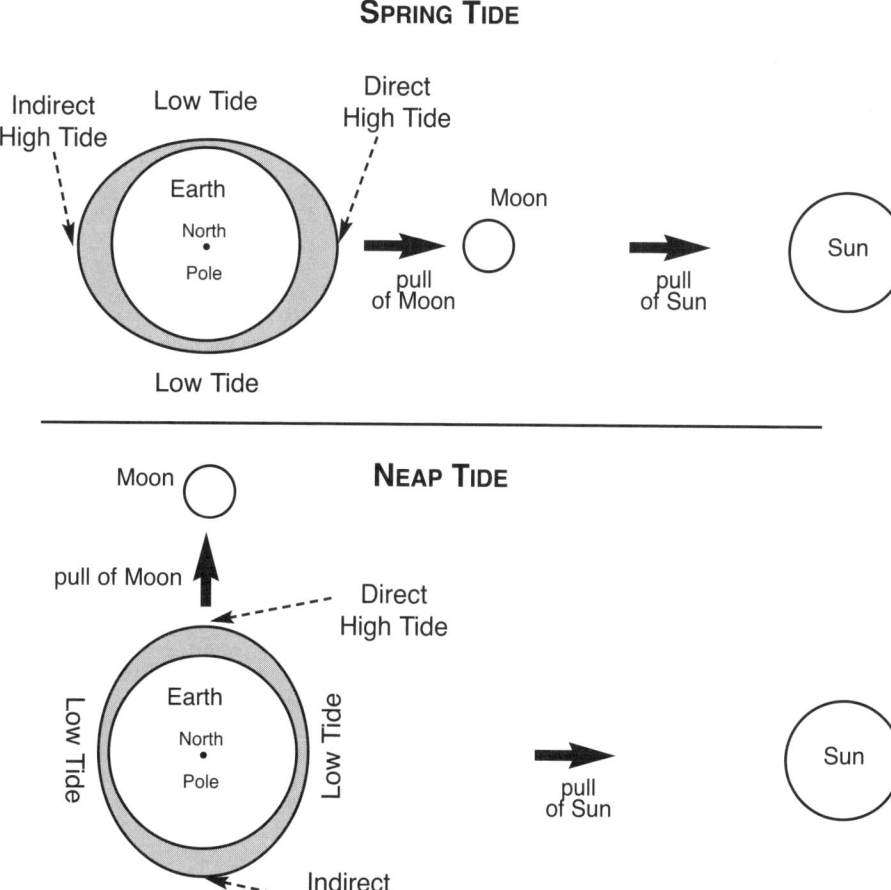

It takes 27.3 days for the Moon to complete one **revolution** around Earth. The Moon also completes one **rotation** on its axis in 27.3 days. Since its period of rotation and revolution are the same, this explains why only one side of the Moon is visible to Earth. Try spinning in place the same time you are moving around a friend, but never taking your eyes off the friend as you move. If you can do this, you are experiencing the motion of the Moon.

The time it takes for the Moon to move from one New Moon to the next New Moon is 29.5 days – the **Lunar Month** (see page 128). Although it takes the Moon only 27.3 days to complete its revolution, the extra two days are due to the changing position of the Sun as Earth revolves around the Sun.

Moon (satellite) Motion Around The Earth
Earth and Moon revolve around their common mass center. Note that the Moon does not revolve around Earth's center.

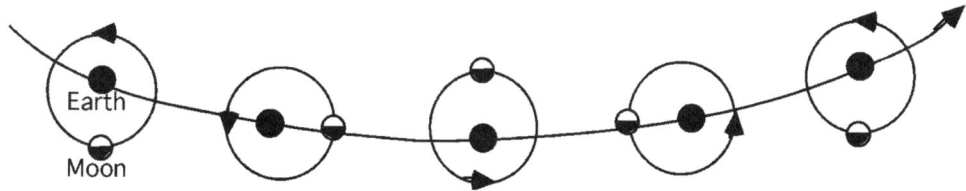

THE MOON'S PHASES

Since your view of the Moon depends on reflected light, the amount of reflected light visible to an observer changes. The change in the amount of light visible to an observer on Earth is called **phases of the Moon**.

- **New Moon** – The Moon is between the Sun and Earth. Only the back side of the Moon is illuminated (the side not visible to a viewer on Earth).
- **New Crescent** – Follows the New Moon.
- **First Quarter** – Appears as a Half Moon after New Crescent. The right side is completely illuminated, but the left side is dark.
- **New Gibbous** – Follows First Quarter looks like part of the Full Moon is missing.
- **Full Moon** – Occurs when the Moon is behind Earth. The entire face of the Moon is illuminated.
- **Old Gibbous** – Occurs as the amount of illumination decreases on the Full Moon.
- **Last Quarter** – Follows the Old Gibbous.
- **Old Crescent** – Last illuminated phase before a New Moon.

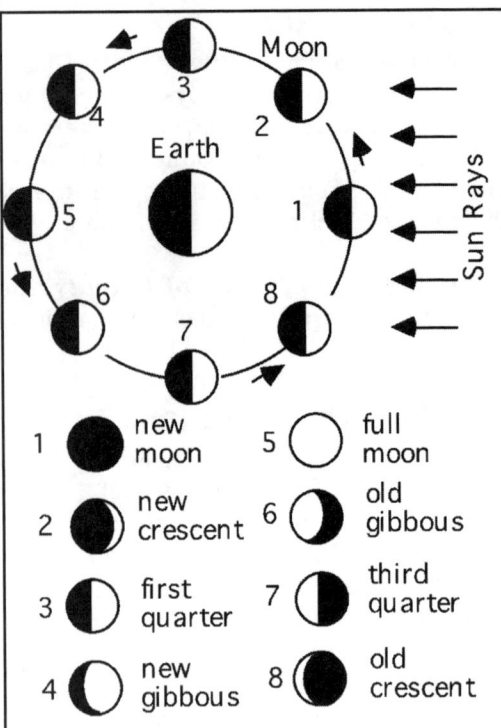

1 new moon
2 new crescent
3 first quarter
4 new gibbous
5 full moon
6 old gibbous
7 third quarter
8 old crescent

9 What causes the phases of the Moon? _____

_____

ECLIPSES

Imagine how a person from ancient times felt when, in the middle of daylight, it suddenly became dark, or in the evening, the Moon seemed to disappear from the sky. Today, it is known that solar and lunar eclipses are caused by the motion of Earth and Moon as they relate to the Sun. Total solar and lunar eclipses are rare events.

In order for a **solar eclipse** to occur, the Moon must be situated at just the right angle between Earth and the Sun. The Moon also must be at just the right distance from Earth so that its shadow strikes Earth.

A **lunar eclipse** occurs only when there is a Full Moon and Earth must be between the Moon and the Sun so that Earth's shadows darkens the Moon.

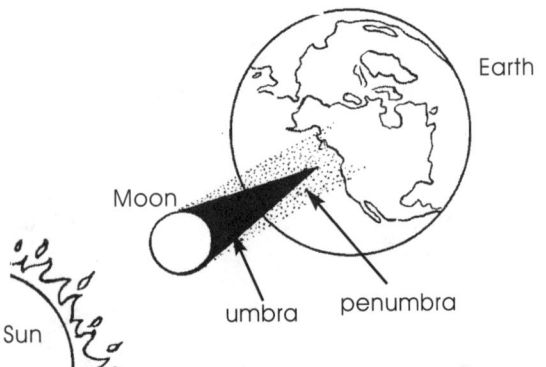

Solar Eclipse
Occurs only at a New Moon. Within the umbra, a total eclipse of the Sun is seen, whereas within the penumbra, only a partial eclipse is seen.

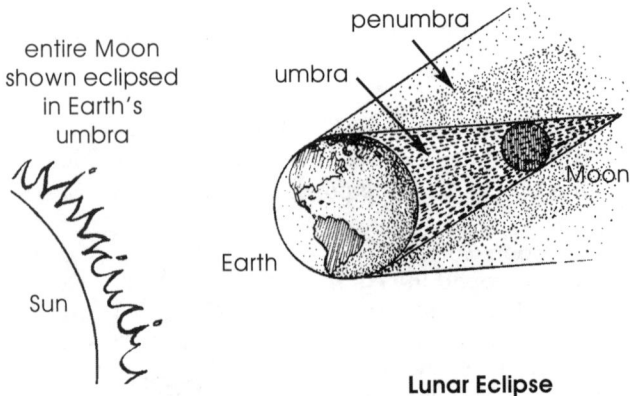

Lunar Eclipse
Occurs only at a Full Moon.

A complete lunar eclipse can last more than an hour, while a total solar eclipse is over in a few minutes. However, lunar eclipses are rarer than solar eclipses. This is because the angles of the Moon's orbit may miss the shadow of Earth completely. When a lunar eclipse occurs, the shadow of Earth on the Moon offers further evidence that Earth is round.

10 What causes solar and lunar eclipses? _____

_____

CELESTIAL DEBRIS

There is considerable evidence that the universe began with a giant explosion, when it was first formed (over 13 billion years ago according to most scientists). This model of the beginning of the universe became known as "***Big Bang Theory***."

Some of the matter from the explosion formed **stars** while other matter, too small to become stars, became planets. A star is composed mostly of hydrogen. Some of the energy

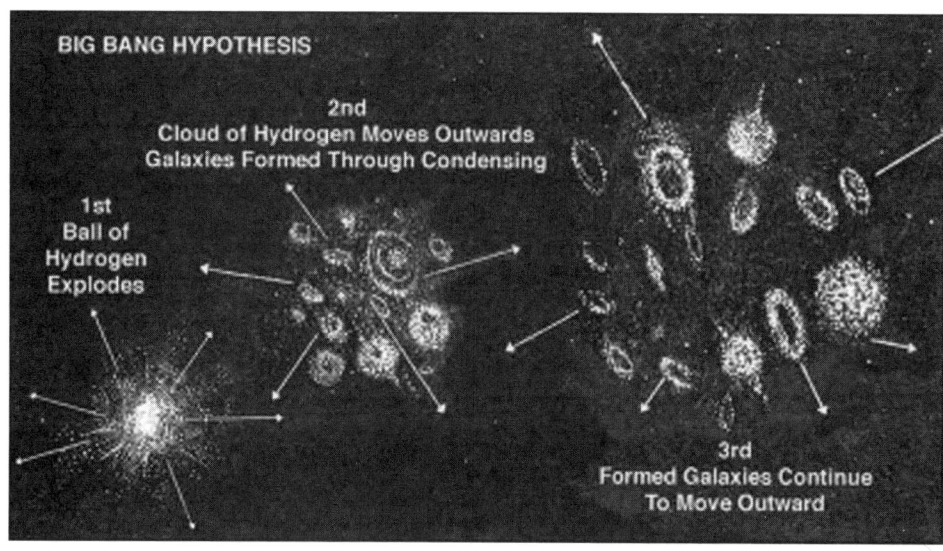

produced by a star is seen as visible light. Some scientists say that the planet Jupiter is actually a "want-to-be" star, but being too small, it became the solar system's largest planet. The star closest to Earth is the Sun.

Some of the smallest pieces of matter became known as **meteors**. The many **craters** on the Moon were most likely formed from the impact of large meteors. Meteors also strike Earth, but when they pass through Earth's atmosphere, the friction causes them to burn up. Most of these meteors are no bigger than a grain of sand. This burning up of meteors is called a **meteor shower** or a **shooting star**. Earth is always being struck by meteors, but large meteor showers can often be predicted because there are times during the year when Earth's orbit passes through meteor swarms (large numbers of meteors). Sometimes meteors do strike the Earth, but that is a rare event today.

In addition to the small debris from the Big Bang, other objects not big enough to be a planet were formed. These are called **asteroids**. The majority of asteroids move between Mars and Jupiter. Many scientists blame the impact of an asteroid on Earth for the extinction of the dinosaurs 65 million years ago.

Comets are relatively small members of the solar system with very elongated elliptical orbits. Comets are characterized by a tail made of fine gases and dust expelled from the head of the comet. Most take tens to thousands of years to travel around the Sun. Comets can travel into and out of the solar system. One of the most famous comets is **Halley's Comet**. Halley's Comet was seen twice during the 20th century.

... talk about your fast track to extinction...

According to the International Astronomical Union (IAU), A planet must do three things: it must orbit a star, it must be big enough to have enough gravity to force a spherical shape, and it must be big enough that its gravity cleared away any objects of a similar size near its orbit.

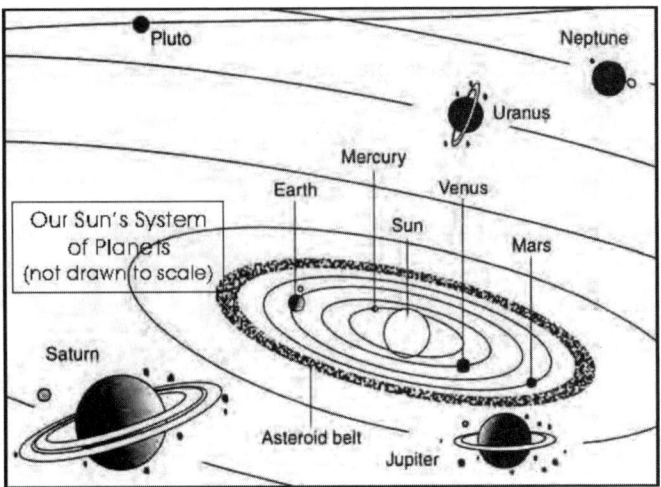

Earth is one of eight planets that make up the solar system. Earth, Venus, Mercury, and Mars, make up the inner rocky planets, while Saturn, Uranus, and Jupiter are the gaseous planets. Pluto, now classified as a dwarf planet, is most distant from the Sun. Some of the planets have one or more natural satellites (moons) that orbit them. Pluto has the most elliptical of orbits. Mercury is the planet closest to the Sun, but Venus is the hottest planet because of the large amount of carbon dioxide in its atmosphere causing a "Greenhouse Effect." Carbon dioxide is considered to be a greenhouse gas because it can trap solar energy. Scientists are constantly gathering new information about the planets. Some scientists think there is evidence for a tenth planet which they have dubbed Planet X. It is hypothesized to be located in the farthest portion of the solar system. It is called Planet X because the X both stands for an unknown and the number ten.

EARTH'S COORDINATE SYSTEM

In order to locate any point on a two dimensional surface such as a map, a **coordinate system** was devised. All locations on a map can be determined using the latitude / longitude system. This is not the only location system, but it is the one most widely used today. The intersection of a line of **latitude** and line of longitude can locate any point on Earth's surface.

The equator is used as a reference in determining latitude. Lines of latitude (or parallels) run east-west and are parallel to one another. The equator is placed at a latitude of 0°, while the latitude at the poles is 90° north or south of the equator. To distinguish between latitudes above and below the equator, degrees of latitude are identified as north or south.

Longitude (meridian): Lines of **longitude** (or **meridians**) run north-south. Since there is no naturally occurring reference point on Earth that runs north-south, a common reference point was needed. In 1884, the United States hosted an international conference at which it was agreed to use a meridian that runs through the Royal Observatory in Greenwich, England. This would be the **Prime Meridian**. The Prime Meridian was given the value of 0° longitude. Since Earth is a sphere, it was divided into two halves, each containing 180° of longitude east and west of the Prime Meridian.

11 How is a point on Earth determined? _____

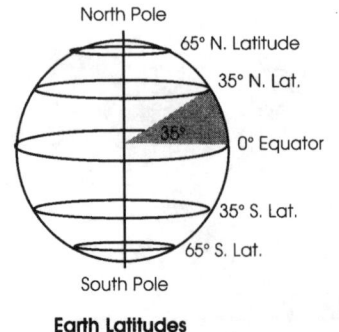

Earth Latitudes

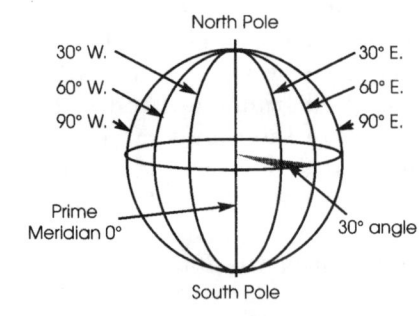

Earth Longitudes

MULTIPLE CHOICE

1 The geocentric model of the universe states that all objects in the universe revolve around the
(1) Earth   (2) Moon   (3) Sun   (4) Milky Way

2 One complete rotation of Earth on its axis makes up a
(1) day   (2) week   (3) month   (4) year

3 Foucault's pendulum helped prove that the
(1) Earth revolves around the Sun   (3) Moon revolves around Earth
(2) Sun revolves around Earth   (4) Earth rotates on its axis

4 The cyclic change in the apparent size of the Sun is due to the Earth's
(1) rotation   (2) tilt   (3) revolution   (4) shape

5 Which diagram best illustrates Earth's rotation?
(1) 1   (3) 3
(2) 2   (4) 4

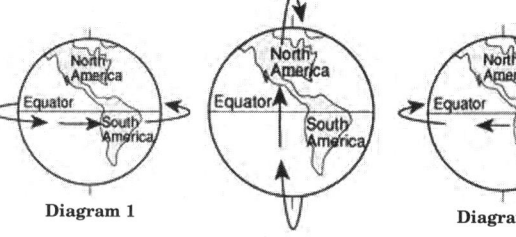
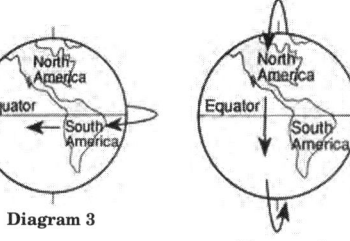

Diagram 1   Diagram 2   Diagram 3   Diagram 4

6 One complete revolution of Earth around the Sun equals a
(1) day   (2) week   (3) month   (4) year

7 Which position of Earth shown in the diagram represents summer in the Northern Hemisphere?
(1) A   (3) C
(2) B   (4) D

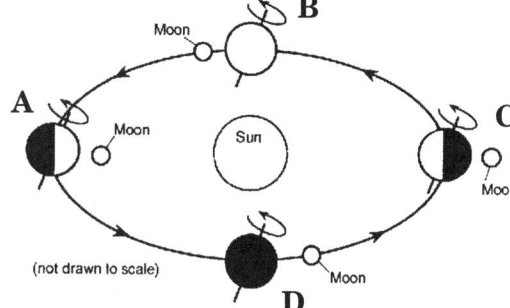

8 Which diagram best illustrates the shape of Earth from

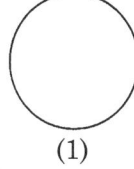   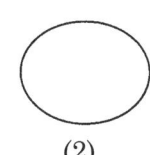   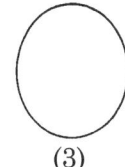   

(1)   (2)   (3)   (4)

9 The Moon shines because it
(1) is very hot   (3) reflects light from Earth
(2) reflects light from the Sun   (4) have special rocks that fluoresce (glow)

10 The length of a month is determined by the
(1) rotation of Earth   (3) revolving of the Moon around Earth
(2) rotation of the Moon   (4) revolving of Earth around the Moon

11 On the diagram at the right, which position of the Moon represents a full Moon to an observer on the Earth?
(1) A   (3) C
(2) B   (4) D

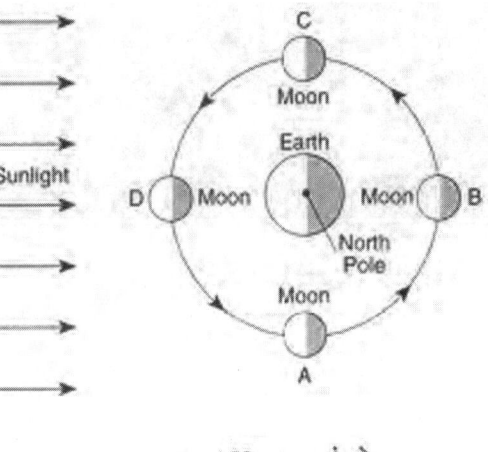

12 Which celestial object has the most elliptical orbit?
(1) Planet   (3) Asteroid
(2) Comet   (4) 4 Meteor

13 Which of the following planets is considered a gaseous planet?
(1) Mars   (3) Venus
(2) Jupiter   (4) Mercury

14 The diagram at the right shows the constellations that can be seen from Earth. Which constellations are visible in the fall?
(1) Hercules and Cygnus
(2) Leo and Gemini
(3) Pisces and Pegasus
(4) Aries and Orion

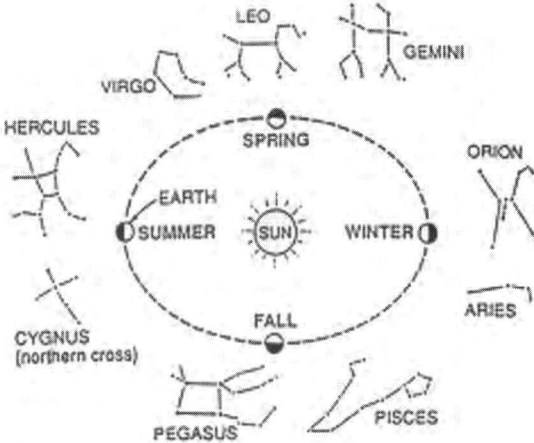

15 The diagram at the right shows four different phases of the Moon. In which phase is the Moon between Earth and the Sun?
(1) A   (3) C
(2) B   (4) D

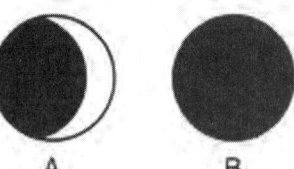 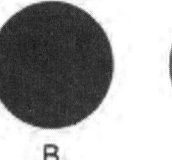 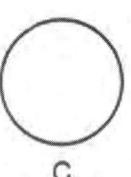 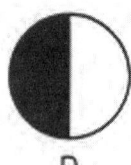

16 The diagram at the right shows a solar eclipse. Which statement is most true about a solar eclipse?
(1) A solar eclipse causes the entire daytime side of Earth to become dark.
(2) Only part of Earth becomes totally dark during a solar eclipse.
(3) A solar eclipse occurs when Earth is between the Moon and the Sun.
(4) During a solar eclipse the side of Earth facing away from the Sun receives sunlight.

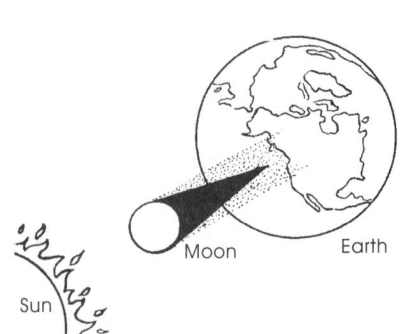

17 Which diagram below most accurately illustrates the relative motion of Earth, the Moon, and the Sun?

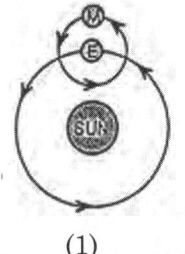 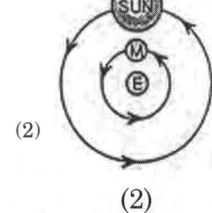 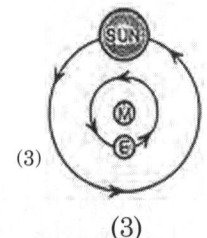 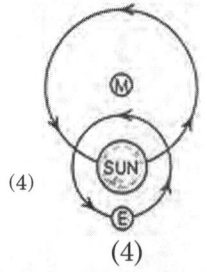

(1)   (2)   (3)   (4)

18 Which diagram below illustrates sunrise?

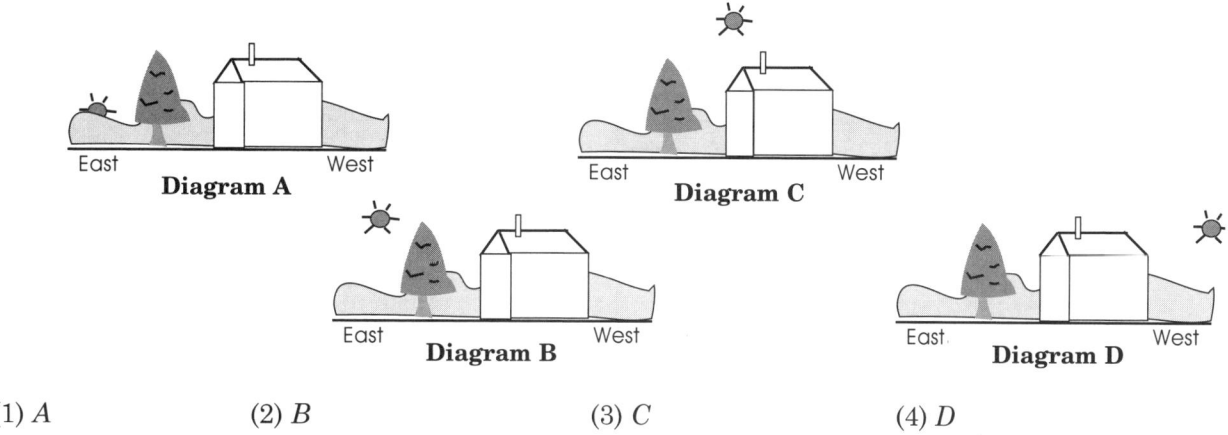

(1) A   (2) B   (3) C   (4) D

19 The position of the Polar Star changes in the evening sky because
   (1) Earth is revolving around the Sun   (3) the Polar Star is rotating on its axis
   (2) the Polar Star is revolving around Earth   (4) Earth is rotating on its axis

20 The force that holds the solar system together is
   (1) magnetism   (2) electricity   (3) gravity   (4) inertiax

THE QUESTIONS THAT FOLLOW TEST YOUR UNDERSTANDING OF SCIENTIFIC PRACTICES, CORES IDEAS IN SCIENCE AND CROSS-CUTTING CONCEPTS SUCH AS PATTERNS AND CAUSE AND EFFECT RELATIONSHIPS."

SHORT ANSWER

1 Directions: Use the diagram of the Moon's phases to answer the questions that follow:

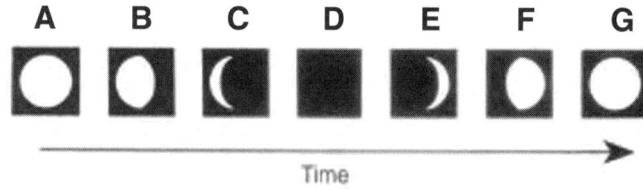

a   Which position of the Moon represents a New Moon? (1 point) _____

b   Explain why you chose that answer (2 points) _____

_____

_____

_____

2   The graph below shows how long it takes the planets Mercury, Venus, Earth, and Mars to revolve around the Sun in Earth days.

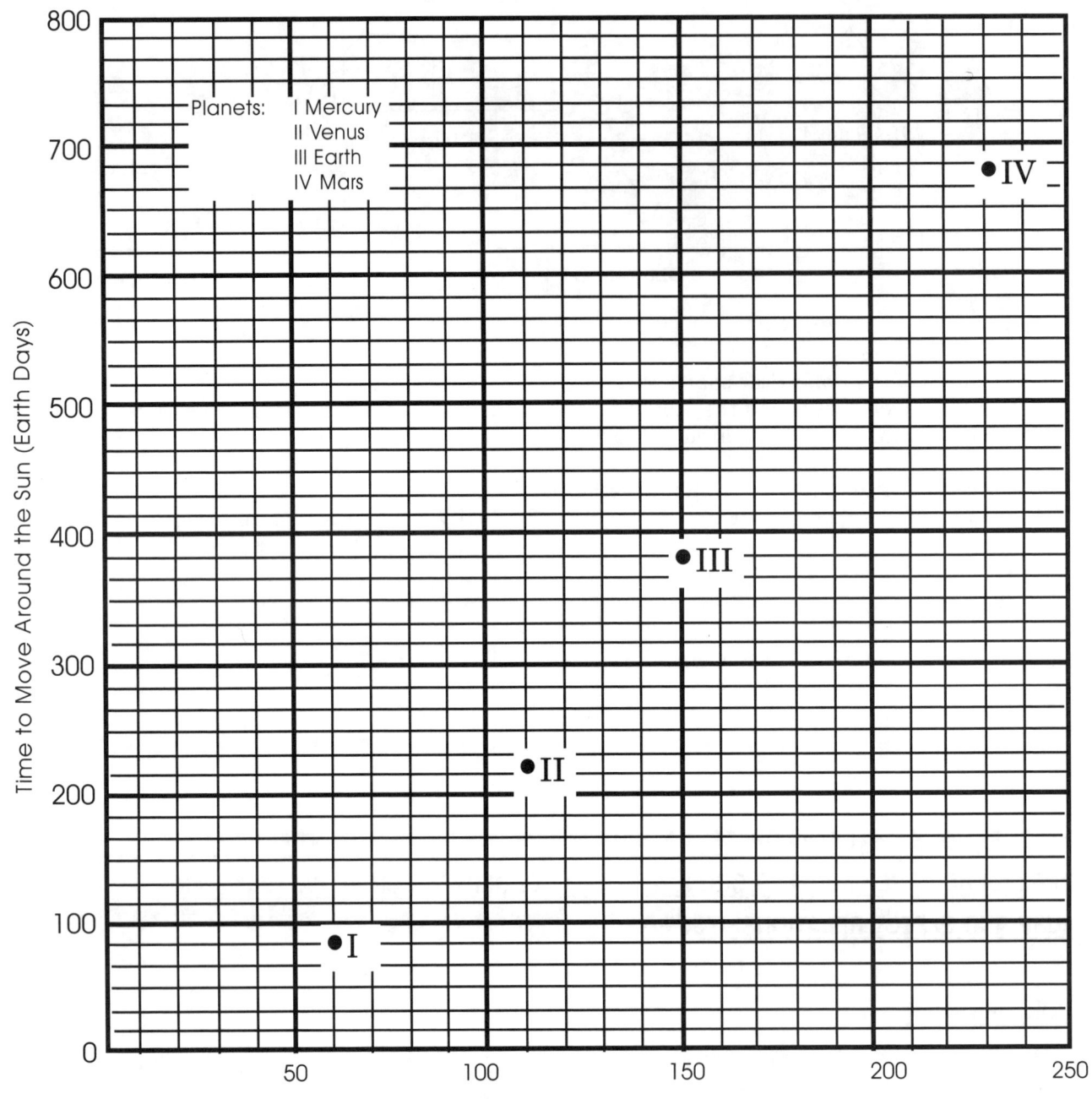

a   According to information in the graph, which planet takes the longest to circle the Sun in Earth days?

(1 point) _____

b   Describe a relationship that can be inferred from the graph and and cite evidence to support this

relationship. (2 points) _____

_____

_____

_____

3   The Earth is tilted on its axis while it revolves around the Sun. Explain how this causes the changing

seasons. (2 points) _____

_____

_____

4   Using the diagrams of solar and lunar eclipses, compare and contrast solar and lunar eclipses. (3 points)

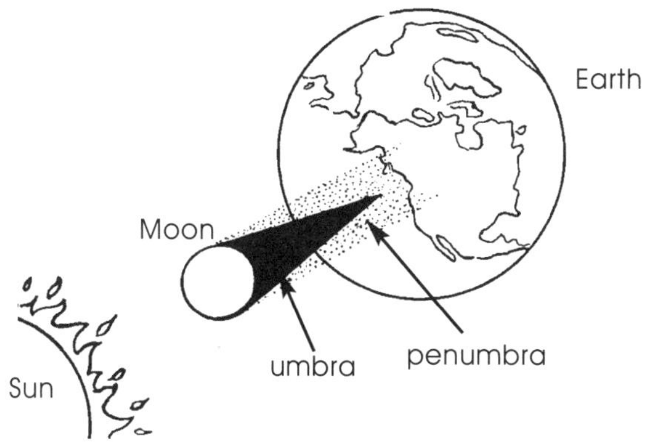

Solar Eclipse

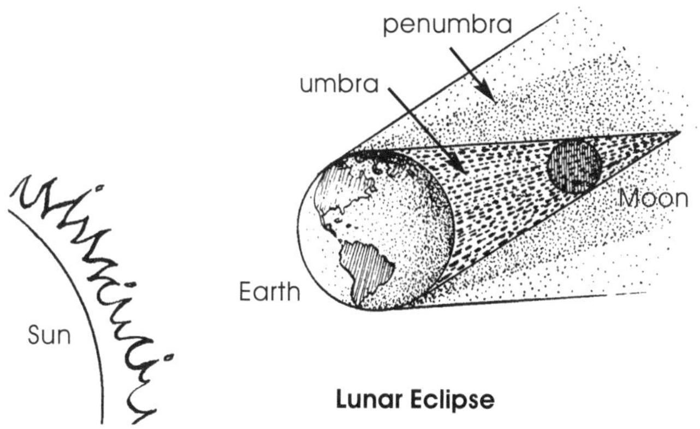

Lunar Eclipse

_____

_____

_____

_____

_____

_____

_____

SHORT ESSAY

1   Use the chart of the bright-
    ness of stars to answer the
    questions that follow.

    Reminder: *Luminosity* as
    used on this chart means
    the same as *brightness*.

Luminosity and Temperature of Stars
(Name in italics refers to star shown by a ⊕)

Luminosity is the brightness of stars compared to the brightness of our Sun as seen from the same distance from the observer.

a   How does the Sun's brightness compare to other stars? _____

_____

_____

b   Generally, what is the relationship between temperature and brightness of a star? Name a star that

      does not follow that pattern? _____

_____

_____

2   Below is a list of information that can be used to support the idea that Earth revolves around the Sun.
    Explain how each piece of information supports that idea.

a   tilt of Earth _____

_____

_____

b   seasons _____

_____

_____

c   Constellations _____

_____

_____

EARTH'S SYSTEMS

DISCIPLINARY CORE IDEAS:

ESS2.A: EARTH'S MATERIALS AND SYSTEMS

ESS2.C: THE ROLES OF WATER IN EARTH'S SURFACE PROCESSES

ESS2.C: HUMAN IMPACT ON EARTH'S SYSTEMS

KNOWING YOUR PLANET!

Today's technology allows meteorologists to make more accurate forecasts about weather and, at times even predict the eruption of a volcano or an earthquake. In order to accomplish all of this, scientists have had to learn how the atmosphere, hydrosphere, and lithosphere interact, evolve, and change.

THE ATMOSPHERE

Our **atmosphere** forms a thin layer around Earth. It is held in place by gravity. It consists mostly of the gases nitrogen (78%) and oxygen (20.9%) with very small amounts of carbon dioxide (0.03%), hydrogen, and the noble gases (helium, neon, argon, krypton, xenon, and radon). The atmosphere also contains water, dust, and other particulates (small particles).

12 What are the main gases that make up the atmosphere? _____

_____

The upper atmosphere is called the **stratosphere**. It contains the gas **ozone**. Ozone helps block ultra-violet radiation. UV radiation can cause skin cancer in humans. Over 99% of the matter that makes up Earth's atmosphere is located below an altitude of 20 miles (32 kilometers). The **troposphere** is the lower part of the atmosphere in which weather occurs and extends about 7.5 miles (12 kilometers) into space.

The temperature and concentration of gases, solids, and liquids changes as you travel upward through the atmosphere. The temperature changes have to do with the composition of the atmosphere's layer. For example, ozone which makes up part of the stratosphere also acts to warm this layer in addition to blocking ultra-violet radiation (UV light).

13 Using the graph of the atmosphere's layers at the right, describe the temperature changes from the lowest to highest part of the atmosphere.

_____

_____

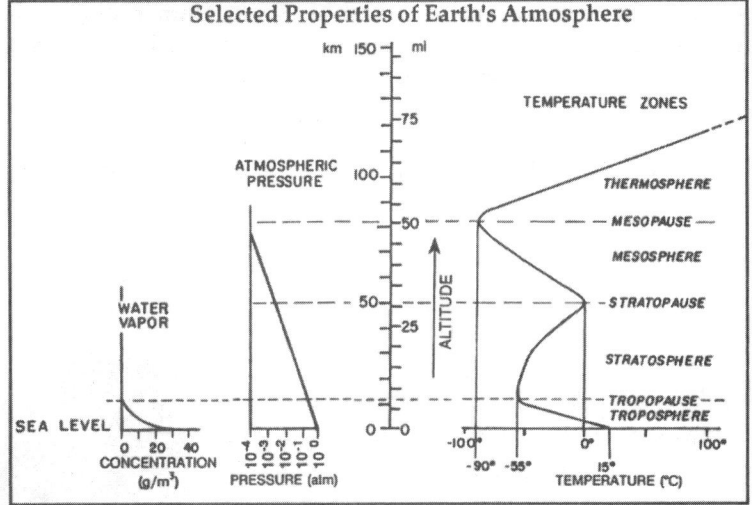

_____

_____

PEOPLE POLLUTERS

Most scientists believe that certain forms of human activity are changing the composition of the atmosphere. For example, global warming refers to the fact that Earth is getting warmer. Many scientists have related this to the massive burning of fossil fuels by humans.

The **Greenhouse Effect** (see below) has led to the hypothesis that Earth is getting warmer due to the build up of heat-absorbing gases in the atmosphere. However, the reason for the increase in temperature is often debated by many scientists.

Greenhouse gases include any gas that prevents heat energy absorbed by Earth from radiating back into space. Carbon dioxide, methane, ozone, nitrogen oxides, and water vapor are all greenhouse gases that enter the atmosphere both naturally and artificially.

- Since the Industrial Revolution, the amount of **carbon dioxide** in the atmosphere has increased by 27%, caused by the burning of fossil fuels such as coal, natural gas, and petroleum.

- The amount of **methane** gas has increased by 100% since the Industrial Revolution, mainly due to farming practices and landfills. For example, cattle burp methane gas and bacteria release methane gas in landfills.

- The amount of **water vapor** in the air is associated with many natural cycles on Earth. However, human activity, such as the draining of swamps and deforestation (cutting down of trees), can affect the amount of water vapor in a region.

- **Ozone** in the lower atmosphere is produced from the exhaust of automobiles as the fumes react to the energy of sunlight. Ozone is associated with a number of respiratory problems as well.

- **Nitrogen oxides** come mostly from the burning of fossil fuels such as coal, petroleum, and natural gas.

GREENHOUSE EFFECT

High energy, short wave radiation is trapped by glass (Earth's greenhouse gases)

Sun

Low energy, long wave radiation heats the greenhouse (Earth's atmosphere)

AN EXPLANATION OF THE GREENHOUSE EFFECT.

When the Sun beats down on a greenhouse, heat energy is transferred into the greenhouse. However, the glass "traps" the heat energy so it cannot escape out of the greenhouse. Greenhouse gases act similar to the glass in a greenhouse. They prevent the excess heat absorbed by Earth from radiating back into space. The concept of global warming is that Earth is getting warmer due to these greenhouse gases.

14 Explain what is meant by the "Greenhouse Effect." _____

_____

_____

Why should you be concerned about the planet getting warmer? The increased temperature will cause glaciers to melt and the warmer water will take up a larger volume (thermal expansion). Water already takes up to three-fourths of Earth. If Earth gets warmer, it will cover more of the lithosphere.

Since many major cities of the world are located along the coasts, many will be under water. The rise in temperature will also affect the climate. The result could be major food shortages.

THERE ARE HOLES IN EARTH'S STRATOSPHERE!

Human activity has even changed the composition of the upper atmosphere called the **stratosphere** which contains the gas ozone. Ozone acts to block UV radiation, which is a cause of skin cancer and does damage to crops. Satellite data has shown the amount of ozone has decreased in the upper atmosphere (particularly over the Antarctic and Arctic Circle). This decrease has formed "holes" in the stratosphere exposing Earth to more UV radiation.

CAUSE OF OZONE DEPLETION

The use of **chlorofluorocarbons** (CFCs) is the major cause of **ozone depletion**. CFCs is a component of manufactured freon (a refrigerant) and has been used as a propellant in aerosol cans and the manufacturing of styrofoam. When this gas reaches the upper atmosphere, it chemically reacts with ozone, breaking it down. Today, CFCs are no longer used in aerosol cans or in the making of styrofoam in the United States and Canada. New refrigerants are also being manufactured.

15 Distinguish between the "Greenhouse Effect" and depletion of the ozone layer in the stratosphere.

_____

_____

_____

THERE IS A GREAT DEAL OF PRESSURE LIVING ON EARTH

The atmosphere exerts a **pressure** on the living and non-living environment. If there was no atmosphere, water would boil at room temperature. The Moon has no atmosphere. When astronauts visit the Moon, they must wear pressurized suits. If not, their blood would boil, because blood is made mostly of water. The pressure of the atmosphere changes with altitude. It is 14.7 pounds per square inch (in the metric system: 10,000 N per square meter) at sea level. It decreases as the altitude increases.

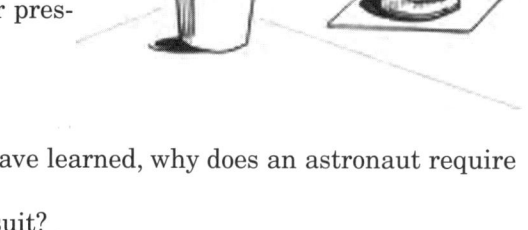

Defying Gravity? Fill a small paper cup $^3/_4$ with water. Holding the cup over the sink, place a 3 x 5 card over the opening, so the top of the cup is *completely* covered. While holding the card tightly over the opening, turn the cup upside down. The card stays on the cup because of air pressure!

16 Applying what you have learned, why does an astronaut require a pressurized space suit? _____

_____

_____

_____

THE EARTH IS LIKE AN ONION

The lithosphere is the solid, rocky outer portion of Earth that extends about 100 km into its interior. It makes up the crust and upper mantle. However, Earth is not solid throughout. Melted rock is released when volcanoes erupt. This melted rock comes from the mantle or lower crust in regions of extreme heat. In fact, if the Earth could be cut in half, it would resemble an onion. Unlike an onion however, the Earth is not made of the same matter throughout.

LAYERS OF THE EARTH

- Crust – upper region
- Mantle – layer under the crust made up of mostly liquid rock
- Outer Core – liquid rock
- Inner Core – solid rock

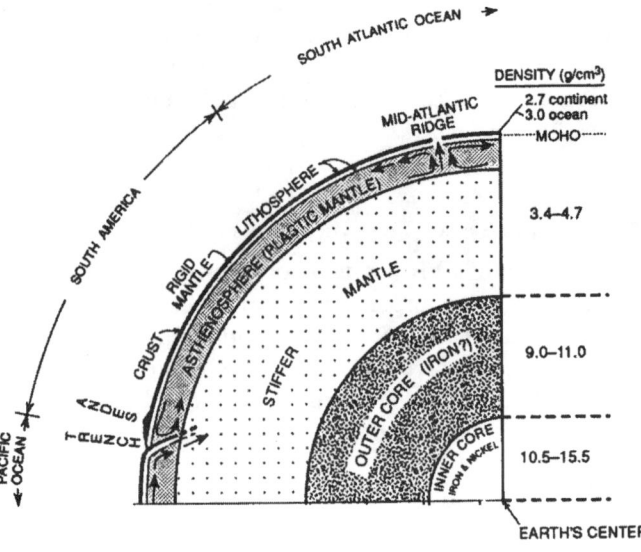

This layering occurred when Earth first formed. The dense matter sank to the center of Earth forming the core, while the less dense materials formed the outer boundary of Earth (the crust).

This model about the structure of Earth was inferred by studying the motion of seismic waves (waves from earthquakes). In 1909, a Yugoslav seismologist, **Andrija Mohorovicic** (mo-ho-ro-vi-chic) studied seismic waves to determine the depth of the crust. He noticed that when seismic waves moved through Earth, they would speed up and slow down. He felt that this can only be explained if the density of the rock changed. Therefore, he hypothesized that Earth must be composed of layers of rock with different properties.

17 Describe Earth's layers and their properties. _____

_____

_____

_____

IT IS ALL WET!

About three-fourths of the lithosphere is covered with water, called the **hydrosphere**. It includes the oceans, lakes, rivers, streams, and ponds. Because the soil and bedrock (rock under the soil) have many cracks and pores, some of the water, called **groundwater**, has seeped underground into these small openings. Some of the water is not in the liquid phase, but exists as ice such as in glaciers and icebergs.

18 About what percentage of the Earth is

covered with water? _____

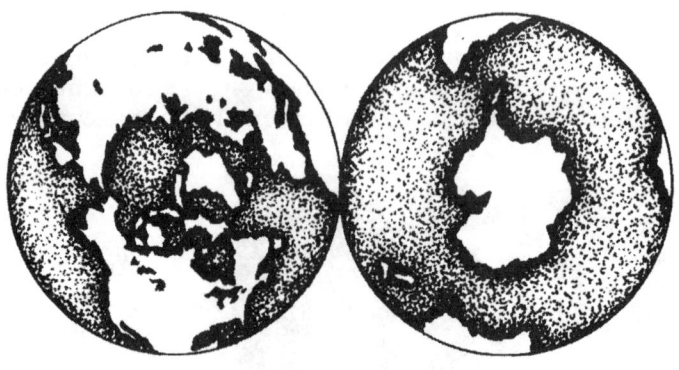

Earth's Hydrosphere
(viewed from North Pole (left), South Pole (right))

THE SHAPING AND RESHAPING OF EARTH'S CRUST BY WEATHERING, EROSION AND DEPOSITION.

WEATHERING AND ITS EFFECTS

The interaction of the lithosphere with the atmosphere and hydrosphere results in **weathering**. Weathering is when rock is changed physically or chemically as it interacts with air, water, and living things. It occurs on or near the surface of the lithosphere.

PHYSICAL WEATHERING

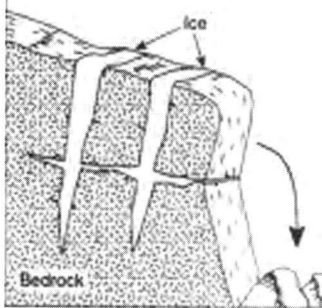

Physical weathering changes the size and shape of rock but does not change its composition.

Cracks and potholes in roads are caused by **frost action** – water that seeped into the asphalt of the road and froze. Since water expands when it freezes below 4°C, the pressure of the expanding ice causes cracks in the road. This can also cause large rocks to break into smaller rocks. Weathering by frost formation can only occur in parts of the United States where freezing and melting occur frequently.

Water and wind carry small particles that act to abrade (grind up) larger rocks into smaller rocks. This is called **erosion**.

As matter is heated, the particles in it move faster, causing it to expand. When heat is removed, the particles slow down and contract. The effects of frequent **temperature** changes places a great deal of stress on solid matter. The heaving of roads (bending upwards) is due to exposure to many temperature changes.

The **water pressure** from the roots of plants can grow into rocks breaking it into smaller pieces. For example, the roots of trees can crack sidewalks and roads and cause rock slides.

19 Describe the effects of two types of physical weathering. _____

_____

CHEMICAL WEATHERING
CHEMICAL WEATHERING CHANGES SIZE, SHAPE, & COMPOSITION OF ROCK

Water is sometimes called the **universal solvent**, because it can dissolve more substances than any other liquid. Certain minerals that make up rock are soluble in water. When rock is covered by water, minerals leave the rock by dissolving. This creates spaces in the rock that can weaken the rock causing it to break into smaller pieces.

Oxygen reacts easily with other substances changing their properties. For example, oxygen readily combines with iron to form iron oxide (rust). This weakens the structure of iron causing it to flake and eventually fall apart.

The gas, **carbon dioxide**, does not contribute to weathering until it dissolves in water. When it dissolves, it forms a weak acid called carbonic acid. This acid readily reacts with a number of minerals in rock. The mineral calcite is dissolved by carbonic acid. The effects of this type of weathering becomes dramatic when the bedrock is composed of limestone. Groundwater containing carbonic acid moves underground dissolving large amounts of the limestone. The result is the formation of spectacular underground **caverns**, such as Howe Caverns in New York State.

America's Caverns © Garnsey, 1976

20 Name two examples of chemical weathering. _____

_____

HUMAN ACTIVITY CAUSES WEATHERING

Acid rain also adds to the weathering of rocks, buildings, automobiles, and statues. Acid rain comes mostly from the burning of coal rich in sulfur compounds by power plants in the Midwest.

When the coal is burned, sulfur oxides form. In the atmosphere, sulfur oxides combine with water droplets forming sulfuric acid. The result is that the precipitation becomes acidic. The jet stream brings this precipitation to other places, such as the Adirondacks so that it literally rains or snows sulfuric acid. Since the Adirondacks does not have much limestone in its bedrock (limestone can neutralize an acid), acid rain has caused much damage to this region of New York State. The life in hundreds of lakes has been destroyed because of acid rain. Some lakes are as acidic as lemon juice.

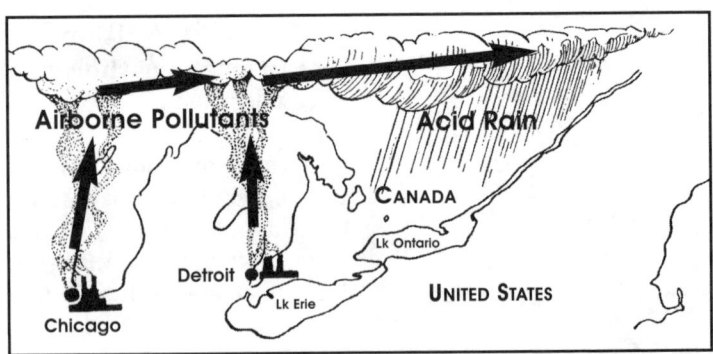

21 Describe the causes and effects of acid rain.

_____

_____

_____

Western Strip Mine © Moreau, 1982

During **strip-mining**, land is literally stripped of its vegetation. Large holes dug into Earth are not covered up. The result is that the exposed land is left to the agents of erosion such as water and wind.

EROSION

The Rocky Mountains are still "growing." Mountain building is caused by the movement of the crustal plates. As they collide, the crust folds and mountains are "born." Erosion continually causes sediments of rock to be deposited in another place. When the forces of erosion are greater than that of mountain building, mountains shrink in size such as what is happening to the Appalachian mountains.

Constructional and Destructional Forces

Mountain Building

precipitation

Erosion

sedimentary layers

ocean

Earth's Crust

magma

bed rock

Gravity is the main force that causes erosion. All sediments on a hill, such as rocks and soil, are affected by gravity. Over time, they may slide down the hill due to Earth's gravitation.

Forces associated with the motion of the hydrosphere and atmosphere also cause erosion. The **agents of erosion** include moving water, wind, and glaciers.

Moving water is one of the main forces that has shaped Earth. When water weathers rock, it smooths it out, eventually changing some rock into soil. Even the splash of a drop of rain striking soil moves it. If an area has received a large amount of rainfall, its groundwater saturates the ground below so that runoff on the surface begins moving from higher to lower regions.

Runoff can also occur during heavy rainfall when water cannot seep into the ground fast enough. This causes a layer of moving water on the surface such as a stream. As a stream moves, it carries sediments. The force of the moving water combined with the sediments in it act together like carving tools shaping Earth as it moves.

..AND A THOUSAND YEARS FROM NOW PEOPLE WILL LOOK AT WHAT WE HAVE BUILT HERE TODAY, AND BE TOTALLY AMAZED...

The movement of the **ocean currents** is constantly reshaping the shores of beaches. Ocean waves are a very powerful erosion agent responsible for many of the features of shores. Sea cliffs, sea caves, and sea terraces are examples of wave erosion on rocky shores.

Wind can also cause erosion, but its impact is not as great as water erosion. The effects of wind erosion are felt mostly in very dry areas with small amounts of vegetation. The rock **buttes** (narrow flat hills) and **mesas** (broad flat hills) are carved out of the softer sediments eroded by the wind.

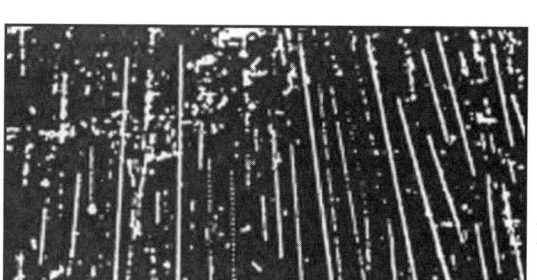

Striations on Limestone Bedrock
(Exposed in New York State)

Glaciers are large bodies of ice formed over many years when snowfall exceeds melting. The ice at the base of the glacier behaves like a fluid causing the glaciers to move from higher to lower elevations. As they move, they carry rocks which act to carve the landscape. **U-shaped valleys** were formed by glaciers. Much of the landscapes of New York and New England were shaped by glaciers. The eroding forces of the glaciers created the beautiful lakes of the Adirondacks such as Lake George and the Finger Lakes. In the photograph on the left, evidence of their motion can be seen in the fine striations (thin parallel lines) on the rock they glided over.

22 What is erosion and what is the main force responsible for erosion? _____

_____

23 Using the Peanuts cartoon, explain why Lucy's brother is concerned. _____

_____

A "ROOT" TO CONTROLLING OF WATER EROSION

Erosion of the land can be checked with **vegetation**. Plants with highly branched roots hold the soil in place. Farming in hilly areas uses a type of **contour farming**, planting into a hill to cut down on the effects of erosion caused by gravity. Along the coast of Long Island, erosion can be checked by stopping the destruction of natural sand dunes that help preserve the beaches.

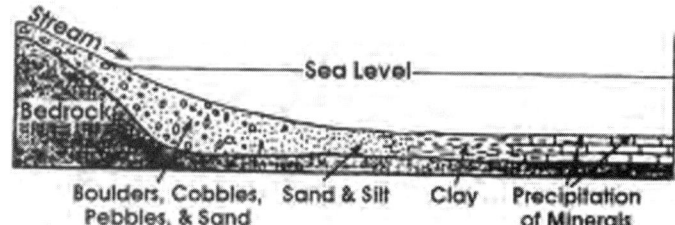

Horizontal Sorting

DEPOSITION OF PARTICLES

Deposition is when the agents of weathering and erosion deposit particles that they can no longer transport. **Sedimentation** is the process by which the particles are deposited. Deposition of sediments varies depending on the agent of erosion. What follows are some examples of deposition and its effects on landscape for some agents of erosion.

Sediments moved through water are eventually deposited in the water or near a shore, such as deltas, at river mouths. Sediments in water are deposited in layers. A stream moves more slowly when the slope of the land flattens. This causes the stream to **meander** or curve. Water moving along the outside curve (*B*) of a stream has greater motion causing erosion of the land on that side of the stream. However, the slower moving water on the inside of the curve (*C*), results in deposition of sediments on that side.

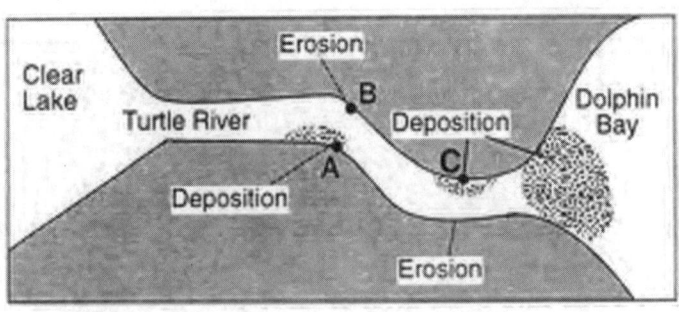

The motion of ocean waves and currents move sediments forming beaches and coastal formations.

Sediments from glaciers sort by size if the melting occurs in the front of the glacier. Northern and southern Long Island, New York shows this type of effect. The hilly landscape of northern Long Island contrasts with the flat landscape of southern Long Island with its beautiful beaches deposited by sand from the melt-water of glaciers.

The spectacular sand dunes of Long Island, New York, were formed by wind. The sand dunes act to cut down the effects of erosion caused by the ocean. Many of the housing developments along the shores of Long Island have destroyed these sand dunes resulting in excessive erosion of the beaches.

Minerals dissolved in water are deposited in the form of crystals after the water **evaporates**. Halite (rock salt) and gypsum are examples of minerals deposited this way.

24 Describe two mechanisms for the deposition of particles. _____

_____

_____

"IT'S A DIRTY BUSINESS!"

Soil is formed from weathered rock. This is a slow and complex process. The length of time for formation depends on the type of bedrock and the climate. It has been estimated that it can take 200-1200 years to form 2.5 cm (1 inch) of topsoil from hard rock. Softer bedrock such as shale and sandstone may take about 20 years to form. The composition of the soil varies depending on its origin. Sometimes it is formed from local bedrock, while at other times it is formed from sediments carried by agents of erosion (wind and water).

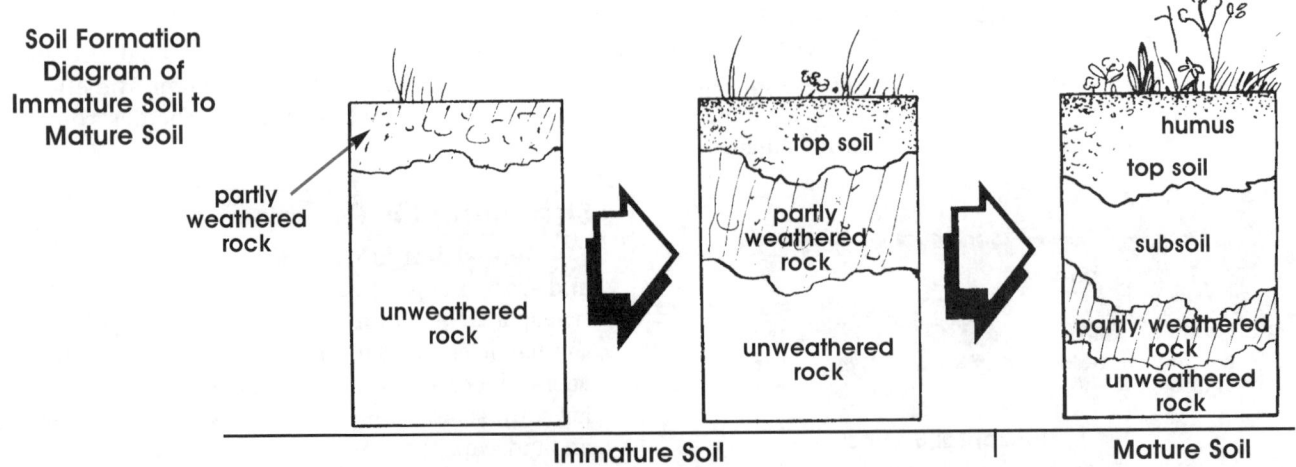

LAYERS OF SOIL

Soil contains a number of layers. Studying these layers tells a lot about the soil and how it can be best used. For example, some soil may be very suitable for agriculture, while other soil would be better used for animal preserves or home building. The soil in which your food is grown is called **top soil**. It makes up the darker component of soil because it contains organic matter (decayed plants and animals and the wastes of organisms). Air and water fills the spaces between sediments in soil.

25 How is soil formed? _____

_____

_____

26 Describe the properties of soil that are good for growing food _____

_____

_____

LIFE IN THE SOIL

Many organisms living in the soil **fertilize** and **aerate** it (add spaces for air to circulate). Earthworms move through soil; they act to aerate it while depositing wastes that make the soil more fertile. Some microorganisms that live in the roots of plants such as clover convert nitrogen from the air to nitrates which add fertility to the soil. It is the activity of bacteria that cycles nitrogen between the lithosphere and atmosphere.

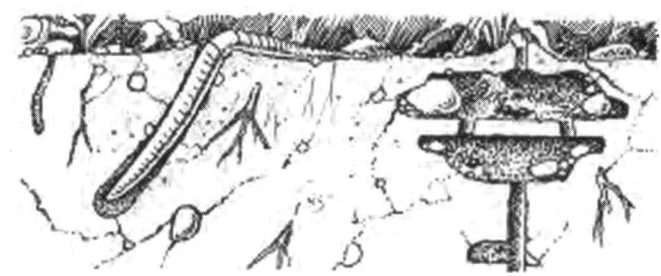

WATER CYCLE

The water above and below the Earth's surface is continually cycled between the lithosphere, atmosphere, and hydrosphere. This movement of water and the change in form is called the **water cycle**.

Water in the form of **precipitation**, enters the lithosphere when the air becomes saturated with water vapor. Depending on the temperature, water falls onto the lithosphere as rain, snow, sleet or hail. If the precipitation is rain, some of the water accumulates in depressions on Earth forming small puddles or streams. The remainder seeps into the ground to form the groundwater. Sometimes, water vapor in the air is deposited on the lithosphere directly as ice. This phase change is called deposition. If you see "snow" on the inside of a window during the wintertime, that is a example of **deposition**. It is also an example of poorly insulated windows!

Liquid water evaporates back into the atmosphere. Water enters the atmosphere when puddles disappear quickly after a rainfall. Streams disappear if there has been a lack of rainfall; rivers move more slowly, their banks narrow; and lakes shrink in size. This means it changes phase from a liquid to a gas becoming a colorless gas called **water vapor**. The warmer and dryer the air is, the faster this phase change takes place. If the water is frozen on Earth, it may return directly to the air by a type of phase change called **sublimation**. This is when the surface of a solid changes directly into a gas. As moisture in the air rises, it often condenses (changes from a gas to liquid) on particles in the air to form clouds. Depending on the temperature, the clouds may contain small water droplets or ice crystals.

Water also enters the atmosphere due to biological activity. Many living organisms exhale water vapor into the air. Plants carry out an activity called **transpiration** in which water moves up a plant and exits through openings in the leaves as water vapor.

Earth's Water Cycle

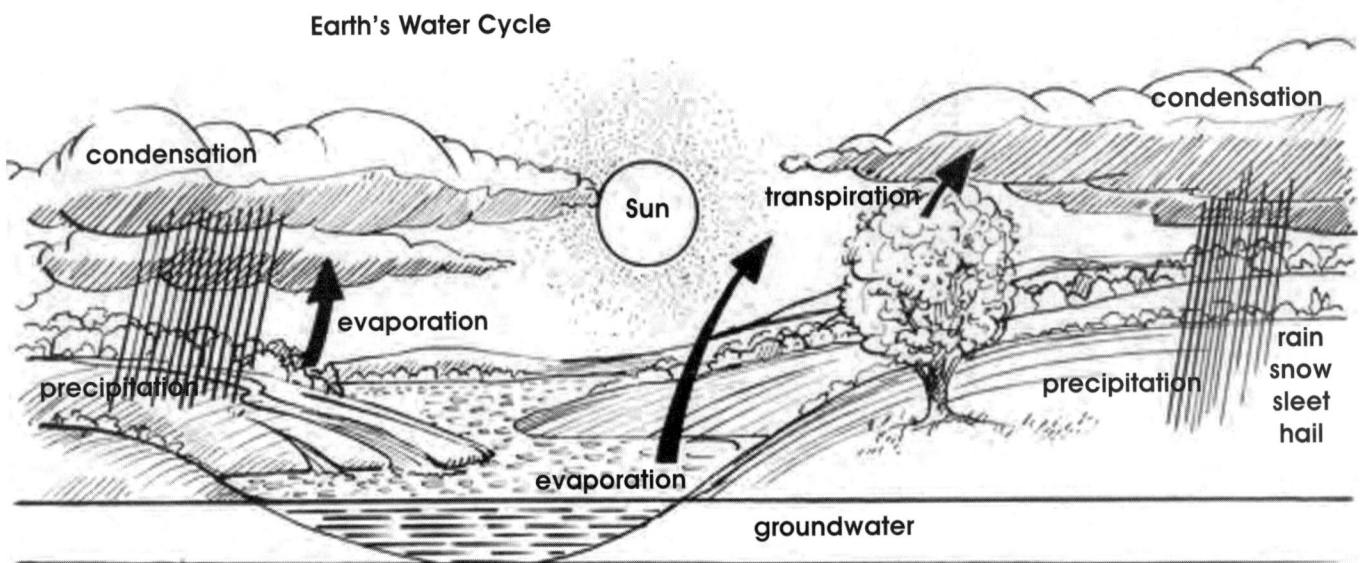

Water enters the hydrosphere during times of precipitation. Most of the water falling to Earth is deposited directly into the oceans, lakes, rivers, and streams. During rainfall, some of the rain restores the **groundwater,** while other becomes **runoff**, eventually reaching a large body of water. In the spring, the snow melts becoming melt water that recharges the rivers and lakes. In the spring, rivers literally roar with moving water.

27 Briefly describe how water moves through the hydrosphere, lithosphere and atmosphere. _____

_____

_____

_____

Disciplinary Core Ideas:

ESS2.B: Tectonic Plates and Large-scale System Interactions

ESS1.C: History of Planet Earth

Rock & Shake, Rattle & Roll!

Rocks And The Rock Cycle

Did you ever notice that when you pick up a rock it does not look the same throughout? The texture and color seems to vary throughout the rock. The assortment of substances that you can identify in a rock is its mineral content. A **mineral** is a pure substance made up of only one type of matter. A pure substance can be an element or a compound. An **element** is the simplest form of matter composed of only one type of **atom**. Gold and uranium are some examples. Usually, elements are chemically combined with other elements to form compounds. Therefore, most minerals are compounds such as fool's gold (pyrite) which is a compound made of the elements iron and sulfur. Only a few rock-forming minerals make up most rocks of Earth.

28 What is a mineral?_____

_____

Minerals are identified on the basis of properties such as streak, hardness, luster, cleavage, and reaction to acid.

- **Streak** is the color of the powdered mineral which can easily be seen by rubbing the mineral on a white or clear smooth surface such as ceramic.

- **Hardness** is the resistance of a mineral to being scratched. The hardest mineral is a diamond and the softest mineral is talc. The hardness of a mineral is usually stated in terms of the **Moh scale** that assigned a numerical value to hardness. Talc, the a the softest is a 1, while a diamond has a value of 10.

- **Cleavage** is the tendency of a mineral to break along regular surfaces in one or more specific directions. For example, the mineral mica separates easily into thin layers along its cleavage surface.

- **Luster** is the ability of the mineral to shine or how it reflects light. It may be dull or shiny like a metal.

- **Reaction to acid.** Some minerals are so similar in appearance that a chemical test is required. For example, halite and calcite are difficult to tell apart without testing with hydrochloric acid. Hydrochloric acid reacts with calcite forming bubbles and releasing the gas, carbon dioxide.

Moh's Hardness Scale		Approximate Hardness of Common Objects
Talc	1	
Gypsum	2	Fingernail (2.5)
Calcite	3	Copper penny (3.5)
Fluorite	4	Iron nail (4.5)
Apatite	5	Glass (5.5)
Feldspar	6	Steel file (6.5)
Quartz	7	Streak plate (7.0)
Topaz	8	
Corundum	9	
Diamond	10	

29 Describe how a mineral can be identified using streak and luster. _____

_____

_____

MAJOR CATEGORIES OF ROCKS AND THEIR CHARACTERISTICS

Scientists study rocks by classifying them into major groups and subgroups. The major categories of rocks are classified according to how they are formed. The mineral content of a rock is also used for subcategories.

As the name implies, **sedimentary rock** is chiefly formed when rock is weathered into small sediments. These sediments settle in water and sometimes, directly on land. Minerals in the water or on the land can cement the sediments together resulting in this type of rock. Rock can also form by **compaction**. The sheer weight of layers of sediment causes them to be compacted (pressed together). Coal is a type of sedimentary rock formed when layers of ancient plants were compacted together by great pressure rearranging the carbon molecules that are part of all living things. Sedimentary rock can also be formed when water containing dissolved salts evaporates, leaving behind salt crystals.

uniform layers

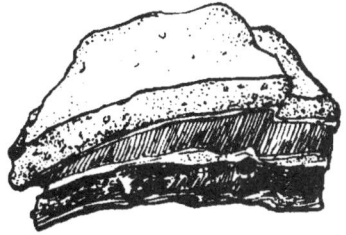
overlying layers

As sediments are deposited, their layers are clearly visible in sedimentary rock. This is because the sediments may vary in thickness, type of mineral, and degree of compacting. The walls of the Grand Canyon are made mostly of sedimentary rock.

ROCKS WITH A HISTORY LESSON

Sedimentary rock also houses a history of life on Earth. When an organism dies, it becomes covered with sediments and may be preserved in the rock. A **fossil** is any remain of past life. It may be the entire organism, an imprint, bones, shells or formed when minerals replace living tissue (**petrification**). Sometimes, sediments can fall into an **imprint** to create a type of fossil called a **cast**. In this case, the sedimentary rock has taken the shape of an ancient organism.

The layers of sedimentary rock can also tell the relative age of a fossil. For example, fossils found at the top of the Grand Canyon are younger than the ones below it with the oldest being in the bottom layer. Sandstone, limestone, shale, and conglomerate are some examples of sedimentary rock.

Using the fossils found in sedimentary rock, scientists can infer certain things about a region's earlier climate and environment. For example, fossil coral found in rocks implies that the area was once covered by seawater.

Fossil Record
When left undisturbed, the oldest fossils are found deeper in the layers of sedimentary rock.

HOT ROCKS

Marble and slate are some examples of **metamorphic rocks**. If you ever have picked up a large piece of slate or marble, it feels fairly heavy. This is because, these type of rocks are very dense compared to other types of rocks. How did this happen? Metamorphic rock is a type of rock that has undergone changes in form. In fact, its name literally means that. "Meta" comes from Greek meaning change while "morph" means form. Metamorphic rocks are formed from exposure to extreme pressure and temperature. This usually happens if the rock is buried and then exposed to hot materials such as magma or the forces of Earth's crust moving. The heat and pressure causes the chemical bonds in minerals to break, melting the minerals, resulting often in the formation of new minerals. When liquefied, minerals of different densities separate. This causes the **banding** which characterizes some metamorphic rock such as gneiss.

Banding in Gneiss
Metamorphic Rock

Rocks that crystallize from molten minerals are called **igneous rocks**. The liquid rock can come from a volcanic eruption or from the magma in the mantle of Earth. The crystal structure of igneous rocks varies depending on whether it cooled quickly or slowly. Igneous rocks that cooled slowly come from the mantle. The slow cooling process results in the large crystals that characterize this type of rock. **Granite** is an example of this type of igneous rock.

Some igneous rocks can float on water! When **lava** flows quickly out of a volcano, some of the lava mixes with the air to form a low density rock called **pumice**. This rock is characterized by many small pores in it.

Scheme for Igneous Rock Identification

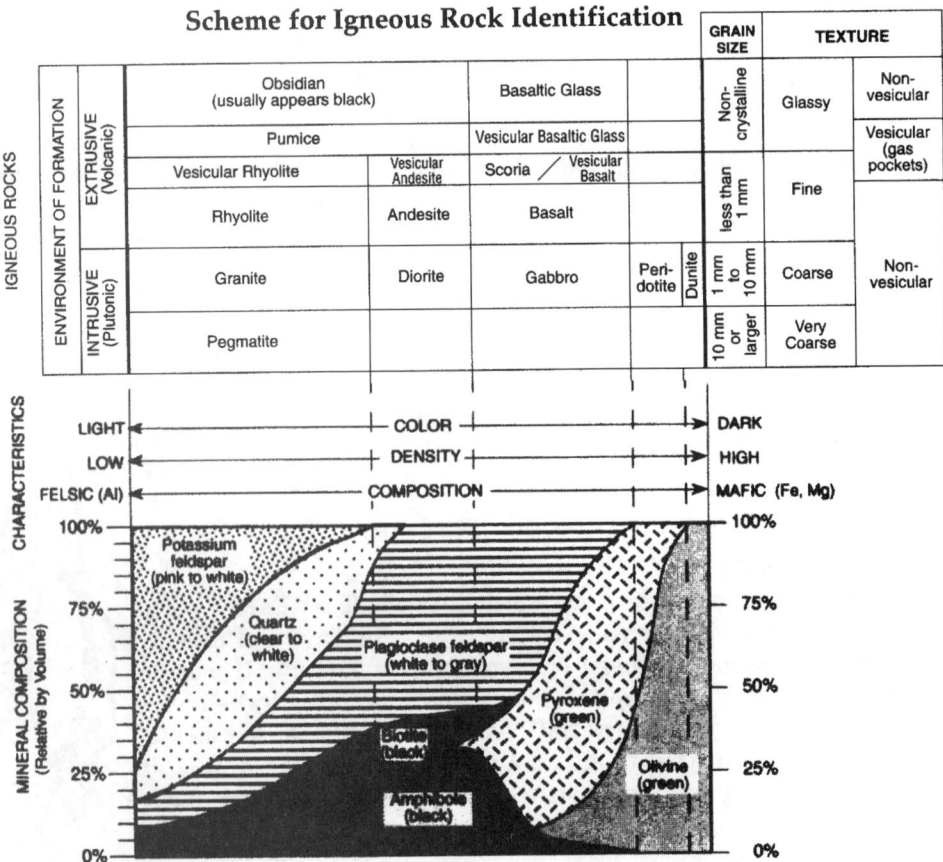

Early humans carved some of their spears out of an igneous rock called **obsidian**, also known as "black glass." This type of rock cooled very quickly. When igneous rocks cool quickly, the crystals are so tiny, it gives the appearance of glass. Igneous rocks are classified by their mineral content and whether they were formed above ground or below ground. The type of minerals present in the rock further help to identify it.

Scheme for Sedimentary Rock Identification

INORGANIC LAND-DERIVED SEDIMENTARY ROCKS

TEXTURE	GRAIN SIZE	COMPOSITION	COMMENTS	ROCK NAME	MAP SYMBOL
Clastic (fragmental)	Pebbles, cobbles, and/or boulders embedded in sand, silt, and/or clay	Mostly quartz, feldspar, and clay minerals; may contain fragments of other rocks and minerals	Rounded fragments	Conglomerate	
			Angular fragments	Breccia	
	Sand (0.2 to 0.006 cm)		Fine to coarse	Sandstone	
	Silt (0.006 to 0.0004 cm)		Very fine grain	Siltstone	
	Clay (less than 0.0004 cm)		Compact; may split easily	Shale	

CHEMICALLY AND/OR ORGANICALLY FORMED SEDIMENTARY ROCKS

TEXTURE	GRAIN SIZE	COMPOSITION	COMMENTS	ROCK NAME	MAP SYMBOL
Crystalline	Varied	Halite	Crystals from chemical precipitates and evaporites	Rock Salt	
	Varied	Gypsum		Rock Gypsum	
	Varied	Dolomite		Dolostone	
Bioclastic	Microscopic to coarse	Calcite	Cemented shell fragments or precipitates of biologic origin	Limestone	
	Varied	Carbon	From plant remains	Coal	

Scheme for Metamorphic Rock Identification

TEXTURE	GRAIN SIZE	COMPOSITION	TYPE OF METAMORPHISM	COMMENTS	ROCK NAME	MAP SYMBOL
FOLIATED — MINERAL ALIGNMENT	Fine	MICA QUARTZ FELDSPAR AMPHIBOLE GARNET PYROXENE	Regional (Heat and pressure increase with depth)	Low-grade metamorphism of shale	Slate	
	Fine to medium			Foliation surfaces shiny from microscopic mica crystals	Phyllite	
				Platy mica crystals visible from metamorphism of clay or feldspars	Schist	
FOLIATED — BANDING	Medium to coarse			High-grade metamorphism; some mica changed to feldspar; segregated by mineral type into bands	Gneiss	
NONFOLIATED	Fine	Variable	Contact (Heat)	Various rocks changed by heat from nearby magma/lava	Hornfels	
	Fine to coarse	Quartz	Regional or Contact	Metamorphism of quartz sandstone	Quartzite	
		Calcite and/or dolomite		Metamorphism of limestone or dolostone	Marble	
	Coarse	Various minerals in particles and matrix		Pebbles may be distorted or stretched	Metaconglomerate	

ROCK CYCLE

The forces acting on Earth interact with rocks causing them to undergo what is called the **rock cycle**. Weathering, erosion, heat, and the movement of the tectonic plates (see below) are constantly changing rocks. Weathered igneous and metamorphic rock can be changed into sedimentary rock. Sedimentary rock or igneous rock, placed under extreme pressure and heat due to the motion of the plates, change into metamorphic rock. Metamorphic and sedimentary rock forced into the mantle by the motion of the **plates** melts and can eventually recrystallize as igneous rock and so the cycle continues.

Rock Cycle in Earth s Crust

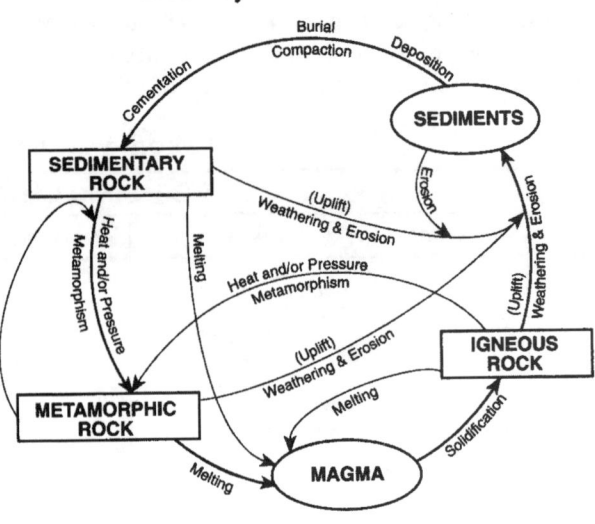

30 Using the rock cycle, explain how sedimentary rock can change to metamorphic and igneous rock.

_____

_____

_____

DISCIPLINARY CORE IDEAS:

ESS1.B: NATURAL HAZARDS

THE RESTLESS EARTH: PLATE TECTONICS, EARTHQUAKES & VOLCANOES

Data collected about earthquakes and volcanoes have provided evidence that the crust of Earth has shifted. Rock formations look like they have been folded, faulted, or tilted by this motion.

A MOVING EARTH STORY – CONTINENTAL DRIFT

Have you ever noticed how the west coast of Africa looks like it can fit into the east coast of South America? Scientists have found that the animal and plant fossils found along these coasts were often identical.

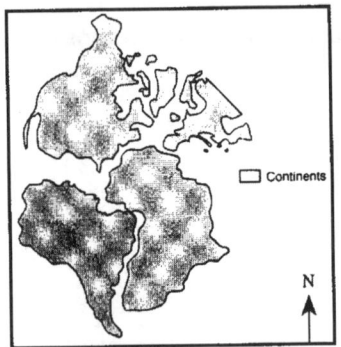

In 1910, German geophysicist **Alfred Wegener** proposed a model to explain these observations. His model came to be called the ***Theory of Continental Drift***. According to this model, there was a time in history when all of the continents were joined together forming one large continent called "Pangaea." Over a period of millions of years, the continents drifted apart into their present position.

AN EXPLANATION FOR CONTINENTAL DRIFT

Wegener was also a meteorologist and the mechanism he gave to explain the motion was tied to how air currents moved. Air currents form when denser, cool air sinks under warmer air. Wegener hypothesized that **convection currents** form in the solid Earth, and that these currents were the force that moved the continents.

31 What is the Theory of Continental Drift? _____

_____

32 What evidence supports the idea that the continents are moving? _____

_____

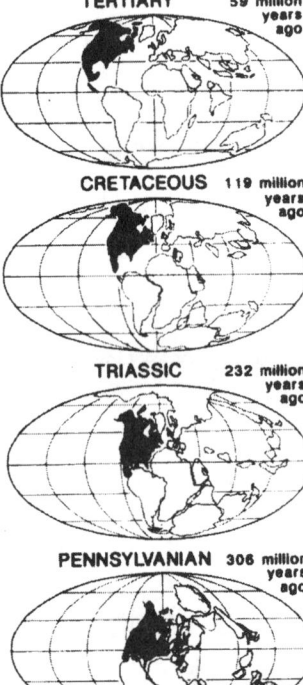

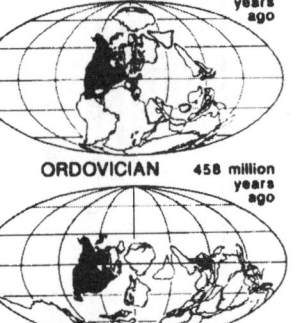

THE JIG-SAW PUZZLE SOLUTION

The Theory of Continental Drift eventually led to the *Theory of Plate Tectonics*. Basically, this theory states that Earth consists of a number of plates. You can think of it as taking the lithosphere and carving it up into a few puzzle pieces. Six major plates have been identified with evidence for numerous smaller ones also existing. In fact, if you marked off on a map where earthquakes and volcanoes occur, you will find that they outline the major plates. Today, it is believed that the crust rides on the upper portion of the mantle powered by convection currents in the mantle.

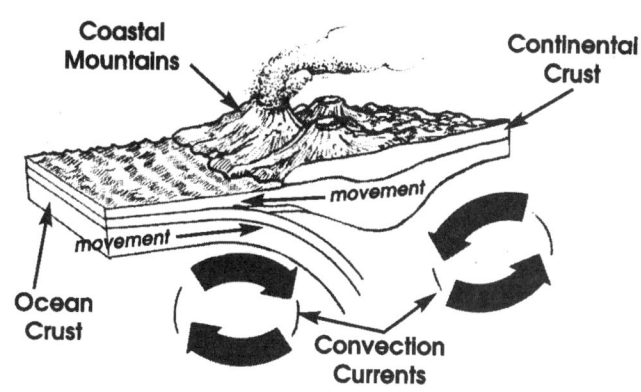

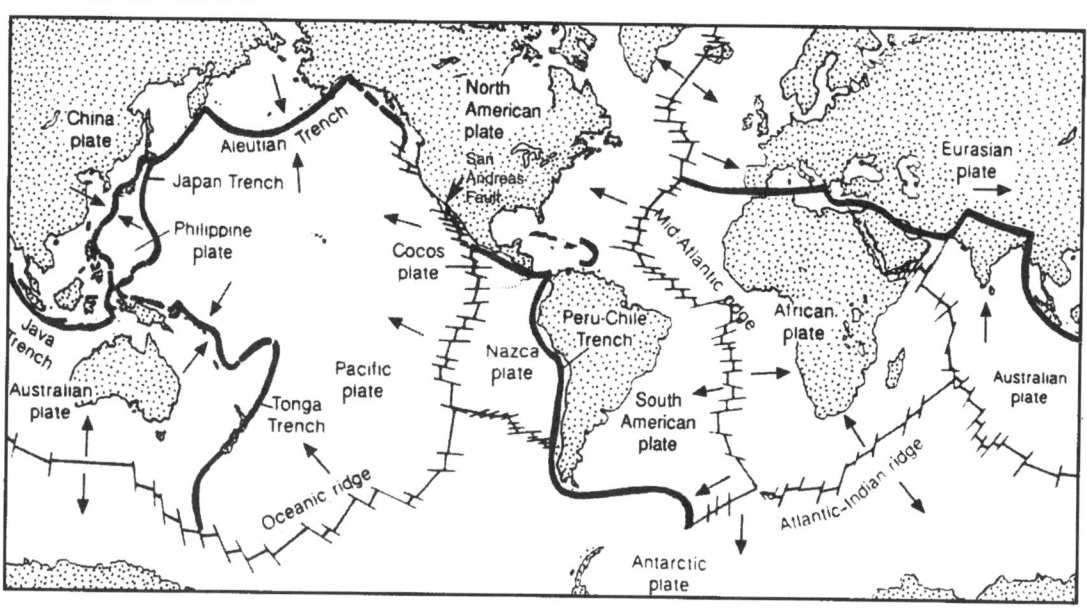

Plate Tectonics

This world map identifies the major plates, trenches, and ridges that are sights of major crustal activity.

33 What is the Theory of Plate Tectonics? _____

_____

_____

_____

EFFECTS OF PLATE TECTONICS

Mountain Building occurs when continental plates collide with one another. The crust folds and mountains, such as the Andes and the Himalayas, are built.

Studies of the ocean basin have shown that when plates collide and one is denser than the other, one plate slides under another plate causing **ocean trenches**. This happens when the denser ocean crust slides under the less dense continental plate. The Pacific Plate is sliding under the North American Plate causing the formation of deep trenches in the Pacific Ocean. This type of motion is is called **subduction**. Volcanic eruptions are also associated with this type of motion. The eruptions of Mount Pinatubo in the Philippines (1991) and Mount St. Helens in Washington (1980) are evidence for this type of motion.

There is so much volcanic activity along the **Pacific Rim** that it is often referred to as the **Ring of Fire**. This area is named the "ring," because it completely surrounds the Pacific Ocean, from North and South America to Asia and Australia. The **San Andreas fault** (fracture in rock) is located where the Pacific Plate is moving past the North American Plate. Since Los Angeles is located on the Pacific Plate and San Francisco is located on the North American Plate, scientists predict that Los Angeles will eventually be north of San Francisco.

Did you know that the distance between North America and Europe is increasing? In other words, the Atlantic Ocean is getting bigger. Studies of the ocean basin have shown there is a giant crack on the floor of the Atlantic Ocean known as the **Mid-Atlantic Ridge**. Through this crack, lava flows onto the ocean floor causing it to

MOVEMENT OF THE MID-ATLANTIC OCEAN RIDGE

spread and get larger. This motion can explain how South America and Africa with coasts that resemble matching puzzle pieces, drifted away from one another. These ridges are found in all of the oceans. It is the lava spewing out of the mid-ocean ridges that form the ocean floors.

Sometimes, the lava forming the mid-ocean ridges reaches the surface, forming islands such as Iceland in the Atlantic Ocean. The Hawaiian Islands are also **volcanic islands** that are still being formed. As the Pacific Plate moves over vents or cracks in the ocean floor, magma flows onto the floor solidifying as lava. Eventually, the lava may pile up high enough to reach the surface of the ocean. When this happens, a volcanic island is formed.

34 Describe two land formations caused by the movement of plates.

_____

_____

_____

_____

_____

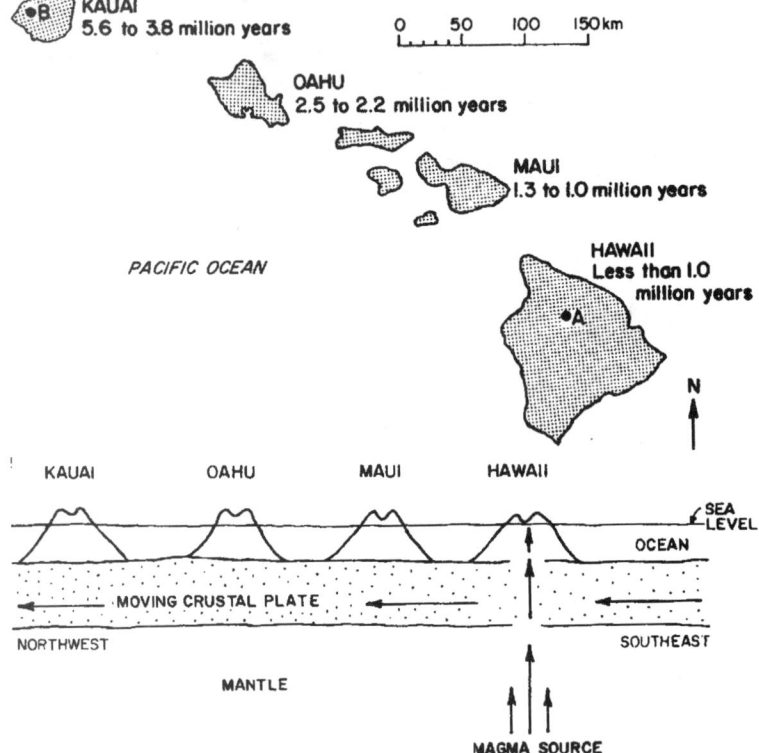

"SMOKY MOUNTAINS" – VOLCANOES

The motion of the crustal plates sliding along the mantle places tremendous stress on the lithosphere. As new ocean floor is being formed at the ridges, it is being destroyed in the trenches. **Volcanoes** and **earthquakes** act to relieve the stress created by the motion of the plates. They occur chiefly near areas of active plate motion and so can be used to show the location of the major plate boundaries.

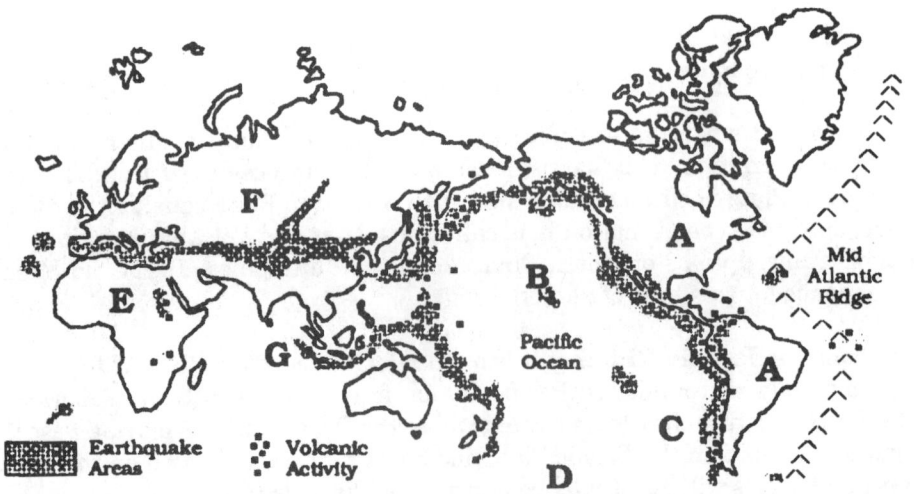

Major Earthquake Areas & Volcanic Activity

Key – Major Plate Areas:

A – American
B – Pacific
C – Nazca
D – Antarctic
E – African
F – Eurasian
G – Indian

When **magma** reaches the surface through a crack in the crust, a **volcano** is formed. The flow of the magma can be gradual, such as the mid-ocean ridges, or it can be very violent, such as the eruption of Mount Pinatubo in the Philippines. The thickness of the magma (viscosity), and the amount of dissolved gases in it, affects how "explosive" the magma is when it erupts. Magma that reaches the surface of Earth is called **lava**. When magma emerges from long cracks or fissures in Earth, a **lava plateau** can form as layers of lava build up over the years. Lava that flows through **vents** creates volcanic mountains. The shape of the mountain depends on how fluid the lava is when it comes out of the vent. These mountains can form above ground, such as Mount St. Helens in Washington, or on the ocean floor, such as the mid-ocean ridges.

AN ISLAND BLOWS UP

One of the most dramatic instances of a volcanic eruption took place on an island called Krakatoa in Southwest Indonesia. Until August 1883, Krakatoa had an area of 47 sq. km (18 sq. mi.). After the volcano erupted, two-thirds of the island disappeared. The eruption, and the tsunami (giant tidal wave) it created, killed thousands of people. The sound of the explosion was one of the loudest noises ever recorded. It was heard 12,875 km (8000 mi.) away! The dust arising from the eruption circulated in the upper atmosphere of the planet for three years.

Volcanic eruptions release sulfur oxides that make their way into the upper atmosphere. Scientists studying the warming of Earth had to factor in the effects of these gases. They do just the opposite of greenhouse gases. Instead of absorbing sunlight, sulfur oxides reflect sunlight resulting in a cooling effect.

A TREMBLING EARTH – EARTHQUAKES

When Earth suddenly trembles, an earthquake is taking place. The magnitude of the **earthquake** depends on the amount of stress placed on the solid rock from the motion of tectonic plates. The energy released during an earthquake travels in waves called **seismic waves**. These waves create vibrations in the solid Earth. Sometimes, the vibrations are so great that giant cracks form on Earth. If the earthquake occurs in the ocean, the vibrational energy is transferred to the water. It can result in giant waves called **tsunamis** if they strike the coast. They are common around the Pacific Ocean. These waves can reach a speed of 935 km/hr (581 mi/hr). As they move from the deeper to shallower water, the height of the wave increases as high as a ten-story building. When they break, they can cause a great deal of destruction.

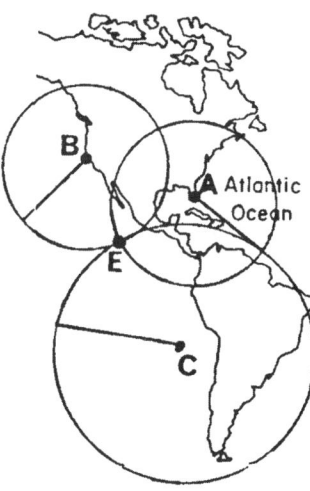

**A, B, & C Recording Stations
Point E – the epicenter**

Rock can crack near the surface of Earth and far below. The location of the crack is called the **focus** of an earthquake. The place on the surface directly above the focus is called the **epicenter** of an earthquake. Using seismic equipment called a **seismograph**, the epicenter and the strength of the energy coming from the wave can be determined.

35 What is the major cause of earthquakes and volcanoes? _____

_____

36 Where do earthquakes and volcanoes most frequently occur? _____

_____

MEASURING EARTHQUAKES

Two scales are used to measure the strength of an earthquake. One scale, called the **Richter Scale**, measures the actual amount of energy coming from the earthquake. The Richter Scale (on the next page) measures magnitudes from 1-9. Each value is a logarithm (10X), therefore, a value of 2 is actually ten times greater than 1; while a value of 3 is 100 times greater than 1, and so on. The **Mercalli Scale** (on the next page) assigns a numerical value to the damage associated with an earthquake.

Table Of Richter Scale

Magnitude	Equivalent Energy by Mass of Exploding TNT	Remarks
0	600 g	enough to blast a stump
1	20 kg	small construction blast
2	600 kg	average quarry blast
3	20,000 kg	large quarry blast
4	600,000 kg	small atom bomb
5	20 million kg	"standard" atom bomb
6	600 million kg	small H bomb
7	20 billion kg	enough energy to heat New York City for one year
8	600 billion kg	enough energy to heat New York City for 30 years
9	20 trillion kg	the energy in the world's production of coal and oil for five years

Table Of Modified Mercalli Scale

Intensity Value	Description of Effects
I	usually detected only by instruments
II	felt by a few persons at most, especially on upper floors
III	hanging objects swing; vibration like passing truck; noticeable indoors
IV	felt indoors by many, outdoors by few; sensation like heavy truck striking building; parked automobiles rock
V	felt by nearly everyone; sleepers awakened; liquids disturbed; unstable objects overturned; some dishes and windows broken
VI	felt by all; many frightened and run outdoors; some heavy furniture moved; glassware broken; books off shelves; damage slight
VII	difficult to stand; noticed in moving automobiles; damage to some masonry; weak chimneys broken at roofline
VIII	partial collapse of masonry; chimneys, factory stacks, columns fall; heavy furniture overturned; frame houses moved on foundations

Disciplinary Core Ideas:

ESS2.D: Weather And Climate

The Long & Short Of It – Climate & Weather

Climate is the average weather conditions over a large geographic area that extends over a long period of time. *The amount of solar radiation and rainfall are the two major factors that determine climate.* These are affected by the relative distribution of land and water, ocean currents, altitude, wind belts, and relief of the land (mountainous or flat). **Weather** refers to local or short term changes in temperature and rainfall, wind, cloud cover, etc. For example, during the summertime, one section of a town may be receiving sunshine and no major clouds, while another section has cloud cover and rain. They have the same climate, but the weather is different in the two sections of the town.

37 What is the main difference between climate and weather? _____

_____

Climate

Classification Of Climate

The type of climate is classified using three criteria: the region's water budget, temperature, and vegetation.

The **water budget** measures the amount of water entering, being stored, and removed from a region. Knowledge about the water budget is one way in which climate can be described. The size of the ratio between water entering (**P**) and leaving (**Ep**) a large region can classify a region as arid, semiarid, subhumid or humid.

Climates Based On The P/Ep Ratio	
P/Ep Ratio Range	Type of Climate
Over 1.2	Humid
0.8 to 1.2	Subhumid
0.4 to 0.8	Semiarid
Less than 0.4	Arid

Maritime m very moist, forms over water **Continental c** dry, forms over land	**Arctic A** forms over ice-covered regions, extremely cold **cA** - dry and frigid, forms north of Canada	**Polar P** forms over high latitudes, relatively low temperature **mP** - cold and moist, forms over N. Atlantic and N. Pacific **cP** - cold and dry, forms over northern and central Canada	**Tropical T** forms over low latitudes, relatively high temperatures **mT** - warm and moist, forms over Gulf of Mexico, middle Atlantic, Caribbean, Pacific south of California **cT** - warm and dry, forms over southwestern U.S. in summer

United States – Climate Zones
(Numbers Represent P/E$_p$ Ratios)

If **temperature** is used to classify a climate, it can be called polar, subpolar (temperate), subtropical, or tropical (see chart on previous page).

The dominant form of **vegetation** in a region can also be used to describe a region. This makes sense since the major limiting factors that determine vegetation are rainfall and temperature. Therefore, large regions can be called **deserts**, **tundra**, **grasslands**, **temperate deciduous forest**, **taiga** (coniferous forests) and **tropical** or **temperate rain forests**. These are also called biomes. A **biome** is a biological term for a large region known by its dominant forms of plant life. Generally, regions with plentiful rainfall support forests. As the rainfall decreases, a grassland may form. A desert is associated with infrequent rainfall.

FACTORS THAT AFFECT RAINFALL IN A REGION

The amount of rainfall a region receives is affected by **natural barriers** such as mountains. For example, weather moves from west to east in the United States. East of the Rocky Mountains is prairie land which has much lower annual rainfall than the region west of the Rockies. As moist air moves across the mountains, it is forced to rise. Air at higher altitudes is cooler and holds less moisture than air at lower altitudes, resulting in frequent precipitation. As the air descends across the eastern side of the Rockies, much of the moisture has been removed from the air. This lower rainfall characterizes the vegetation found on prairie land.

The uneven heating of Earth creates high and low pressure belts. The result of this is the formation of **air currents** or **wind belts**. Air that is warm holds more moisture than cool air. The trade winds or tropical easterlies carry warm, moist air to regions such as the Virgin Islands that are very popular for sailing holidays because of the prevalent trade winds.

Planetary Wind & Moisture Belts in the Troposphere

38 Describe two factors that affect rainfall in a region. _____

FACTORS AFFECTING TEMPERATURE

Because of the tilt of Earth, regions located near the Equator receive the most **insolation** (solar radiation) and have the highest average annual temperature. Therefore, as the **latitude** increases, north or south, there is a decrease in the amount of solar radiation. This results in lower average annual temperatures.

The temperature in higher **altitudes** is cooler because the air is thinner and so cannot retain as much heat energy as denser air. Temperature decreases about 1°C for every 100 meter increase in elevation. A hike up the Rocky Mountain National Park during the summertime illustrates this. As the altitude increases, the temperature decreases and the vegetation changes, creating mini-climate zones or biomes.

The temperature of a region can be affected by **ocean currents**. For example, the Gulf Stream is a warm ocean current that originates from the Straits of Florida and travels in a general NE direction up along the

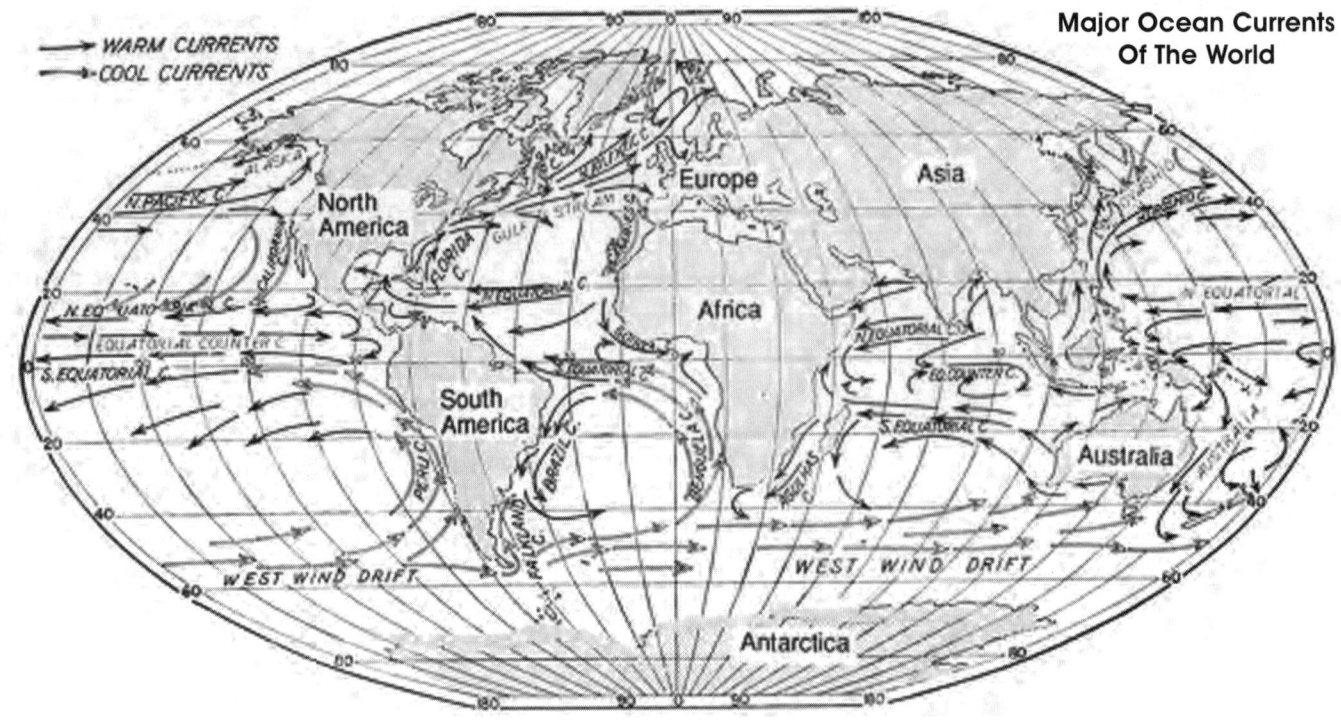

coast of Europe. This section of the Gulf Stream is also called the **North Atlantic Drift**. This warm water alters the air so that cities such as London, which is at a higher latitude than New York, are characterized by a warmer climate.

Water absorbs and loses heat relatively slowly. On the other hand, land regions absorb and release heat much faster. This causes the temperature of the water and land to be always different from one another in a particular region. This also affects the temperature of the air above land and water. During the summertime, the temperature of large bodies of water increases. This causes the air above the water to also become warmer. When winter approaches a coastal region, the warmer air over the water moves toward the land, causing the average temperature of the land to be higher.

A desert is characterized by extreme temperature changes because of its lack of water to moderate the air temperature. Inland cities have a broader range of temperature between summer and winter because there are no large bodies of water to moderate the temperature of the air.

The amount of heat absorbed and released by land depends on the properties of the land. For example, rocky land absorbs heat more quickly than land that is highly vegetated. Tundra (ice fields) tend to reflect a large amount of solar radiation.

39 Describe two factors that affect temperature in a region. _____

_____

WEATHER

The temperature, cloud cover, precipitation, wind speed, and humidity in any particular location represent its weather. The major cause of weather is the unequal heating of Earth. This creates **density** differences in the air. Warm air is less dense than cold air. Gravity causes the denser cold air to sink below the warmer air. This results in continuous air circulation or **convection currents**. Add to this the air brought to a region by the prevailing and local winds, and the properties of air are constantly changing.

40 What is the major cause of weather? _____

_____

Weather Map Symbols

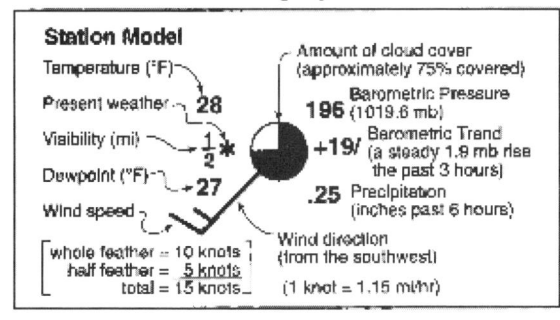

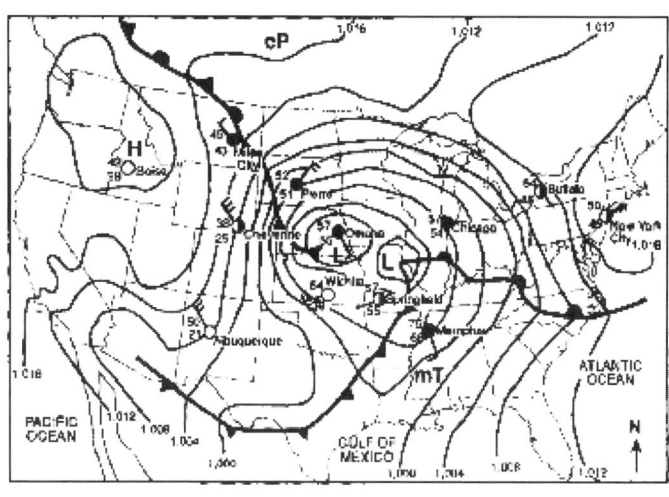

FORECASTING WEATHER

Weather Stations located in thousands of regions throughout the world collect numerous data associated with weather. **Weather satellites** (space craft orbiting the Earth) can measure temperature at various altitudes, cloud cover, and even the temperature of ocean currents.

Weather maps graphically show temperature, barometric pressure, wind speed, and relative humidity. When this information is recorded on a series of weather maps, **weather patterns** can be determined. These weather patterns become very useful in forecasting weather. A **synoptic weather map** is a summary created from data collected from numerous **weather stations**.

The information from a series of weather stations can be used to create **field maps**. Field maps of temperature and air pressure are very useful because they can show weather patterns. A field map is created by connecting points of "equal value" called **isolines**. If the points are connecting barometric pressure, they are called **isobars**. If temperature values are connected, the lines are called **isotherms**.

Isobars on a synoptic weather map show areas of **low** and **high pressure**. Since warm, moist air is less dense than cool, dry air, low pressure areas are very humid. If a cold front is moving toward a low on a map, rain will take place in that region. This is because the cooler air cannot hold as much moisture as the warmer air. As a result, the excess moisture is removed as precipitation.

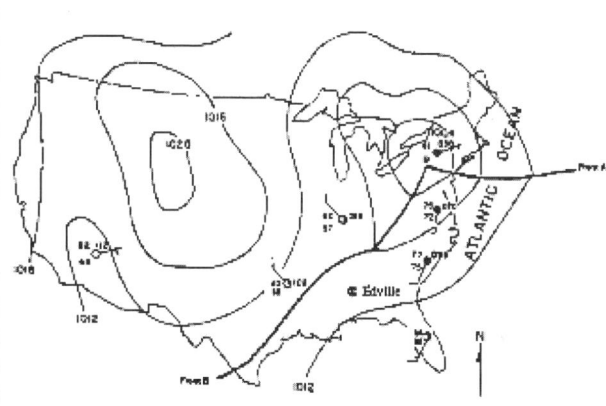

Example Of Isobar Field Map

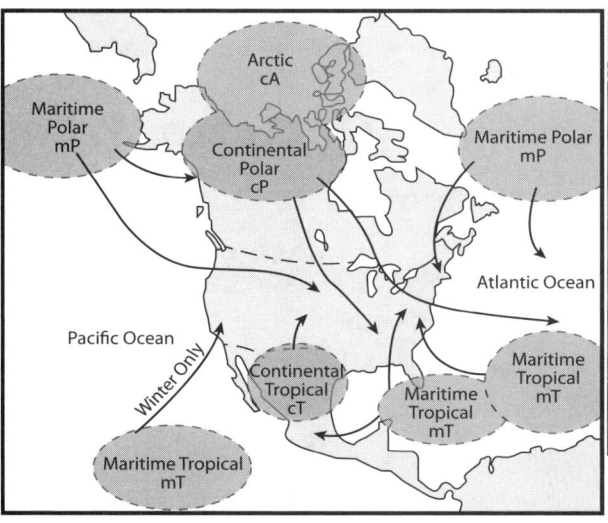

Meteorologists (scientists who study weather) study the motion of **air masses** to help forecast the weather. An air mass represents a large areas of air with similar characteristics. For example, if the air was over the ocean, it will be moist, while air over a desert will be dry

41 Using the air mass map at the left, describe the air mass

most affecting the weather in New York State. _____

_____

_____

_____

In the United States, planetary winds and the **jet stream** (a large stream of air moving from west to east in the upper atmosphere) move air masses.

WEATHER FRONTS

When two air masses meet, the region is called a **weather front**. It is often characterized by rapid weather changes accompanied by precipitation.

A **cold front** forms when a cold air mass meets a warm air mass. In the summer time, cold fronts are often associated with thunderstorms. Since the cold mass is more dense than the warm mass, it pushes the warmer air up. The moisture in the warmer air quickly condenses in the cooler upper air, forming **cumulus clouds** (towering thick white clouds) that characterize cold fronts.

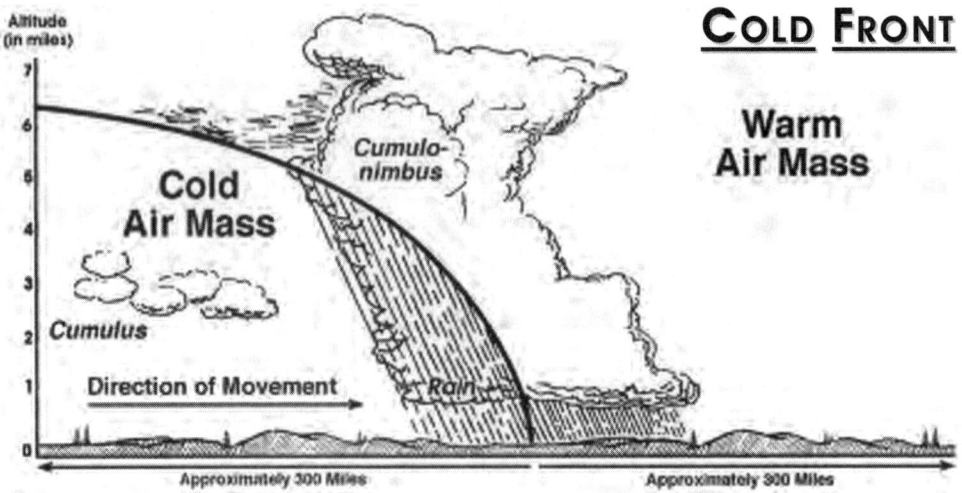

A **warm front** occurs when warm air catches up with cold air. The less dense warm air glides over the denser cold air. The slow rising air also condenses in the cooler air as it rises. However, since it does not rise as rapidly as when a cold front collides with a warm front, it is characterized by the formation of feathery **cirrus clouds** followed by sheet-like **stratus clouds**. If the air had significant moisture in it, there would be widespread precipitation.

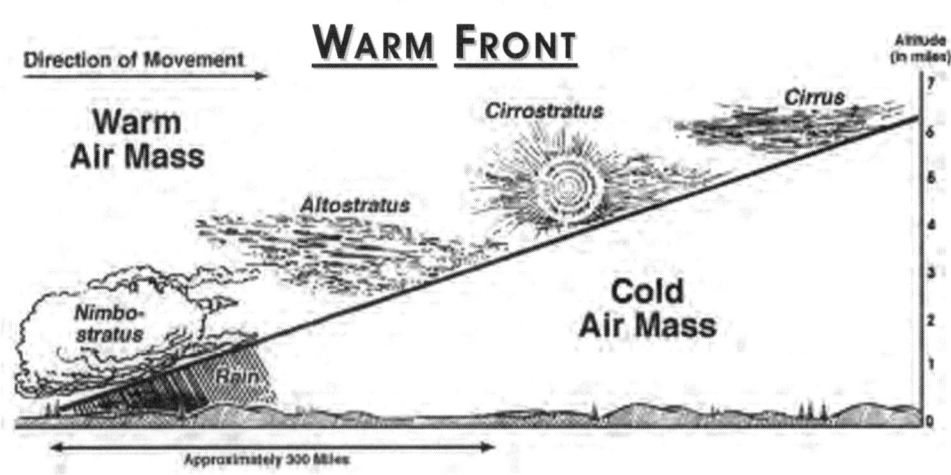

An **occluded front** takes place when a fast moving cold front overtakes a warm front, lifting it quickly, and causing large scale condensation and precipitation.

When a cold front and warm front meet, but move very slowly relative to one another, a **stationery front** has formed. The denser cold air sinks slowly under the warmer air mass. Clouds form, but they form very slowly. They can last for a number of days until another air mass moves in with enough force to move the stalled air masses.

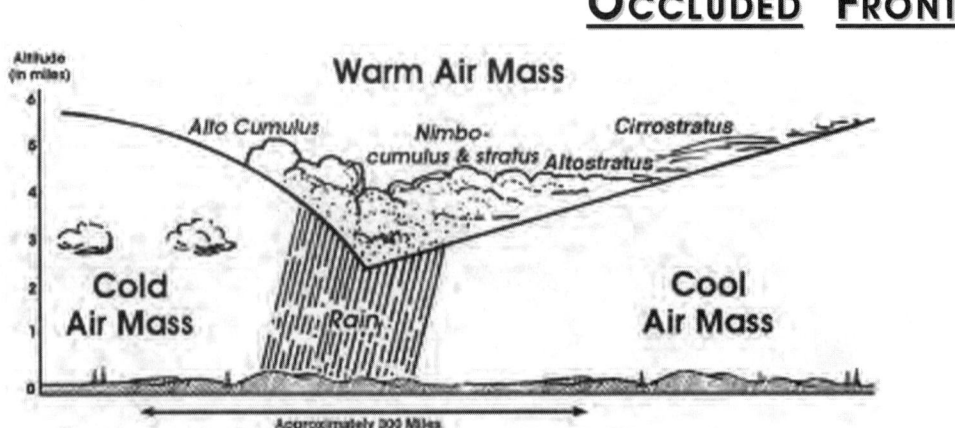

Present Weather						Air Masses	Front Symbols		Hurricane

Present Weather

Drizzle · Rain · Smog · Hail · Thunder-storms · Rain Showers

Snow · Sleet · Freezing Rain · Fog · Haze · Snow Showers

Air Masses

cA continental arctic
cP continental polar
cT continental tropical
mT maritime tropical
mP maritime polar

Front Symbols

Cold
Warm
Stationary
Occluded

Hurricane

Front Symbols: Synoptic weather maps use symbols to represent the motion of fronts, because fronts can be used to predict weather. For example, a warm front carrying moist air is followed by a cold front. This often results in rain. This is because cool air cannot hold as much moisture as warm air. Cloud formation and precipitation follows as the colder air causes the temperature to drop to the dew point temperature. The **dew point temperature** is the temperature in which water vapor in the air condenses.

42 What is a cold front? What type of clouds and weather is usually associated with a cold front?

a _____

b _____

43 What is a warm front? What type of clouds and weather is usually associated with a warm front?

a _____

b _____

RELATIVE HUMIDITY

The amount of water vapor in the atmosphere varies with the temperature. Warm air can hold more water vapor than cold air. **Relative humidity** measures the amount of water vapor in a given sample of air and compares it to the maximum amount of water vapor that the air sample can hold.

Using a chart developed by meteorologists that compares a wet bulb thermometer with a dry bulb thermometer, the relative humidity of a given sample of air can be calculated.

SAMPLE PROBLEM

What is the relative humidity if the dry bulb temperature is 10°C and the wet bulb temperature is 5°C?

Relative Humidity (%)

Dry-Bulb Tempera-ture (°C)	Difference Between Wet-Bulb and Dry-Bulb Temperatures (C°)															
	0	1	2	3	4	5	6	7	8	9	10	11	12	13	14	15
−20	100	28														
−18	100	40														
−16	100	48														
−14	100	55	11													
−12	100	61	23													
−10	100	66	33													
−8	100	71	41	13												
−6	100	73	48	20												
−4	100	77	54	32	11											
−2	100	79	58	37	20	1										
0	100	81	63	45	28	11										
2	100	83	67	51	36	20	6									
4	100	85	70	56	42	27	14									
6	100	86	72	59	46	35	22	10								
8	100	87	74	62	51	39	28	17	6							
10	100	88	76	65	54	43	33	24	13	4						
12	100	88	78	67	57	48	38	28	19	10	2					
14	100	89	79	69	60	50	41	33	25	16	8	1				
16	100	90	80	71	62	54	45	37	29	21	14	7	1			
18	100	91	81	72	64	56	48	40	33	26	19	12	6			
20	100	91	82	74	66	58	51	44	36	30	23	17	11	5		
22	100	92	83	75	68	60	53	46	40	33	27	21	15	10	4	
24	100	92	84	76	69	62	55	49	42	36	30	25	20	14	9	4
26	100	92	85	77	70	64	57	51	45	39	34	28	23	18	13	9
28	100	93	86	78	71	65	59	53	47	42	36	31	26	21	17	12
30	100	93	86	79	72	66	61	55	49	44	39	34	29	25	20	16

In order to use the chart, the difference between the wet bulb temperature and the dry bulb temperature must be determined. In this situation the difference between the dry bulb temperature and the wet bulb temperature is 10°C - 5°C = 5. Using the Relative Humidity chart, the relative humidity is found at the intersection between the dry bulb temperature (10°C) and 5. As can be seen the relative humidity is 43%.

44 Determine the relative humidity when the dry bulb reading is 24°C and the wet bulb reading is 20°C.

The relative humidity is _____ %

HAZARDOUS WEATHER CONDITIONS

Hazardous weather has caused great human suffering and death. It has been responsible for severe damage to the environment, the infrastructure of cities and towns, and private property. Millions and millions of dollars have been spent to repair the damage from **hazardous storms**. These storms include **thunderstorms**, **hurricanes**, **tornadoes**, **ice storms**, and **blizzards**. Knowledge about storm formation allows meteorologists to better predict the hazardous storms and prepare people to take necessary precautions.

Thunderstorms normally form after severe heating of Earth's surface. This results in numerous powerful convection cells. A convection cell forms a circular motion of air due to differences in densities. As cooler air sinks, it pushes up warmer air which then cools and sinks. The process repeats itself with great frequency. However, when this occurs very rapidly, a certain amount of internal friction results creating clouds that are electrically charged. The discharging of a cloud is called **lightning**. It produces **thunder**, the booming sound produced by rapidly expanding air along the path of the lightning. Thunderstorms are also associated with heavy rain and flooding.

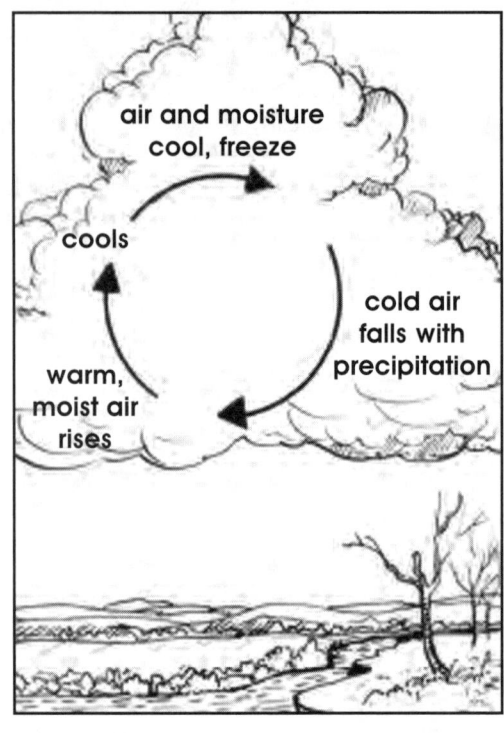

Tornadoes often form late in the day when Earth's surface is the warmest. These funnel shaped clouds are small in size (average diameter, 100 meters). They are brief in duration and develop over land from severe thunderstorms. The wind speeds near the center of a tornado may reach 500 km an hour or more. When these funnel-shaped, spinning clouds touch ground, they cause extreme damage.

Hurricanes form over warm oceans. Many convection cells merge into huge convection cells. Winds begin to blow into the low pressure area and the Coriolis Effect causes the wind to deflect to the right in the Northern Hemisphere, resulting in a counterclockwise spinning. When these cyclones (regions of low pressure air surrounded by high pressure air) form, and the wind speeds exceeds 119 km/hr, the cyclone is called a hurricane. These cyclones often have a diameter of over 500 km.

Blizzards are storms with high winds and frozen precipitation that form in the upper atmosphere. The combination of heavy snow and high winds can cause considerable damage to trees, homes, and infrastructure.

courtesy of NOAA/National Climatic Data Center

Ice Storms occur when the temperature of the upper atmosphere is warmer than the lower atmosphere. Since the temperature of the lower atmosphere is at or below the freezing point of water, frozen precipitation falls. These storms are often associated with many traffic accidents.

45 Describe one type of hazardous storm and how early warning can be helpful.

SEA BREEZES AND LAND BREEZES

Summer vacationers like to be near the water because during the daytime, a breeze often comes off the water. This is because the air over the water is cooler than the air over the land. The denser cool air moves toward the land forming a **sea breeze**. In the evening, the opposite is true. The air over the land is cooler than the air over the water. Air moves out toward the water forming a **land breeze**. Vacationers with sailboats can take advantage of a land breeze for an evening sail.

sea breeze

Land Breeze

46 *a* What causes a land breeze? _____

_____

_____

b What causes a sea breeze? _____

_____

_____

The most watched section of a TV newscast is the weather report. Knowing the weather allows people to plan events and dress appropriately. There is a little meteorologist in all of us, so knowledge about the factors that affect weather and how to read a weather map is obviously very useful.

MULTIPLE CHOICE

1 Most of the atmosphere is made up of the gases
(1) hydrogen and oxygen   (3) carbon dioxide and oxygen
(2) nitrogen and oxygen   (4) ozone and oxygen

2 The gas that can block UV radiation in the upper atmosphere is
(1) oxygen   (2) nitrogen   (3) ozone   (4) carbon dioxide

3 Weather occurs in the part of the atmosphere called the
(1) troposphere   (2) stratosphere   (3) hydrosphere   (4) lithosphere

4 According to the graph of temperature changes in the atmosphere (on page 137), as the altitude increases in the troposphere, the temperature
(1) increases then decreases   (3) stays the same
(2) decreases   (4) decreases then increases

5 The "Greenhouse Effect" refers to the
(1) warming of Earth by certain gases   (3) warming of Earth due to volcanic activity
(2) cooling of Earth by certain gases   (4) cooling of Earth due to volcanic activity

6 The least dense layer of the Earth is the
(1) inner core   (2) outer core   (3) mantle   (4) crust

7 The idea that Earth is composed of layers was inferred from
(1) volcanic activity   (3) seismic waves
(2) electromagnetic waves   (4) sedimentary rock

8 The percent of water covering the surface of Earth is about
(1) 90%   (2) 75%   (3) 50%   (4) 25%

9 The liquid part of Earth is called the
(1) atmosphere   (2) hydrosphere   (3) lithosphere   (4) stratosphere

10 As the altitude increases, air pressure
(1) decreases   (2) increases   (3) remains the same   (4) goes up, then increases again

Base you answer to question 11 on the three stream tables to the right. Table 1 has bare soil banks, Table 2 has grassy banks, and Table 3 has rocky banks.

11 Rank the boxes illustrating erosion along a river bank from experiencing the least to most erosion.
(1) 1, 2, 3   (3) 3, 2, 1
(2) 2, 3, 1   (4) 3, 1, 2

Table 1   **Table 2**   **Table 3**

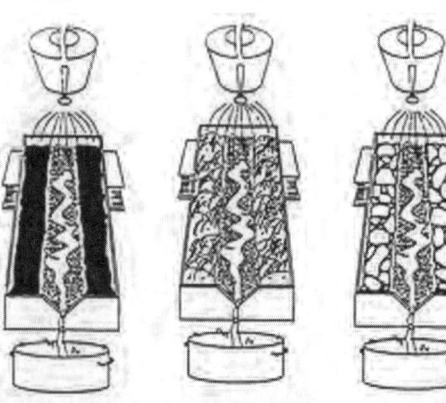

12 Chemical weathering of rock occurs when
   (1) water freezes in the cracks of rocks
   (2) moving water abrades rocks
   (3) carbonic acid combines with rocks
   (4) roots grow into rocks

13 The Appalachian Mountains are shrinking in height because
   (1) the forces of mountain building and erosion are in balance
   (2) the force of mountain building is greater than erosion
   (3) the force of erosion is greater than mountain building
   (4) the force of mountain building has stopped

14 What is the main force responsible for change in the rock formations illustrated at the right?
   (1) erosion
   (2) folding
   (3) faulting
   (4) deposition

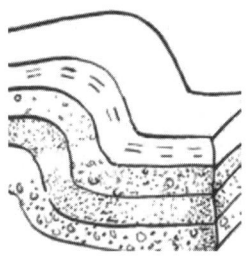

15 The diagram at the right illustrates bedrock being weathered. The main cause of weathering in this diagram is due to
   (1) gravity
   (2) chemical changes
   (3) acid rain
   (4) frost action

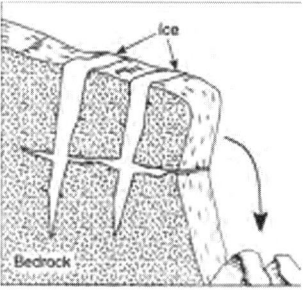

16 Which rock(s) was (were) transported the greatest distance in water?

 (1)
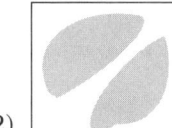 (2)
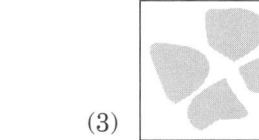 (3)
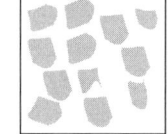 (4)

17 At the right is a diagram of a meandering stream. Which section of the stream will experience the most deposition?
   (1) A   (3) C
   (2) B   (4) D

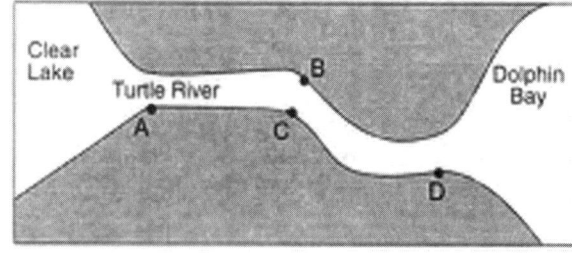

18 Study the diagram of the water cycle at the right. The path of the water cycle from A to B to C to D, illustrates
   (1) evaporation, run-off, precipitation, condensation
   (2) condensation, precipitation, run-off, evaporation
   (3) evaporation, condensation, precipitation, run-off
   (4) run-off, evaporation, condensation, precipitation

19 Rocks are classified as igneous, metamorphic, and sedimentary because of differences in their
   (1) formation   (3) number of minerals
   (2) chemical composition (4) size

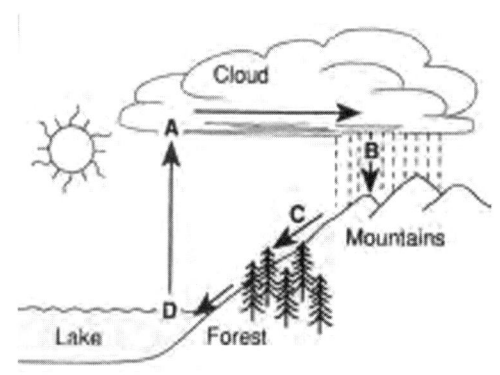

20 Which rock is most likely an example of a metamorphic rock?

(1)   (2)    (3) 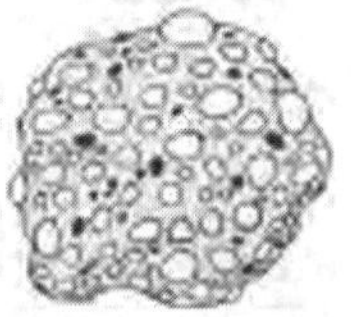   (4)

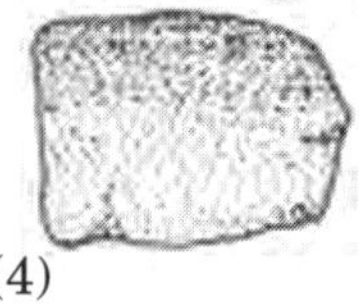

21 Rocks that contain fossils are classified as
   (1) igneous   (2) sedimentary   (3) metamorphic   (4) fossilized

22 Which layer of the rock shown at the right
   would contain the oldest fossils?
   (1) A
   (2) B
   (3) C
   (4) D

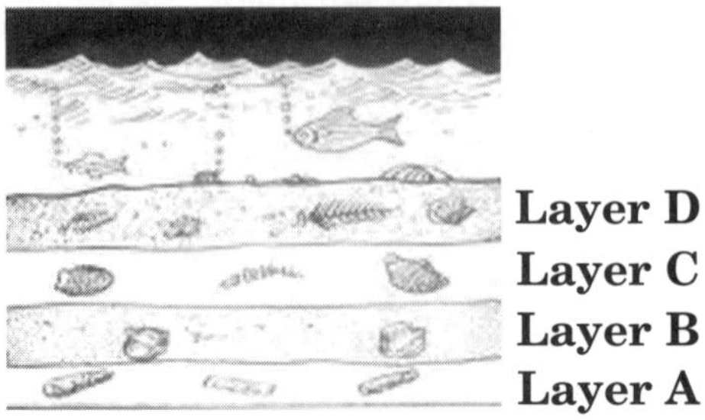

Layer D
Layer C
Layer B
Layer A

23 The landscape to the right has been affected by the
   motion of plates.

   Which diagram below best illustrates the direction
   of the plates?

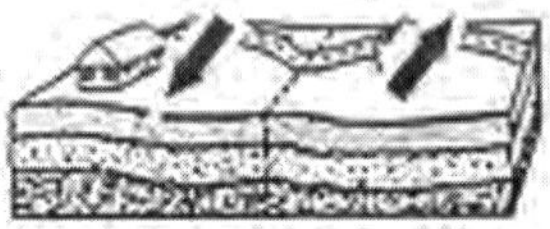

   (1) 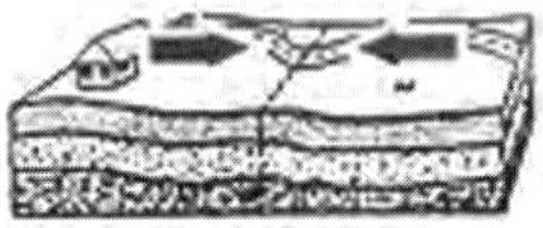   (2) 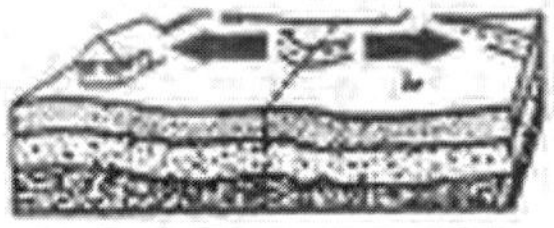   (3) 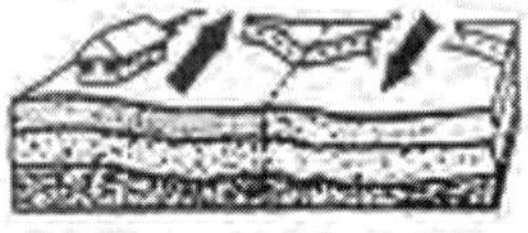   (4)

24 The landscape illustrated at the right best illustrates
   (1) faulting and deposition
   (2) folding and erosion
   (3) erosion and faulting
   (4) erosion and deposition

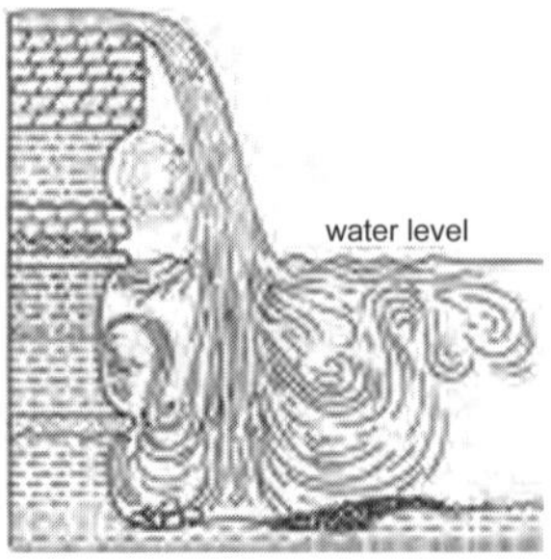

water level

25 Which landscape formation supports the theory of Plate Tectonics?
   (1) deposition   (2) erosion   (3) folding   (4) buttes and mesas

26 What effect does a volcanic eruption have on the atmosphere? The released gases
   (1) reflect sunlight, having a warming effect   (3) absorb sunlight, having a warming effect
   (2) reflect sunlight, having a cooling effect   (4) absorb sunlight, having a cooling effect

27 An earthquake was measured to be 3 on the Richter Scale in California. An earthquake in Turkey was
   measured to be 5 on the Richter Scale. The Richter Scale goes from 1-9. Each number represents a
   logarithm (10X). How many times stronger is the earthquake in Turkey than the one in California?
   (1) 2   (2) 10   (3) 100   (4) 1000

28 The main cause of weather is
   (1) the amount of rainfall a region receives   (3) the latitude of a region
   (2) the type of bedrock in a region   (4) the uneven heating of Earth

29 The two main factors that determine climate are
   (1) elevation of land and solar radiation   (3) rainfall and elevation of land
   (2) solar radiation and rainfall   (4) rainfall and latitude

30 The map at the right shows the location of four air masses. Which air mass would hold the coldest and driest air?
(1) A
(2) B
(3) C
(4) D

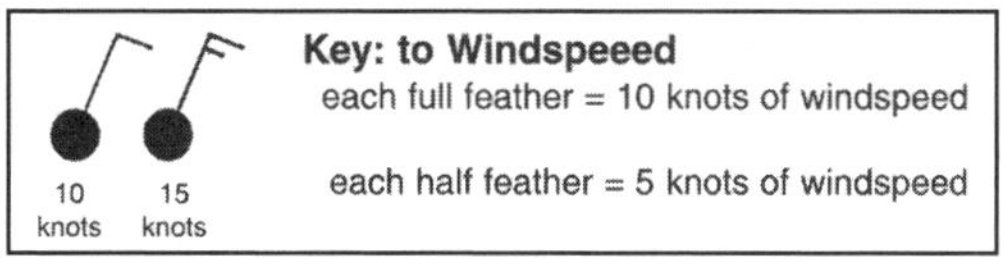

Key: to Windspeeed
each full feather = 10 knots of windspeed

each half feather = 5 knots of windspeed

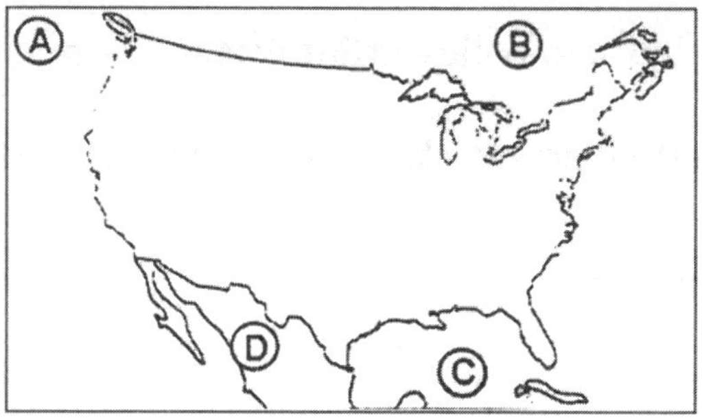

10 knots   15 knots

31 Below are four stations indicating the speed of the wind. Which symbol shows the highest wind speed?

(1)    (2)    (3)    (4)

32 Which point has a barometric pressure of 1012 mb?
(1) A
(2) B
(3) C
(4) D

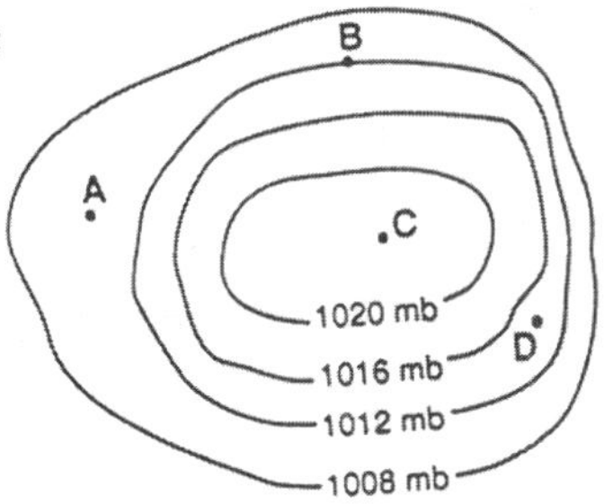

1020 mb
1016 mb
1012 mb
1008 mb

33 Refer to the weather map at the right, what type of weather will City A experience after the cold front moves through?
(1) cold and rainy   (3) warm and rainy
(2) cold and dry   (4) warm and dry

34 Which direction is the warm front moving?
(1) West   (3) Northeast
(2) East   (4) Northwest

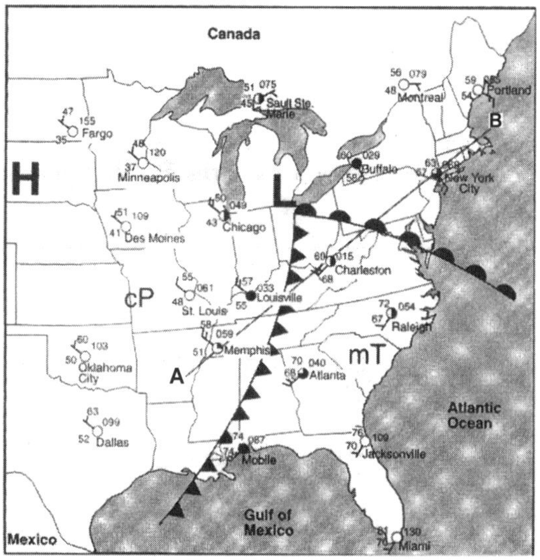

THE QUESTIONS THAT FOLLOW TEST YOUR UNDERSTANDING OF SCIENTIFIC PRACTICES, CORES IDEAS IN SCIENCE AND CROSS-CUTTING CONCEPTS SUCH AS PATTERNS AND CAUSE AND EFFECT RELATIONSHIPS.

SHORT ANSWER

1   Directions: Use the graph at the right that shows the temperature changes in the Earth's atmosphere.

a   Approximately, how many kilometers of atmosphere makes up the stratosphere?

(1 point) _____

_____

b   Describe the temperature changes from 0 km to 100 km. (2 points)

_____

_____

_____

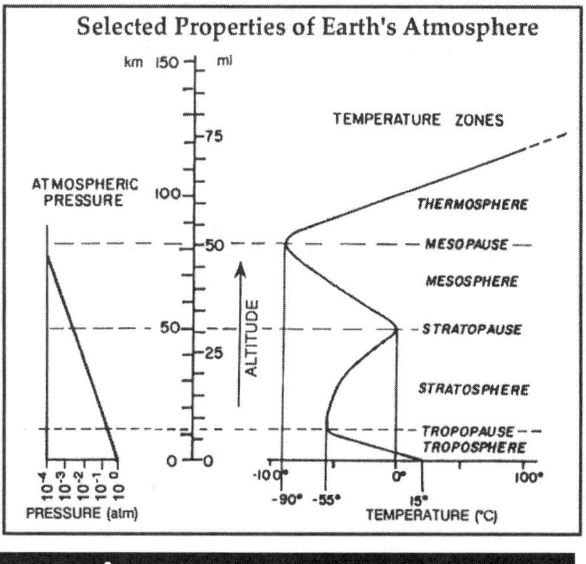

Selected Properties of Earth's Atmosphere

2   Most scientists believe that certain human activity is having a "Greenhouse Effect" on Earth.

     a   What do scientists mean by the "Greenhouse Effect?" (1 point) _____

_____

_____

     b   Describe two types of human activity that are believed to have contributed to the "Greenhouse Effect." (2 point)

_____

_____

SHORT ESSAY

1   A show called "The Restless Earth" was telecast on educational TV.

     a   What do you think the show was about? (1 point) _____

     b   Describe the main reason why the Earth is "restless?" (1 point) _____

_____

     c   Describe two ways the Earth's "restlessness" affects the shape of the land? (2 points) _____

_____

_____

2   Design an experiment to show that water retains heat longer than soil. Your response should be well organized and include the following components. (each component is 1 point)

     a   State your hypothesis. _____

_____

     b   Identify the dependent variable. _____

     c   Identify the independent variable. _____

     d   Discuss factors to be controlled. _____

_____

_____

     e   Construct a table to use for collection of data.

Section Four
PHYSICAL SCIENCE

MATTER AND INTERACTIONS

DISCIPLINARY CORE IDEAS:

PS1.A: STRUCTURE AND PROPERTIES OF MATTER

PS1.B: CHEMICAL REACTIONS

PROPERTIES OF MATTER

Matter is anything that has mass and occupies space. The space that matter occupies is its **volume**. The amount of matter is its mass. Properties that can be determined without changing the matter are called **physical properties. Chemical properties** can only be identified by changing the matter. Each pure substance has characteristic physical and chemical properties that can be used to identify it.

PHYSICAL PROPERTIES OF MATTER

Some physical properties of matter can be described by using your **senses**. These properties include color, odor, taste, texture, luster, and hardness. **Luster** refers to how well light is reflected. Most metals have high luster. **Hardness** is a relative term based on the matter's resistance to scratching. For example, diamonds are considered the hardest matter because nothing can scratch a diamond except another diamond. **Texture** refers to the matter's relative smoothness. Some types of matter have a characteristic **smell**, such as the oil of wintergreen found in wintergreen candy. **Color** depends on the light reflected by the matter. For example, matter that is red absorbs all colors of the spectrum except red (which it reflects).

1 How would you describe the properties of aluminum? _____

_____

TESTING FOR PHYSICAL PROPERTIES

Some physical properties of matter can only be determined by carrying out a test on the matter. These properties include boiling and freezing points, solubility, conductivity of heat and electricity, density, and weight. Other physical properties include whether the matter is malleable, ductile, flexible, or elastic.

The **boiling point** and **freezing point** of matter can be used to help identify the matter. For example, water freezes at 0°C and boils at 100°C. The temperature in which matter changes phase from a liquid to a solid is called its freezing point. The temperature in which matter changes phase from a liquid to a gas is called its boiling point.

Solubility refers to how well a type of matter can dissolve in a solvent. The most common solvent is water. The temperature of the solvent also affects its solubility. The substance being dissolved is called the **solute** while the substance doing the dissolving is called the **solvent**. In a **solution** of salt water, salt would be the solute, and water would be the solvent. Air is a solution of gases. Solid solutions are called **alloys**, such as steel and bronze. All solutions appear the same throughout.

FACTORS THAT AFFECT RATE OF SOLUBILITY

- Generally, the higher the **temperature** of the solvent, the more soluble the matter will be. However, if the matter happens to be a gas, its solubility will decrease with an increase in temperature.

- **Pressure** only has a significant effect on solubility when a gas is being dissolved in a liquid. The greater the pressure, the more soluble the gas will be. This explains why when you open a bottle of soda, it begins to bubble. The opening reduces the pressure and the bubbling is the gas (CO_2) coming out of solution.

- Have you ever noticed how granulated sugar dissolves faster in water than a cube of sugar? This is because granulated sugar has more **surface area** (surface in contact with the solvent water) helping it dissolve faster.

- When a solution is stirred, it helps the solute mix faster with the solvent. This helps it dissolve faster.

SOLUBILITY GRAPHS

Solubility graphs have been developed by scientists to summarize the **solubility** of various substances in a given mass of water at changing temperatures. The x-axis of the graph gives the temperature, while the y-axis tells the grams of solute that can be dissolved in 100 grams of water. The curves on the graph show the number of grams needed to create a saturated solution in 100 grams of water. Solutions can be considered unsaturated, saturated, or supersaturated.

An **unsaturated solution** is a solution that can hold more solute for a given temperature.

A **saturated solution** is a solution that holds the maximum amount of solute it can for a given temperature. If too much solute is added, the extra solute will settle out of the solution.

A **supersaturated solution** is a solution that is holding more solute than it normally can for a given temperature. Have you ever eaten rock candy? You can make rock candy by bringing to a boil 250 ml (about one cup) of water. Add 500 mL (2 cups) of sugar to the boiling water. Let it cool to room temperature. If the sugar is still dissolved, the solution is called a supersaturated solution. It is very unstable and easily **crystallizes** (the sugar comes out of solution). If a string is added to the solution, it will act as a seed on which the crystals form. As you can see in the diagram at the right, there is a little science to everything, even in the making of rock candy.

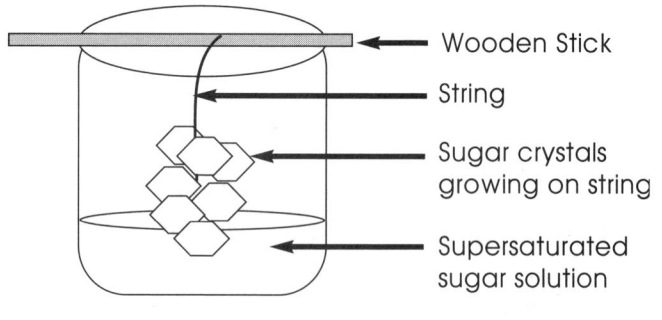

- Wooden Stick
- String
- Sugar crystals growing on string
- Supersaturated sugar solution

2   Based on the graph at the right, how would you describe the solubility of NaCl as the temperature changes? _____

_____

_____

_____

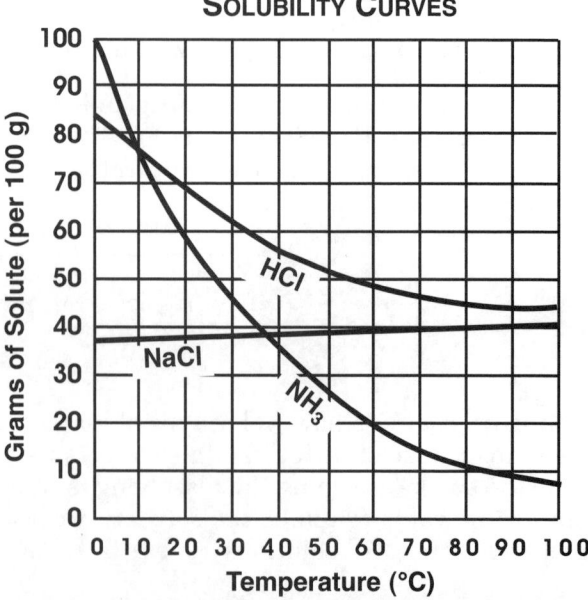

SOLUBILITY CURVES

3   What is the solubility of NH_3 at 20°C? _____

Conductivity refers to how well the matter can conduct heat or electricity. Generally, metals are excellent **conductors** of heat and electricity. Some solutions can conduct electricity. Solutions that conduct electricity are called electrolytes. Batteries contain **electrolytes**. The contents of living cells contain electrolytes. The water in swimming pools contains electrolytes, which is one reason you must get out of a pool during a lightning storm. Because the pool water conducts electricity, you can be seriously injured or killed by electrical shock.

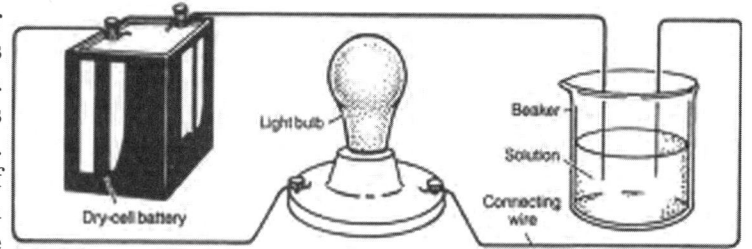

4   Explain why the handle of a silver spoon gets very hot when placed in a cup of hot chocolate. _____

_____

Matter that does not conduct heat or electricity very well is said to be an **insulator**. The plastic or rubber coating of an electrical wire is an insulator. The plastic or wood handles on pots and pans insulate against the transfer of heat.

5   What materials make good insulators? _____

DENSITY & WEIGHT

Have you ever noticed that the oil in salad dressing always floats to the top of the vinegar no matter how many times you shake the dressing bottle? This is because the density of oil is less than vinegar (which is mostly water). **Density** refers to the amount of matter (mass) that occupies a given volume.

Mathematically, it is expressed as $$\textbf{Density (D)} = \frac{\textbf{Mass (grams)}}{\textbf{Volume (cm}^3\text{ or mL)}}$$

Weight is due to the pull of gravity on mass. The more mass, the greater its response to the pull of gravity. The gravitational force exerted by Earth is greater than the Moon's because Earth has more mass than the Moon.

6   How would your mass and weight change if you were an astronaut and

traveled to the Moon? _____

_____

7   An astronaut is, for all practical purposes, weightless on the space

shuttle. *a* Does the astronaut have mass? _____ *b* Explain. _____

_____

_____

A WEIGHTY CONCEPT

The weight of something does not determine its density. Density is a relationship between mass and the volume it occupies. If you look at the formula for density, you should notice that the numerator is the mass and the denominator is the volume. Two types of matter may have the same mass, and so the same weight, but if one occupies a larger volume, it will be less dense. For example, a gold bar weighing 500 grams is much smaller in size (volume) than a silver bar of the same weight, because gold is more dense than silver.

Also, it is important to note that the mass of the silver and gold are the same, so they have the same weight. However, the volume of silver is greater than gold. If you do the calculations for density, you will find that the density of the gold is greater. The density of gold is 19.3 g/mL while the density of silver is 10.5 g/mL. The density of an element is a property that can be used to identify an element.

SAMPLE DENSITY PROBLEM

What is the density of a substance that has a mass of 2.5 g and a volume of 0.93 cm³? In order to solve this problem, write the formula for density and plug in the values.

$$D = \frac{Mass\ (grams)}{Volume\ (cm^3\ or\ mL)} = \frac{2.5\ g}{0.93\ cm^3} = 2.7\ g/cm^3$$

8   The density of water is **1 g/mL at 4°C**. Matter with a density less than water will float while matter with a density greater than water will sink.

Explain why ice floats on water. _____

_____

9   The density of wood is 0.5 g/mL.   *a*   Will it sink or float in water? _____   *b*   Explain. _____

_____

_____

EFFECTS OF DENSITY

BUOYANCY

Some objects sink while others others float in water. This is referred to as the object's **buoyancy**. All fluids (liquids and gases) exert a buoyant force on matter. While gravity exerts a downward force on objects, the buoyant force is an upward force. In order for an object to float, the buoyant force must be equal to or greater than the pull of gravity. Since a force is the "push" or "pull" on matter, you can think of the buoyant force as the push and gravity as the pull on an object. The buoyant force is equal to the weight of the fluid displaced by the object.

- **Floating Objects**: If an object weighs 50 Newtons (a Newton (N) is a measurement of metric weight and 1 kg = 9.8 Newtons) and displaces 50 Newtons of water, it will float on the water.

- **Sinking Objects**: If an object weighs 150 Newtons and displaces 100 Newtons of water, it will sink.

56 *a*   Will an object that weighs 125 N and

displaces 100 N of water float? _____

b   Explain. _____

_____

_____

EARTH'S DENSITY

Scientists believe Earth is layered because when Earth first formed, it was all liquid. The densest matter sank to the center of Earth. The least dense matter made up the lithosphere (crust) which floats on the denser mantle. Weather occurs because of the uneven heating of the planet. Temperature differences in the atmosphere create density currents (convection) with the denser cold air sinking under the warmer, less dense air. The property of density is responsible for many of your observations about the natural world.

MALLEABILITY & DUCTILITY

Metals are examples of matter that are both malleable and ductile. **Malleable** matter can be hammered into thin sheets. In ancient times, a gold beater was a person who hammered gold into thin sheets to be used for decorative purposes. **Ductile** matter is matter than can be pulled into wires. Metals can also be pulled into wires when heated. Fiber optics are made of very pure glass that are pulled into "wires" to conduct light energy.

FLEXIBILITY & ELASTICITY

Matter that can bend without breaking is considered **flexible**. Matter that can stretch and return to its original shape without breaking is called **elastic**.

11 Name some examples of matter that are flexible, elastic, or both. _____

_____

CHEMICAL PROPERTIES

These are properties that can only be determined by changing the properties of the matter. Flammability and reactivity are examples of chemical properties. For example, **petroleum** is very **flammable**. This means that it burns readily. Some substances are relatively stable while others **react** readily with other substances. Hydrogen is no longer used in balloons because it is very reactive and flammable. Helium is a very stable substance so it is used instead. The element sodium is so reactive that it must be stored under oil to prevent it from reacting with moisture in the air. If it reacts with the moisture in the air, sodium releases hydrogen gas from the water. The reaction produces so much heat that the hydrogen gas bursts into flame.

12 Describe one chemical and one physical property of matter. _____

_____

_____

DISTINGUISH BETWEEN CHEMICAL AND PHYSICAL REACTIONS

Both physical and chemical reactions involve the absorption or release of energy. For example, the physical reaction of ice melting involves the absorption of energy and when it freezes, the release of energy. The chemical reaction of the burning of wood involves the release of energy while the cooking of an egg involves the absorption or storage of energy.

PHYSICAL CHANGE

The heat from the Sun melts your snowman into a puddle. Is the puddle of water and the snowman made of the same matter? If a new type of matter was created, the answer would be yes. However, upon close examination, you should find that the puddle and the snowman are both made from water because they share the identical properties. The chemical properties of the matter are not changed; therefore, this is a physical (not chemical) change.

- A **phase change** is always a physical change (see phase changes on page 177).

- When one substance dissolves into another substance, a physical change has taken place. A sugar cube is still sugar after it **dissolves** in water. Drink the water and you will notice that it tastes sweet because the sugar is still there. If the water is evaporated, the sugar can be recovered.

- Wood can be **chopped** into smaller pieces and paper can be **cut** into sections. If the density of each cut piece of paper or chopped piece of wood was measured, it would be found to be the same as the larger piece, because only a physical change had taken place.

- When substances are combined, but keep their own individual properties, a **mixture** forms. Mixtures can be separated by physical means.

SEPARATING MIXTURES

The physical properties of matter can be used to separate a mixture of different types of matter.

- A **magnet** separates a mixture of magnetic and nonmagnetic matter (e.g., sawdust and nails).
- **Filtration** separates a liquid and a solid or two solids with different particle size. The colander that separates pasta from water is an example. A filter is used to separate coffee grinds from the coffee pot.
- **Evaporation** separates substances in a solution. After the solvent evaporates, the solute is left behind.
- **Distillation** separates two liquids by differences in their boiling points. The liquid with the lower boiling point vaporizes first from the liquid mixture leaving behind the liquid with the higher boiling point.
- **Decantation** (**filtration**) separates the liquid portion of an insoluble mixture from the solid part by simply pouring off the liquid part.
- **Chromatography paper**, a type of filter paper, separates mixtures by differences in their solubility and how they adhere to the paper.
- A **centrifuge** separates materials of different densities (in solution) by spinning them about a central axis.

13 Describe two ways to separate a mixture. _____

_____

_____

CHEMICAL CHANGE

The type of change in which the physical and chemical properties of matter are changed is called a **chemical change**. A nail rusts, silver tarnishes, food is digested, deep caverns are formed in Earth, dead organisms decay, milk sours, wood burns, and a metal bridge corrodes. All of these are examples of chemical reactions because the matter created has new chemical compositions and different properties than the matter from which it was formed.

THE SALT CONNECTION

Do you put salt on your food? You might think twice about eating the elements that make up the compound called **table salt**. The element sodium is a highly reactive metal. It is silver in color, has luster, and is so soft, it can be cut with a knife. The element chlorine is a green pungent smelling, highly reactive gas. However, when they **chemically react** with one another, they form a white, cubic, crystal shaped, salty tasting matter: common table salt. The properties of sodium and chlorine are very different than the properties of table salt. Salt is necessary for cell function. However, too much salt can be harmful and is linked to high blood pressure.

THE CHEMICAL EQUATION

A **chemical equation** summarizes a chemical reaction. The chemicals that begin the reaction are called **reactants**, while the substances produced from the reaction are called **products**. An arrow means the same thing as an equal sign in a mathematics formula. It stands for "**yields**."

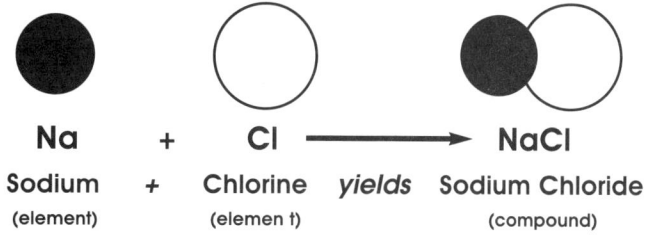

Na	+	**Cl**	⟶	**NaCl**
Sodium	+	Chlorine	*yields*	Sodium Chloride
(element)		(elemen t)		(compound)

CLASSIFYING MATTER BY REACTIVITY

Some types of matter are more reactive than other types. The **Noble gases** are a group of elements that are highly inert (stable). Hydrogen is a very reactive gas. It is the least dense of gases and so was used in balloons and dirigibles (blimps). However, since the tragic 1937 explosion of the German airship *Hindenberg* (a dirigible filled with hydrogen), helium, another noble gas, has replaced hydrogen in balloons and blimps. Helium is the second least dense gas.

A group of metals that include sodium and potassium are considered so reactive that they are stored under oil to keep air (oxygen) away. Fluorine and chlorine are examples of nonmetals and gases that are highly reactive. Metals can corrode or rust when they chemically combine with a nonmetal such as oxygen or sulfur. A chemical reaction between a metal and a nonmetal produces a chemical called a **salt**. Another way of describing the reaction is: the reaction between a base (NaOH) and an acid (HCl) yields a salt (NaCl) and water (H₂O).

NaOH	+	**HCL**	⟶	**NaCl**	+	**H₂O**
Sodium Hydroxide		Hydrochloric Acid		Sodium Chloride (salt)		Water

LAW OF CONSERVATION OF MASS

During a chemical reaction, the mass of the reactants (substances that begin the reaction) must equal the mass of the product (what is formed). Basically, this law states that during a chemical reaction, matter cannot be destroyed or created, only rearranged. For example, if a total of 5 grams of Hydrogen and Oxygen gases were combined, 5 grams of water would result.

14   Why is the mass of the water equal to the combined mass of the hydrogen and oxygen? _____

_____

DEVELOP MENTAL MODELS TO EXPLAIN COMMON REACTIONS AND CHANGES IN STATES (PHASES) OF MATTER.

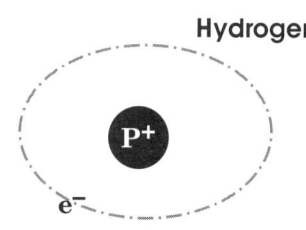

ATOMS

An **atom** is the smallest part of an element that still has the properties of that element. Atoms of the same element are alike and atoms of different elements are different. All atoms have a neutral charge and are in constant motion. An atom consists of a central **nucleus** in which most of the mass is located. In this nucleus are two main subatomic particles called neutrons and protons. **Protons** (P^+) are positively charged while **neutrons** (N^0) have no charge. Orbiting the nucleus are negatively charged particles called **electrons** (e^-). Electrons are located in the electron cloud, a region located around the nucleus of an atom. The mass of an electron is very minute (1/1836 a.m.u.s or **atomic mass units**). The mass of neutrons and protons are very similar, each having a mass of about 1 atomic mass unit (a.m.u.). The number of protons an element has determines what element it is. For example, hydrogen is hydrogen because it has 1 proton, while helium is helium because it has 2 protons, lithium is lithium, because it has 3 protons and so on. (Additional atoms illustrated on the next page.)

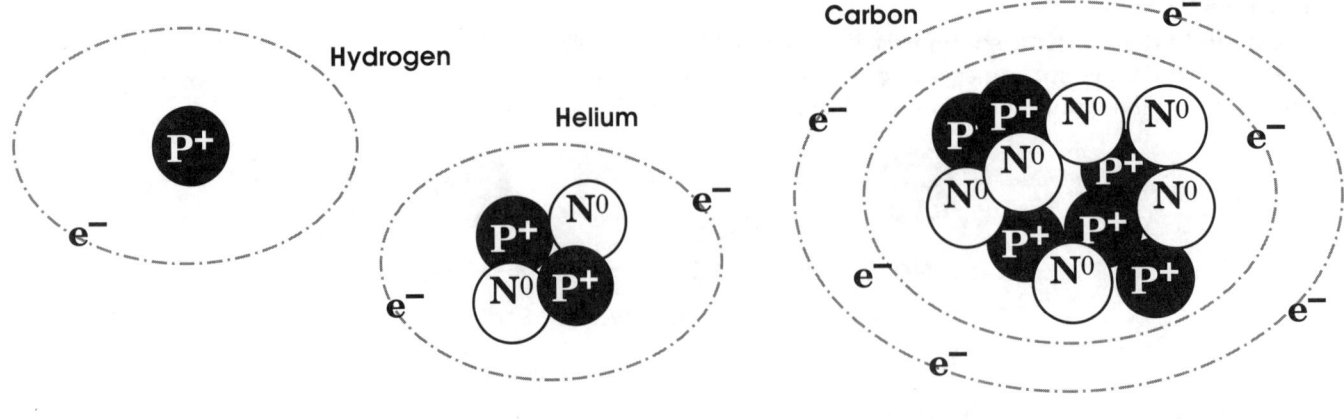

15 Describe the charge, mass, and location of the three main particles in an atom. _____

_____

16 What makes one atom different from another atom? _____

_____

ELEMENTS

An **element** is a pure substance made up of only one type of atom. Some of the atoms of an element have a different number of neutrons. These atoms are called **isotopes** of one another. For example, hydrogen has three isotopes. Hydrogen with no neutrons is called **protium**. Hydrogen with one neutron is called **deuterium**, while hydrogen with two neutrons is called **tritium**. Most elements are chemically combined with other elements and so are not found in their pure form. Atoms form molecules that range in size from two to thousands of atoms.

ISOTOPES OF HYDROGEN

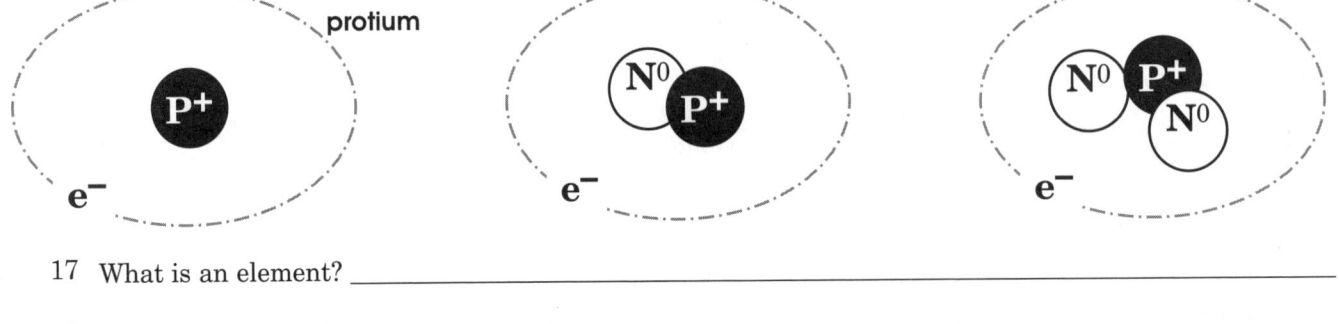

17 What is an element? _____

18 What is an isotope? _____

PERIODIC TABLE OF ELEMENTS

All elements are organized on the ***Periodic Table of Elements***. In the mid-eighteen hundreds, the Russian scientist **Dmitri Mendeleev** developed the first Periodic Table of Elements. The elements were arranged by increasing **atomic mass** (the number of protons plus neutrons). This early table was useful, because it was the first attempt to organize the elements in a manner that grouped elements by similar properties. However, sometimes, elements seemed to be placed incorrectly on the table. Fifty years later, the English scientist **Henry Moseley** discovered the atomic number. The **atomic number** is the number of protons an element has. Remember, it is the number of protons an element has that determines what element it is. When Moseley arranged all of the elements by increasing atomic number, numerous patterns became apparent. Elements with similar properties would always fall into the same columns. Each element is represented by a one or two letter symbol. The first letter is capitalized while the second letter (if used) is small case, for example: Helium is He.

19 What is the main difference between Mendeleev's and Moseley's Periodic Table of Elements? _____

_____

Periodic Table of the Elements

KEY

Atomic Mass → 12.011
Selected Oxidation States → -4 +2 +4

Symbol → **C**

Atomic Number → 6

Electron Configuration → 2-4

Relative atomic masses are based on $^{12}C = 12.000$

Note: Mass numbers in parentheses are mass numbers of the most stable or common isotope.

*The systematic names and symbols for elements of atomic numbers above 109 will be used until the approval of trivial names by IUPAC.

**Denotes the presence of (2-8-) for elements 72 and above

Arrangement Of The Modern Periodic Table

The columns in the table represent **families** or **groups** of elements. All members of the same group share similar properties. For example, sodium (Na), and potassium (K), located in the first column are both very reactive metals.

Each row on the Periodic Table represents a **period**. As you read from left to right on the periodic table the properties of elements change in a cyclic pattern. Elements to the far left are the most reactive metals and as you move to the right, the elements become less and less metallic and more nonmetallic. The elements on the far right are considered nonmetals. The reactivity of the nonmetals increases as you move to the right, except for the column farthest to the right. These elements are non-reactive, or **inert**. They are the **noble gases** (helium, neon, argon, krypton, xenon, and radon). The elements become more metallic on the left side of the Periodic Table. As you move from row to row, this pattern of elements grouped by their properties repeats, hence the name periodic.

Metalloids are elements with both metal and nonmetal properties. They are located along either side of the zigzag line on the periodic table.

20 Name a reactive and nonreactive element on the Periodic Table. _____

_____

21 What is the mass and atomic number of oxygen (O_2)? _____

_____

22 Identify a metal (metallic element) by its symbol on the Periodic Table. _____

_____

Molecule

A molecule is composed of more than one atom chemically combined. The smallest particle of water that can still be identified as water is a molecule of water. H_2O is the formula for one molecule of water. Subscripts tell the number of atoms of each atom in a compound. No subscript means 1. The subscript 2 below hydrogen shows that the hydrogen and oxygen combine in a 2:1 ratio.

Pure Substance

A pure substance is one composed of one type of element, molecule or compound. Examples of this are distilled water, which is composed of only water molecules, and the graphite in your pencil composed of only the element carbon.

Compound

A compound is composed of two or more elements chemically combined. A chemical formula is used to represent a compound. NaCl (sodium chloride) and Fe_2O_3 (iron oxide) are examples of chemical formulas. Sodium chloride units are packed together in large arrays held together by opposite charges while the iron oxide is organized in well defined molecules.

Chemical Bond

The forces that hold atoms together are called **chemical bonds**. There are several kinds: ionic, covalent, and metallic.

The Rate Of A Chemical Reaction

Collision Theory

According to the **Collision Theory**, all atoms and molecules are in constant motion. Since atoms are moving, there is a chance that they may **collide**, and if they collide with sufficient force, a chemical reaction may take place.

- Chemical reactions will occur more rapidly at higher **temperatures**. The internal particles in matter move more rapidly at higher temperatures. This increases the chance for a collision to take place.

- As the **concentration** of the reactants is increased, the rate of a chemical reaction increases. Again, this is because there is a greater chance of a collision among the particles reacting.

In the human body, chemical reactions take place when substances are dissolved or immersed in water. When substances dissolve, they break apart into their smallest components which may be a molecule or **ion** (charged atom). This improves the chance for a collision, so that a reaction will take place. It would be impossible to digest food if water was not present. The dissolving capacity of water also makes chemical reactions easier to take place in the nonliving environment.

PHASES OF MATTER

Four phases of matter have been identified by scientists. They are solid, liquid, gas, and plasma. Only solid, liquid, and gas occur naturally on this planet. The phase of matter is related to how close together the particles are to one another and their relative motion.

If the temperature remains the same for a particular type of matter:

- **Solids** have a definite shape and volume. The particles in a solid are moving the slowest and are closest together, compared to other phases of that matter.
- **Liquids** have a definite volume but take the shape of their container. The movement and position of the particles are faster than solid, but slower than gas.
- **Gases** have no definite shape or volume. The particles in a gas are moving faster and are separated more than either solids or liquids.
- **Plasma** particles of matter are not only moving very fast, but have been stripped of their electrons. When the temperature reaches millions of degrees Celsius, plasma forms. Because of the extreme temperatures on the Sun and all other stars, plasma is the normal phase of matter for these celestial bodies.

PHASE CHANGES

A **phase change** is when matter changes from one phase to another. Phase changes are always accompanied by the transfer of heat. Heat energy is either absorbed or released.

Have you ever walked by the refrigerator and felt the heat coming out from its bottom. This is because the water in food is undergoing a phase change. As heat is removed from it, the particles in the matter slow down and eventually the water in the food freezes. This is because the amount of heat that can be transferred from matter is related to the motion and position of particles in matter. When the particles move closer together, the particles have slowed down so heat is released. When the particles speed up and move farther apart from one another, they have absorbed heat energy. Types of phase changes include:

- **Melting**: This is a change from a solid to a liquid. Melting occurs at the freezing point of matter. For example, ice melts at 0°C which is also its freezing point.
- **Freezing**: This is a change from a liquid to solid. Water freezes at 0°C.
- **Vaporization**: This is a change from a liquid to gas. It occurs at the boiling point of matter. It occurs at 100°C for water.
- **Evaporation**: This is a change from a liquid to gas on the surface of a liquid.
- **Condensation**: This is a change from a gas to a liquid.
- **Sublimation**: This is a change from a solid to a gas.
- **Deposition:** This is a change from a gas to a solid.

23 Using the information above as a reference, identify whether heat is absorbed or released for each of the phase changes in the table at the right.

Type of Phase Change	Heat Change (absorbed or released)
Freezing (liquid to solid)	
Melting (solid to liquid)	
Vaporization (liquid to gas)	
Evaporation (liquid to gas)	
Sublimation (solid to gas)	
Condensation (gas to liquid)	
Deposition (gas to solid)	

At the right is a diagram of the relative position of molecules in a solid, liquid, and a gas.

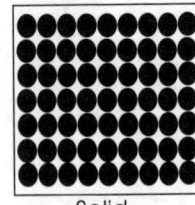
Solid

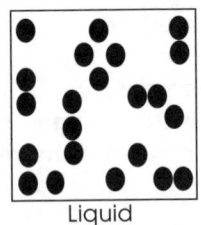
Liquid

Gas

24 Water is continually being cycled through the atmosphere, lithosphere, and hydrosphere.

What type of phase change takes place in a cloud when ice crystals form from water vapor?

_____

25 *a* Does the above phase change add or remove heat from the atmosphere? _____

b Explain. _____

GRAPHING A PHASE CHANGE

If the temperature of the matter undergoing a phase change is recorded over a period of time and graphed, a plateau (flattened area) appears on the graph. This plateau appears because the temperature does not change when matter is going through a phase change. This is because heat energy is being absorbed or released during a phase change resulting in no net change in temperature until the phase change is complete.

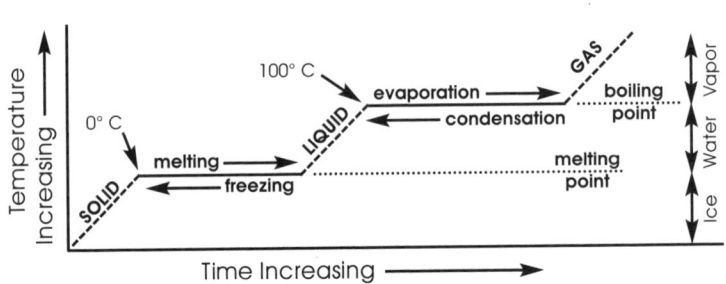

Phase Change Graph for Water

26 Refer to the graph at the right, explain why the temperature does not change during a

phase change. _____

_____

_____

_____

_____

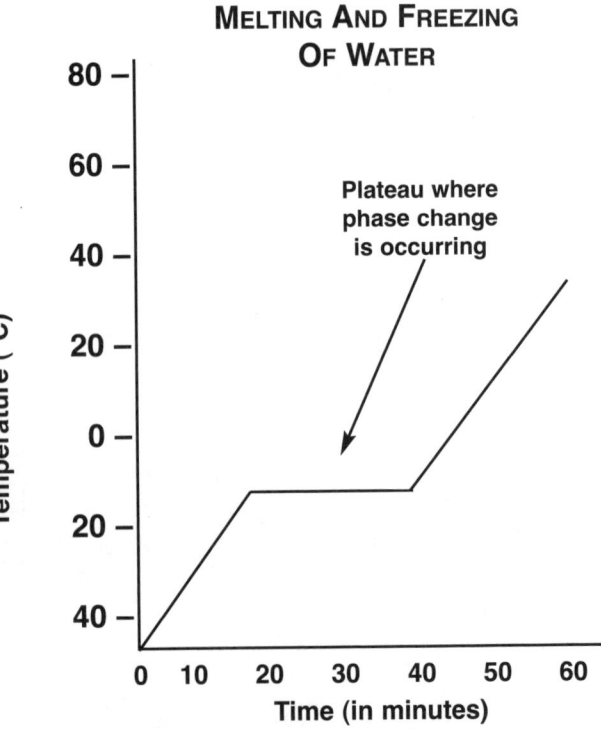

MELTING AND FREEZING OF WATER

Plateau where phase change is occurring

MULTIPLE CHOICE

1   Which property of matter would not be considered a physical property?
    (1) color   (2) reactivity   (3) luster   (4) texture

2   The type of matter that can conduct heat the best is
    (1) wood   (2) plastic   (3) metal   (4) rubber

3   The boiling point of matter is the temperature in which matter
    (1) changes from a gas to a liquid   (3) changes from a solid to a gas
    (2) changes from a liquid to a gas   (4) changes from a gas to a solid

Use the graph at the right to answer questions 4 & 5.

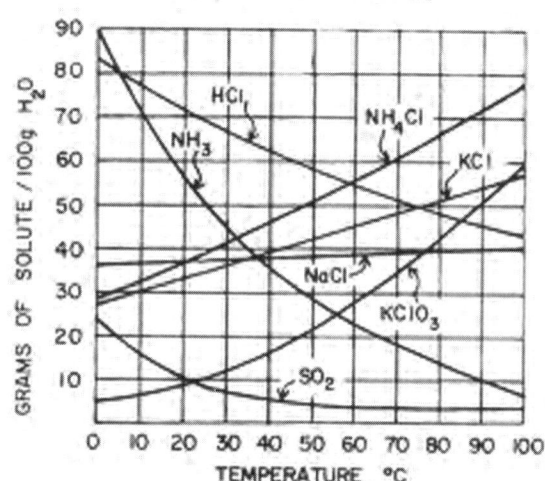

SOLUBILITY CURVES

4   How many grams of NH_3 are needed to create a saturated solution at 20°C?
    (1) 30 g
    (2) 55 g
    (3) 70 g
    (4) 90 g

5   What temperature is needed to create a saturated solution of 50 g of NH_4Cl?
    (1) 50°C
    (2) 60°C
    (3) 70°C
    (4) 80°C

6   Pots and pans use wood or plastic on the handles of the pot because they are
    (1) insulators   (2) conductors   (3) malleable   (4) flexible

7   The pull of gravity on matter is its
    (1) density   (2) volume   (3) weight   (4) luster

8   What is the density of matter that has mass of 15 grams and a volume of 3 cm³?
    (1) 5 g/cm³   (2) 10 g/cm³   (3) 15 g/cm³   (4) 20 g/cm³

9   A mixture that can be separated wi   th a magnet is
    (1) salt and water   (2) sugar and salt   (3) sugar and lead   (4) salt and iron

10 Which of the following is an example of a chemical change?
    (1) dissolving   (2) melting   (3) rusting   (4) evaporating

11 Hydrogen is no longer used in balloons because it is too
    (1) malleable   (2) reactive   (3) stable   (4) metallic

12 If 5 grams of substance A reacts with 10 grams of substance B, how many grams of product are formed?
    (1) 5 grams   (2) 10 grams   (3) 15 grams   (4) 20 grams

13 Which of the following substances is not an element?
    (1) water   (2) calcium   (3) carbon   (4) hydrogen

Periodic Table of the Elements

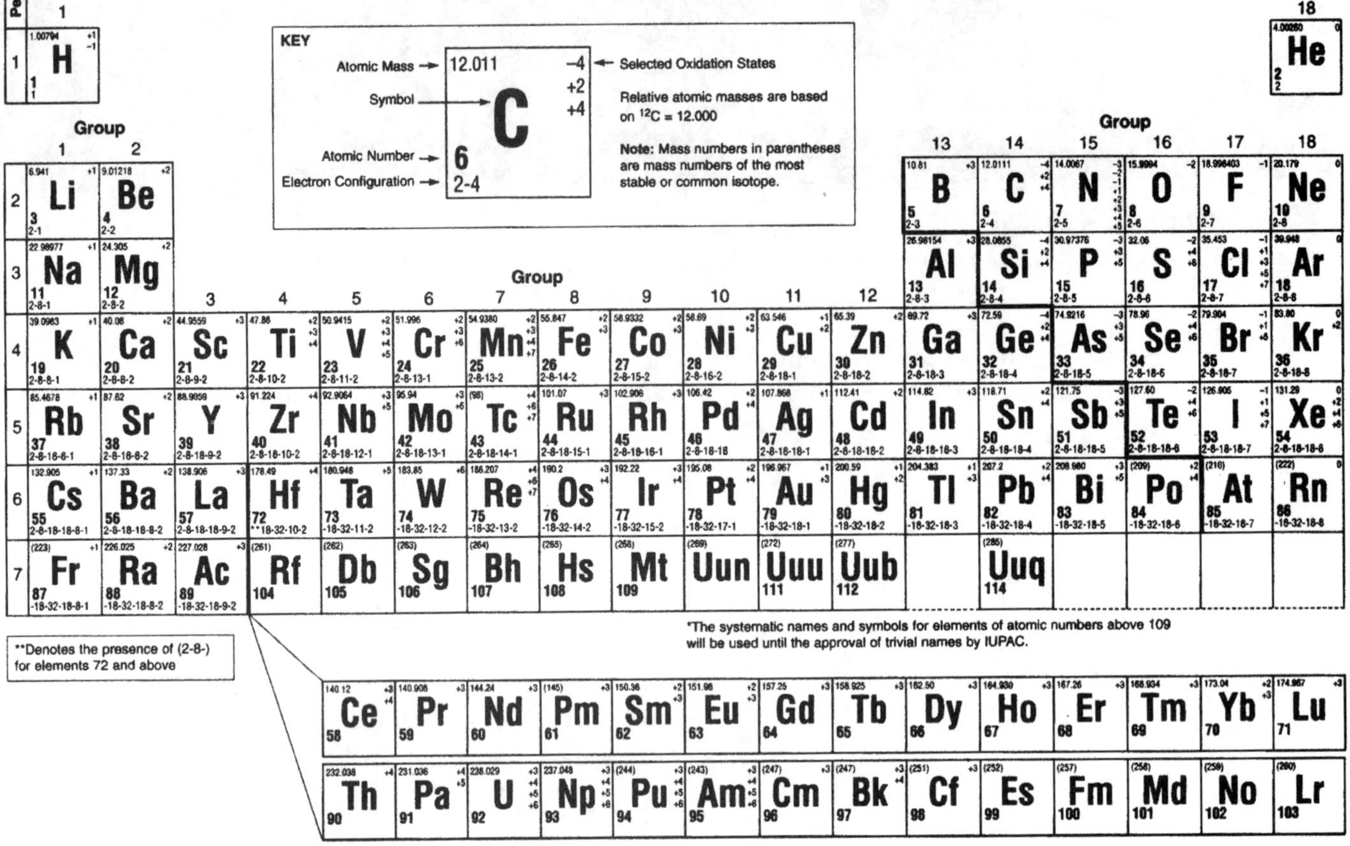

Period
Group
KEY
Atomic Mass
Symbol
Atomic Number
Electron Configuration
12.011
C
6
2-4
-4 +2 +4
Selected Oxidation States
Relative atomic masses are based on ¹²C = 12.000
Note: Mass numbers in parentheses are mass numbers of the most stable or common isotope.
**Denotes the presence of (2-8-) for elements 72 and above
*The systematic names and symbols for elements of atomic numbers above 109 will be used until the approval of trivial names by IUPAC.

Directions: Use the periodic table above (or refer to page 175) to answer questions 14-16.

14 Which group has the most reactive metals?
(1) 1 or IA (2) 2 or IIA (3) 3 or IIIB (4) 4 or IVB

15 In which period is potassium (K) found?
(1) 1 (3) 2 (3) 3 (4) 4

16 What is the atomic number of sodium (Na)?
(1) 3 (2) 11 (3) 19 (4) 37

17 The formula for carbon dioxide is CO,. What is the ratio of carbon to oxygen in this compound?
(1) 1:1 (2) 2:1 (3) 1:2 (4) 2:2

18 As the temperature increases, the rate of a chemical reaction
(1) increases
(2) decreases, then increases
(3) decreases
(4) increases, then decreases

19 The phase of matter in which the particles are moving the slowest is a
(1) liquid (2) solid (3) gas (4) plasma

Directions: use the diagrams at the right illustrating phase changes in water to answer questions 20 and 21.

ICE FORMING
STEAM FORMING
LIQUID WATER
COLD SURFACE
DROPS OF LIQUID WATER FORMING
ICE
LIQUID WATER FORMING
1
2
3
4

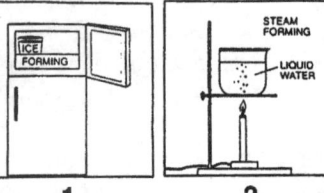

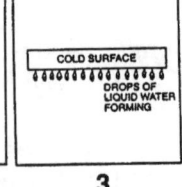

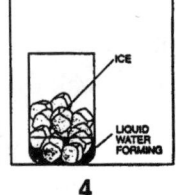

20 Which diagram illustrates condensation?
(1) 1 (3) 3
(2) 2 (4) 4

21 Which two illustrations show a phase change that requires the absorption of heat?
(1) 1 & 3 (2) 2 & 4 (3) 3 & 4 (4) 1 & 4

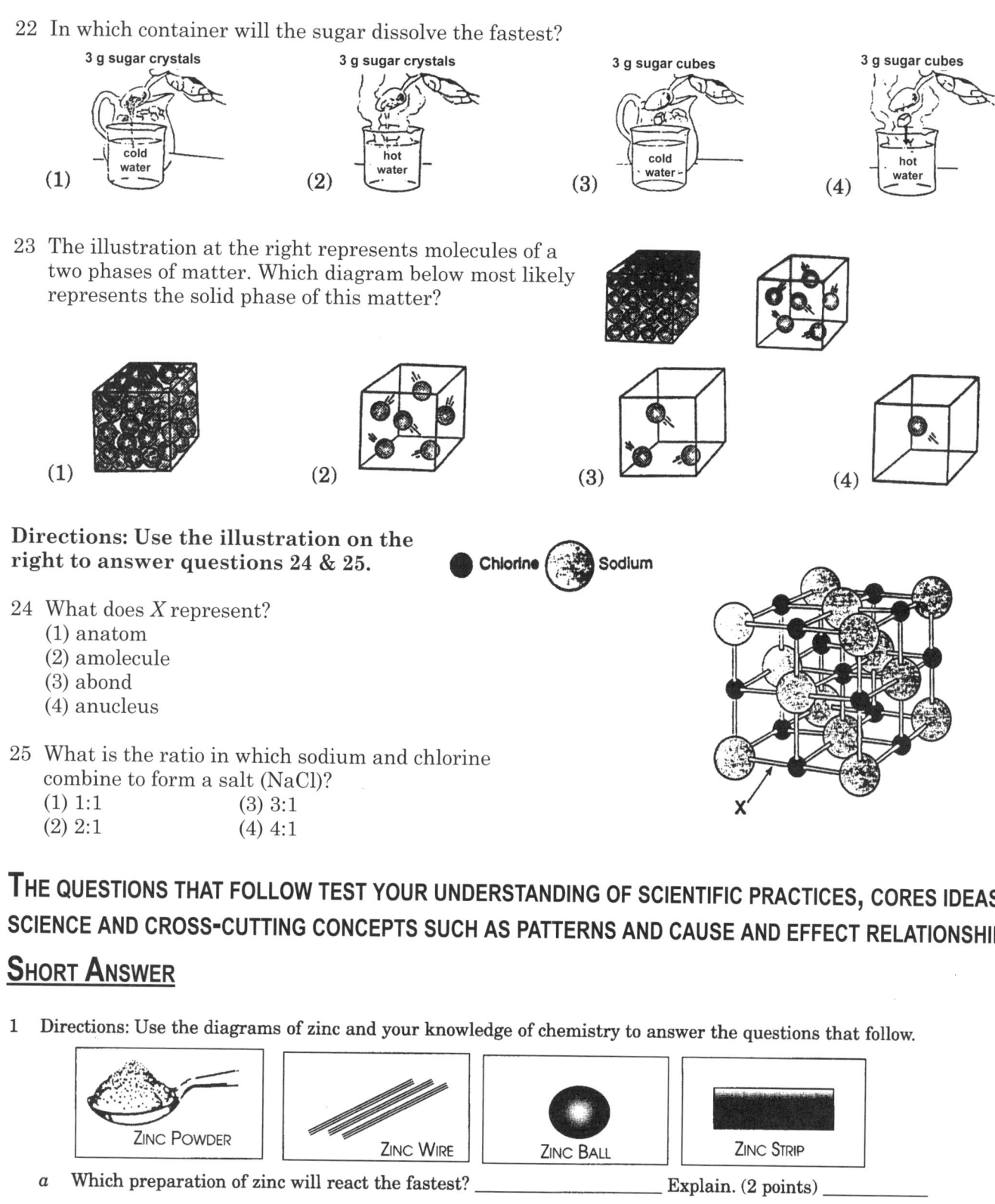

22 In which container will the sugar dissolve the fastest?

3 g sugar crystals (1) cold water

3 g sugar crystals (2) hot water

3 g sugar cubes (3) cold water

3 g sugar cubes (4) hot water

23 The illustration at the right represents molecules of a two phases of matter. Which diagram below most likely represents the solid phase of this matter?

(1)   (2)   (3)   (4)

Directions: Use the illustration on the right to answer questions 24 & 25.

● Chlorine   ◯ Sodium

24 What does X represent?
(1) anatom
(2) amolecule
(3) abond
(4) anucleus

25 What is the ratio in which sodium and chlorine combine to form a salt (NaCl)?
(1) 1:1   (3) 3:1
(2) 2:1   (4) 4:1

X

THE QUESTIONS THAT FOLLOW TEST YOUR UNDERSTANDING OF SCIENTIFIC PRACTICES, CORES IDEAS IN SCIENCE AND CROSS-CUTTING CONCEPTS SUCH AS PATTERNS AND CAUSE AND EFFECT RELATIONSHIPS.

SHORT ANSWER

1 Directions: Use the diagrams of zinc and your knowledge of chemistry to answer the questions that follow.

ZINC POWDER   ZINC WIRE   ZINC BALL   ZINC STRIP

a Which preparation of zinc will react the fastest? _____ Explain. (2 points) _____

_____

_____

b Describe two properties of zinc. (3 points) _____

_____

_____

2 The illustration shows water droplets forming on the outside of a pitcher of water that contains ice cubes.

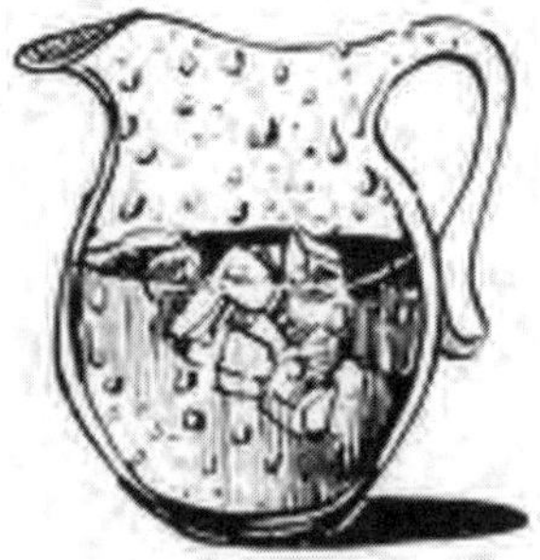

a Name and describe the type of phase change that took place on the outside of the glass. (2 points)

____

____

b Using your understanding that heat is absorbed or released during a phase change, explain why this phase change took place. (2 points)

____

____

3 All matter has physical and chemical properties. Name and describe two physical properties of metals.

(4 points) ____

____

4 Use the solubility graph at the right to answer the questions that follow.

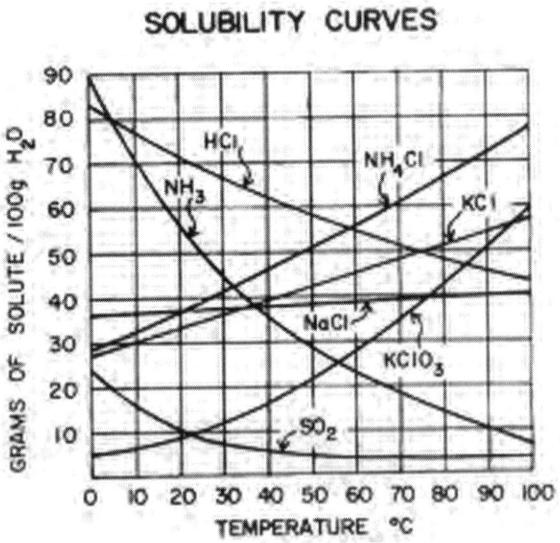
SOLUBILITY CURVES
GRAMS OF SOLUTE / 100g H2O
90
80
70
60
50
40
30
20
10
0
10
20
30
40
50
60
70
80
90
100
TEMPERATURE °C
HCl
NH4Cl
NH3
KCl
NaCl
KClO3
SO2

a What is the solubility of NaCl at 20°C? (1 point)

____

b 60 g of NH_3 are dissolved in 100 g of water at 25°C. Is this an unsaturated, saturated, or supersaturated solution? ____________________ Explain. (3 points)

____

____

____

5 Mixtures can be separated by physical means. Explain and give your reasoning as to how to separate a mixture of each of the following substances.

a Mixture of alcohol and water (2 points) ____

____

____

b Mixture of sand and iron filings (2 points) ____

____

____

____

SHORT ESSAY

1 The experiment below is testing the solubility of a tablet at various temperatures.

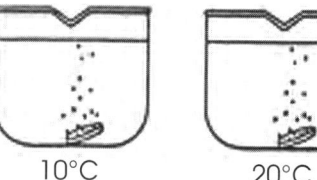 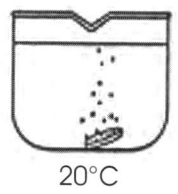 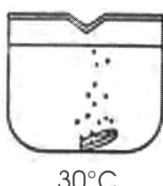 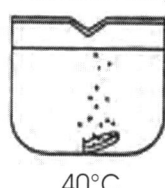 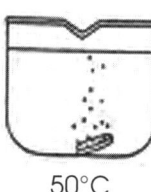

    10°C   20°C   30°C   40°C   50°C

a   What can you conclude about the solubility of the tablet? (1 point) _____

_____

_____

b   Describe another experiment you could do to test the solubility of the tablet? Include both a dependent and an independent variable, identify your control group, and construct a data table. (4 points)

_____

_____

_____

_____

_____

2   The chart to the right shows the density of 4 liquids. Using the letter of the liquid, place the liquids in the correct order in the beaker on the blank before each arrow. Explain your decision. (4 points)

LIQUID DENSITY	
Letter	Density (g/mL)
A	2.3
B	4
C	1.6
D	3.4

_____

_____

_____

_____

_____

Letter _____ →

Letter _____ →

Letter _____ →

Letter _____ →

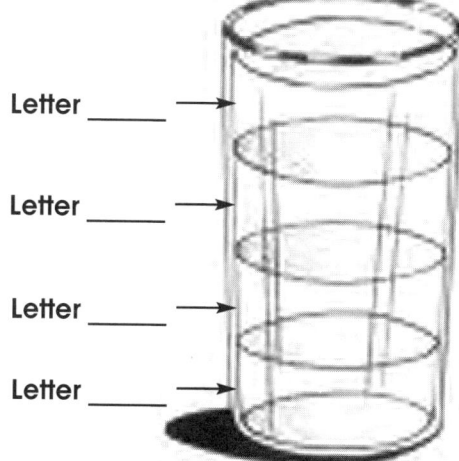

3   The matter in the diagram below has undergone a phase change

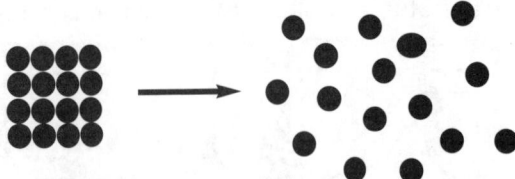

a   Name a type of phase change this matter may have undergone. (1 point)

_____

_____

b   State whether heat is absorbed or released during this phase change and explain your choice. (3 points)

_____

_____

_____

_____

ENERGY EXISTS IN MANY FORMS
AND WHEN THESE FORMS CHANGE, ENERGY IS CONSERVED.

DISCIPLINARY CORE IDEA:

PS3.A: DEFINITIONS OF ENERGY

SOME ENERGIZING INFORMATION: WHAT IS ENERGY?

The **universe** is composed of **matter** and **energy**. Remember, matter is anything that has mass and occupies space. A pizza pie is matter because it occupies space and has mass. Can you order out energy? No, because energy is something associated with matter, but it does not occupy space, nor does it have mass. The pizza delivery person cannot arrive at your door with the message that your energy has arrived. You are waiting for a box with a pie inside of it. Since energy is not matter, it must be described by what it does. **Energy** allows work to be done.

ENERGY COMES IN MANY FORMS

A pizza pie has an assortment of energy associated with it. When you digest pizza, chemical bonds are broken and **chemical energy** is released. A pizza pie coming straight out of the oven is transferring a lot of **heat energy** because the particles that make up the pizza are moving relatively fast. If you place your hand over the pizza pie without touching it, you feel heat. This is because some of the heat energy is converted to **radiant energy** which is what you feel on your hand. As the pizza cooks in the oven, you may hear it sizzle because **sound energy** is being transmitted to your ears. As the pizza sizzles, cheese may be bobbing up and down. The motion of the cheese can be called **mechanical energy**, a type of energy associated with the motion of matter. Some of the atoms that make up your pizza pie release energy from their nuclei. Energy coming from the nucleus of an atom is called **nuclear energy**. If an electric oven cooked your pizza, then **electrical energy** was used in the preparation of the pizza.

The major forms of energy are summarized in the following chart:

ENERGY CHARACTERISTICS

TYPE OF ENERGY	CHARACTERISTICS
HEAT	This is the energy transferred from warmer to cooler objects. It is due to the internal motion and position of particles in matter (thermal energy). No heat can be transferred from objects at absolute zero (equal to 0 K, -273.15°C, or -459.67°F).
MECHANICAL sound	Any energy associated with the motion of matter. Sound is a type of vibrational mechanical energy that is transmitted by waves moving through matter. Its speed is about 1200 km/hr (750 mi/hr). All vibrational energy is wavelike, spreading away from its source.
ELECTROMAGNETIC	Travels as a transverse wave and does not require a medium through which to travel. Light and other forms of radiant energy make up the electromagnetic spectrum. Electromagnetic energy travels through a vacuum at 300,000 Km/s (186,000 mi/sec). Examples include: radio-waves, short wave, AM, FM, TV, radar, infrared, visible light, ultraviolet, x-rays, and gamma rays.
NUCLEAR	The energy produced from the inside of an atom is called nuclear energy. Some of the energy is part of the electromagnetic spectrum that produced ionizing radiation (gamma rays & cosmic rays are examples). Ionizing radiation can break chemical bonds and has been associated with causing cancer.
CHEMICAL ENERGY	This is the energy found in chemical bonds. It is released when chemical bonds are broken.

IT IS THE LAW: LAW OF CONSERVATION OF ENERGY

The interesting thing about energy is that it never gets lost. It just changes into another form of energy. Since this is always true, it can be expressed as a scientific law. The *Law of Conservation of Energy* states that energy cannot be created or destroyed.

27 What are the major forms of energy associated with the making, cooking, and eating of a pizza pie?

_____

_____

28 How is energy different than matter? _____

_____

_____

29 *a* Can energy be lost? _____ *b* Explain _____

_____

_____

KINETIC & POTENTIAL ENERGY

Did you ever carefully watch a ball bounce? If not, drop any ball and watch what happens as it bounces. First of all, you must have noticed that the ball bounces lower and lower after each bounce. It also makes a noise when it makes contact with the ground. A bouncing ball undergoes a number of energy transfers.

The **potential energy** of the ball is **stored energy** and is determined by its position. In the case of the bouncing ball, the highest level from which it was dropped represents its greatest amount of potential energy.

Kinetic energy is energy of **motion**. As soon as the ball begins to drop, some of the potential energy is converted to kinetic energy. As the ball drops, more and more potential energy is converted to kinetic energy until there is only kinetic energy. This happens at the moment the ball makes contact with the ground. As the ball rises back into the air again, kinetic energy is being converted back into potential energy, but it is never as much as the starting potential energy.

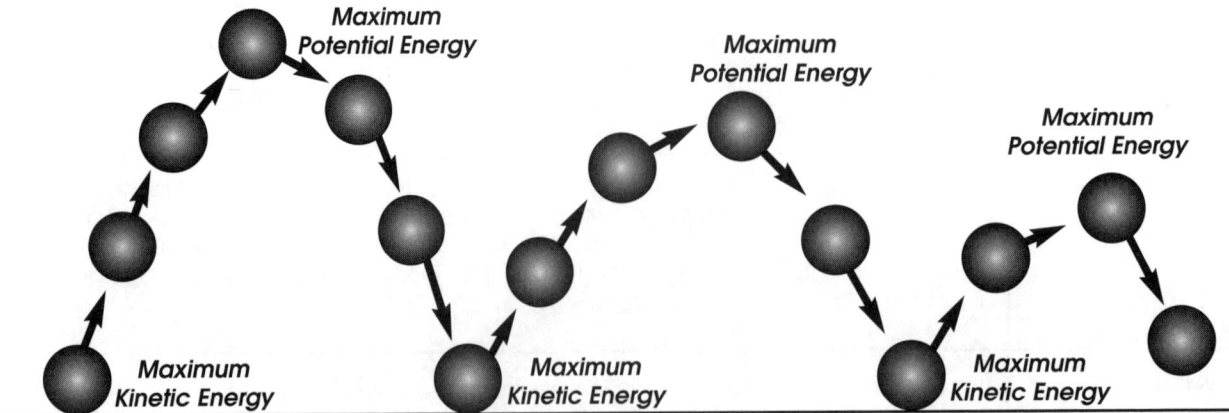

The question still remains as to why a ball bounces lower with each bounce. The answer lies in energy transfers. The mechanical energy of the ball is transferred into energy that does not support bouncing. As the ball makes contact with the ground, some of the energy is converted into heat and sound energy.

MEASURING GRAVITATIONAL POTENTIAL ENERGY

If a bowling ball and tennis ball were both dropped from the same height, which has more potential energy? A ball falls because of the pull of gravity. Therefore, this type of potential energy is called **gravitational potential energy**. Since the bowling ball has more mass than the tennis ball, its weight will be greater. The total amount of potential energy depends on the weight and height of the object. Mathematically, it can be expressed with the following equation:

$$\text{Gravitational Potential Energy} = \text{Weight} \times \text{Height}$$

It can be easily seen from this formula that there is a direct relationship between the amount of gravitational energy and weight and height. As the weight and height increases, so does the potential energy of an object.

MEASURING KINETIC ENERGY

The amount of kinetic energy depends on the mass of an object and its velocity. Objects dropped from the same height will have the same velocity while falling, because of the **Law of Gravity**. In comparing a tennis ball and a bowling ball, the bowling ball has more kinetic energy than the tennis ball, because it has more mass. The amount of kinetic energy associated with matter can be expressed with the following equation:

$$Kinetic\ Energy = \frac{mass \times velocity^2}{2} \qquad KE = \frac{m \times v^2}{2}$$

This formula shows that the amount of kinetic energy is equal to one-half the mass times the square of the velocity of an object. According to this formula, an object can have a relatively small mass, but if it is moving at a high velocity, will have a great deal of kinetic energy. For example, NASA has found that pock marks on the glass of the space shuttle were caused by paint chips from old satellites. A paint chip traveling at 17,5000 mi/hr (the average speed of the space shuttle) can exert enough force to damage the glass of the space shuttle. In fact, NASA is greatly concerned about the large amount of litter that orbits Earth from old satellites.

30 Why does a ball bounce lower after each bounce? _____

_____

31 When does a bouncing ball have the most amount of potential energy? _____

_____

32 Why is NASA concerned about small pieces of orbiting litter? _____

_____

A LIGHT SHOW FROM WINTERGREEN CANDY: ENERGY AS A PROPERTY OF MATTER

All forms of energy are associated with the structure of matter. Did you know that when you chew on wintergreen Life Savers, you create micro-lightning? Try it! Place about 3-4 pieces of wintergreen Life Savers in your mouth, obtain a mirror, and then go into a pitch, dark room. Start chewing on the Life Savers, and you should see sparks coming out of your mouth. Wintergreen Life Savers are made mostly of sugar and wintergreen oil. Sugar is a crystal. When crystals are crushed by your teeth, electrons in atoms jump up to higher energy levels. When they fall back, they release energy in the form of light. However, the light energy they release is not visible light. When wintergreen is present with sugar, the light energy released by the crushing of the sugar crystal is absorbed by the wintergreen oil. This excites its electrons to jump up to higher energy levels in the electron cloud. When the electrons fall back to lower energy levels, they release light energy in the visible part of the electromagnetic spectrum. The light you see is the **micro-lightning**.

RENEWABLE & NONRENEWABLE ENERGY

Since human activity is using fossil fuels faster than they can be replaced, fossil fuels are considered a nonrenewable energy resource. The United States has 5% of the world population, but uses over 25% of the world's energy resources. Seventy percent of the electrical energy generated in the United States comes from the burning of coal and other fossil fuels. Fossil fuels are considered nonrenewable energy resources because there is a fixed amount of it on Earth and it takes millions of years for nature to form fossil fuels. They are called fossil fuels because they come from the remains of ancient forms of life. Fossil fuels are nonrenewable energy resources that have trapped solar energy captured by living organisms. Generally, scientists consider renewable energy resources as those energy resources that will be around for millions of years.

How are fossil fuels formed? Over a period of millions of years sediments from erosion cover the dead remains of these organisms. Molecules in these organisms are placed under tremendous amounts of pressure from the piling up of sediments. This causes chemical bonds to break and new ones to form. Over time, this process left Earth with deposits of coal, petroleum, and **natural gas**. Petroleum was formed mostly from the remains of plants and animals living in shallow oceans. Shallow oceans that once covered Northern Africa left behind the huge petroleum deposits found in the Middle East. Millions of years ago, fern trees, thriving in wetlands created present day **coal deposits**. Natural gas is usually found in association with petroleum because it is formed in the same way.

dead plants covered by
swamp water

over years, great pressure is
applied

after a long, long time, coal
deposits form

RENEWABLE ENERGY RESOURCES

Sunlight, wind, falling water, tides, geothermal (heat from Earth's interior), and biomass are considered renewable or alternative energy sources because they are, for all practical purposes, everlasting. Wood can also be considered a renewable resource as long as there is careful forest management.

- Natural ecosystems use only a fraction of 1% of the energy from the Sun. Scientists have estimated that the energy needs of all humans can be met with a similar percentage of the total solar radiation striking Earth. The heat absorbed from the Sun by **solar collectors** can heat water and homes. Photovoltaic cells convert solar energy directly into electrical energy such as the one you use in your solar calculator. The heat from the Sun can also be used to boil water to create steam to turn a turbine.

- As the Sun strikes Earth, the atmosphere is unevenly heated. This creates **wind. Wind energy** can be used to turn the turbines of a generator producing electricity.

- Under the lithosphere of Earth, lies hot liquid rock. The heat from inside Earth, **geothermal energy**, can be used to heat homes or boil water to create steam that drives turbines.

- **Wood** can be considered a renewable resource as long as trees are replanted. Wood-burning stoves have been very popular for heating homes, but the burning of wood also releases a greenhouse gas — carbon dioxide.

- Falling water can turn a turbine to generate electricity, called hydroelectricity. Power plants are built next to many naturally occurring falls, such as Niagara Falls and artificial falls created by dams.

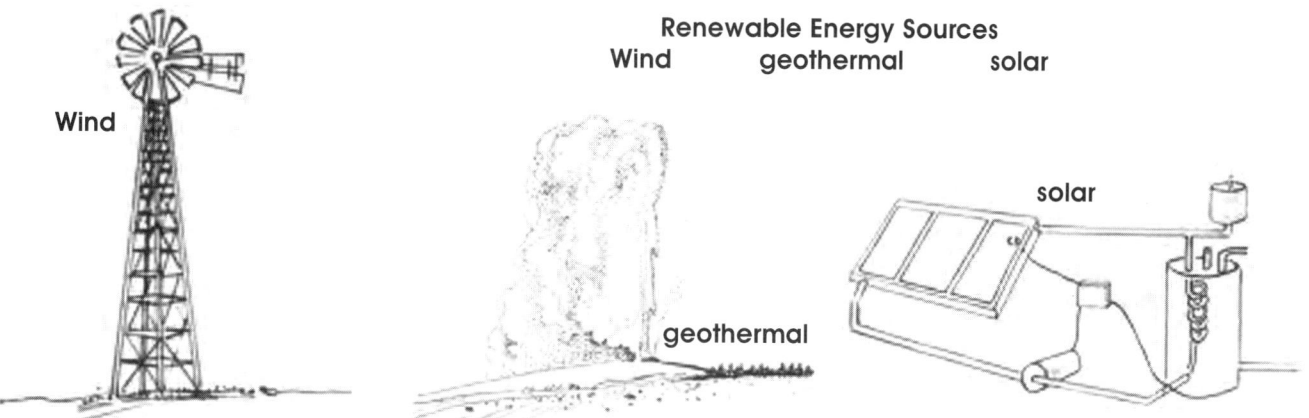

Renewable Energy Sources
Wind   geothermal   solar

Wind

solar

geothermal

ENVIRONMENTAL ISSUES ASSOCIATED WITH THE BURNING OF FOSSIL FUELS

When fossil fuels burn, they release the gas carbon dioxide. This is one of the major suspected causes of the "Greenhouse Effect," which is associated with the warming of Earth. Furthermore, the burning of coal contributes to acid rain. Transport of fossil fuels has also been associated with oil spills into the waterways. These oil spills have caused much damage to wildlife directly or by damaging the wildlife's environment.

NUCLEAR ENERGY

Some scientists promote the use of nuclear energy as an alternative to fossil fuels. Energy that comes from the nucleus of an atom is considered **nuclear energy**. There are no oil spills or release of greenhouse gases when nuclear energy is used to generate electricity. The element uranium is the fuel for nuclear energy. A nuclear power plant produces energy from a nuclear reaction called fission. During **fission**, the heavy nucleus of uranium breaks into two smaller nuclei. The energy released during this reaction can be used to boil water to create steam to turn a turbine.

The energy coming from the Sun is also a nuclear reaction. The Sun is composed of mostly hydrogen. Hydrogen atoms have the smallest of nuclei. When hydrogen nuclei join together a type of nuclear reaction called **fusion** occurs. The heat produced from this reaction warms Earth. Presently, fusion requires extremely high temperatures (millions of degrees Celsius). If scientists can figure out a way to make this happen at low temperatures, it could be a future source of energy for Earth.

ENVIRONMENTAL PROBLEMS ASSOCIATED WITH NUCLEAR ENERGY

A major problem associated with nuclear energy is disposing of the radioactive wastes. When uranium can no longer be used to produce nuclear energy, its wastes are still radioactive. Where to safely bury these wastes is of great concern to many. If a nuclear power plant should break down, there must be a plan to evacuate the surrounding region. This too has created another problem, particularly for densely populated areas such as Long Island.

33 What is the difference between renewable and nonrenewable energy? _____

_____

_____

34 What are some examples of nonrenewable energy sources? _____

_____

35 What are some examples of renewable energy sources? _____

_____

_____

Nuclear Energy & The Law Of Conservation Of Mass

When a nuclear reaction takes place, some of the mass seems to disappear. According to the Law of Conservation of Mass, this should not happen. The answer was figured out by German-American physicist **Albert Einstein** between 1905 and 1915. He said that when a nuclear reaction takes place, some of the energy is converted into mass. This is summarized in the equation: $E = mc^2$. *E* stands for energy, *m* stands for mass, while *c* stands for the constant of the speed of light. Basically, Einstein showed that energy and matter are opposite sides of the same coin. Each can be converted into the other. This knowledge led to the beginning of the **Nuclear Age**.

Generation of Electricity

Fossil fuels and some forms of alternative energy can be used to produce electricity from a generator. Basically, electricity is created when a metal rod turns between a U-shaped magnet (at right). Placing a turbine on the rod can facilitate the turning of the rod. Steam, wind or falling water can turn the turbine to generate electricity.

Disciplinary Core Idea:

PS3.B: Conservation of Energy and Energy Transfer

Heat Energy

Heat was once believed to be a fluid that moved through matter. The fluid was called **caloric**. In 1798, the American scientist **Benjamin Thompson** noticed that heat was produced whenever holes were drilled in cannon barrels. He decided to perform an experiment and drilled holes in a barrel under water. After several hours, he observed that the water began to boil. The heat from the drilling did **work**, because it caused the water to boil. This led Thompson to conclude that heat must be a form of energy and not a fluid.

About 40 years later, British scientist **James Prescott Joule** performed a number of experiments and concluded that heat energy is related to the internal motion and position of particles in matter. As heat is added to matter, the internal particles speed up and when heat is removed, the motion of the particles slow down. Heat is a form of energy that always moves from areas of higher temperature to areas of lower temperature.

36 What contributions were made by Thompson and Joule to your understanding of heat energy? _____

_____

_____

Specific Heat

Did you ever notice that some types of matter feel colder than other types of matter? For example, if you touch pieces of wood and metal that have been sitting on your desk for a while, the metal feels colder than the wood. This is because the metal tends to lose and gain heat more quickly than wood. The ability of matter to gain or lose heat is called its **specific heat**. Specific heat is a property of matter that can be used to identify the matter. For example, the specific heat of water is 1, which is the value of the amount of heat needed to raise the temperature of 1 gram of water by 1°C. The specific heat of other matter is compared to water, since water has a relatively high specific heat and a value of 1.

Measuring heat

Heat is measured in a unit called a **calorie**. A calorie is the amount of heat needed to raise the temperature of 1 gram of water by 1°C. A calorie is also the specific heat of water. The measurement of the amount of heat that can be transferred from a sample of matter depends on three things: (1) the amount of matter; (2) the change in temperature of the matter after heat is transferred from it, and (3) the specific heat of the matter.

GETTING THE HEAT & COLD OF IT:
WHAT IS THE DIFFERENCE BETWEEN TEMPERATURE AND HEAT?

Temperature measures the average kinetic energy of the particles in matter. Suppose a cup of water and a quart of water are both brought to a boil. Since water boils at 100°C, they would both have the same temperature. However, it would take longer for the quart of water to come to a boil. More heat would have to be added to the quart of water in order to bring it to a boil because it has more mass. Therefore, more heat can be transferred out of the quart of water than the cup of water. Temperature measures the average kinetic energy in matter. Heat is the energy transferred from an area of higher temperature to one of lower temperature. There is no heat energy until it is transferred. Thermal energy refers to the amount of heat energy that can be transferred due to the kinetic energy of the particles in the matter.

HEAT TRANSFER

Heat is transferred from areas of higher temperature to areas of lower temperature. When the heat is evenly distributed, the temperature becomes the same for the two areas and heat is no longer transferred. How does the transfer of heat take place? During a summer day, you can feel the heat of the Sun beating down on you, while during the winter, you may feel the warmth of a fireplace. Though the Sun is 93 million miles away, its energy can reach you. You do not have to touch the fire to feel the warmth from it.

Energy from the Sun and fireplace reach you by a type of heat transfer called **radiation**. In order to travel through space, heat from the Sun and fireplace is first converted into electromagnetic energy (radiant energy) which travels as waves. When the wave energy strikes you, it becomes heat energy again.

Convection is heat transfer in a fluid such as the water in the pot or the air in a room. Convection only can occur when there are density differences in the matter. It occurs best when the phase of matter is a gas or liquid (fluids), because fluids can flow which creates the **density currents** associated with heat transfer. Density currents form when the cooler, denser fluid sinks under and displaces the less dense warmer fluid. In the case of boiling water, the water closest to the bottom of the pot is receiving more heat than the water at the top. Since warm water is less dense than cooler water, the cooler water sinks down displacing the warming water. As the water rises in the pot, it loses heat, becoming denser and so sinks down again. This constant circling of the warmer and cooler water forms a **convection cell**.

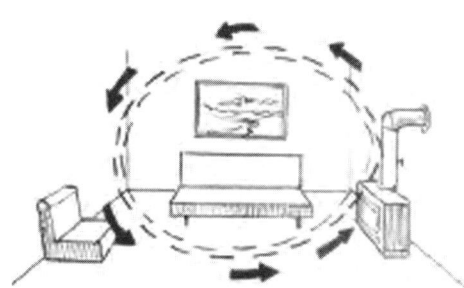

Air is circulated in a room by differences in the density of the air. The air near a radiator is warmer and so less dense than the air farthest from the heat source. Eventually, the denser, cooler air sinks under the warmer air, moving it away from the radiator. As the air cools again, it becomes denser, and will eventually sink under the warm air by the radiator. The continual rising and sinking of the air creates a convection cell in the room that circulates the warm air.

The handle of a spoon placed in hot soup soon feels warm. Heat transfer by contact is called **conduction**. When particles of matter collide with one another, heat energy is transferred. This heat transfer is associated with solids, because the particles of matter for a given sample are packed closest together. The heat from the spoon travels up to the handle.

37 Describe three ways in which heat can be transferred. _____

_____

_____

EFFECTS OF COLOR & TEXTURE ON HEAT TRANSFER

Generally, the darker the color, the more heat it will absorb. Black surfaces absorb heat faster than silver or white surfaces which tend to reflect heat energy. Surfaces with texture tend to absorb heat better than smooth, shiny surfaces which reflect some of the heat energy striking them. Obviously, wearing a dark, textured shirt on a hot summer day can be very uncomfortable.

38 Equal amounts of water are placed in two cans, one black and one silver.

    a   When placed in the Sun, which water sample will get hot first? _____

    _____

    b   Explain. _____

    _____

DISCIPLINARY CORE IDEAS:

PS4.A: WAVE PROPERTIES

PS4.B: ELECTROMAGNETIC RADIATION

PS3.B: CONSERVATION OF ENERGY AND ENERGY TRANSFER

OBSERVE AND DESCRIBE ENERGY CHANGES AS RELATED TO CHEMICAL REACTIONS.

Humans can maintain an average internal body temperature of 37°C. How is your body able to do this? The answer lies in your **metabolism** of food. When food is digested, chemical bonds are broken and energy is released. This chemical energy is quickly converted to heat energy. Whenever a chemical reaction takes place, energy transfers are involved. All chemical reactions involve the absorption (storage) or release of heat energy. For example, when an acid and a base are combined together in a test tube, it becomes very hot because heat is released. Have you ever used a cold pack? This is a type of reaction in which heat is absorbed by the pack. This causes the pack to become cold.

When the gasoline in a family's automobile combines with oxygen, it explodes, providing the force to move the pistons down which drive the axle and propel the car forward. In this case, chemical energy was converted into mechanical energy. Batteries contain **electrolytes**. These are chemicals that form charged particles when dissolved. Electrolytes can conduct electricity. Batteries produce energy by converting chemical energy into electrical energy.

39 **Describe the energy transfer involved when a campfire burns.** _____

WAVES AND THEIR APPLICATIONS IN TECHNOLOGIES FOR INFORMATION TRANSFER

ELECTROMAGNETIC RADIATION (SPECTRUM)

Radiant energy makes up the **Electromagnetic Spectrum.** Light energy is part of the spectrum as well. Sometimes, after a rainstorm, a beautiful rainbow appears in the sky. The rainbow makes up the **visible light** of the spectrum. However, there is some light energy you cannot see, such as ultraviolet or infrared light. Ultraviolet light can tan or burn the skin, while infrared light is felt as heat. Visible light makes up a very small band of the electromagnetic spectrum. It can be red, orange, yellow, green, blue, and violet.

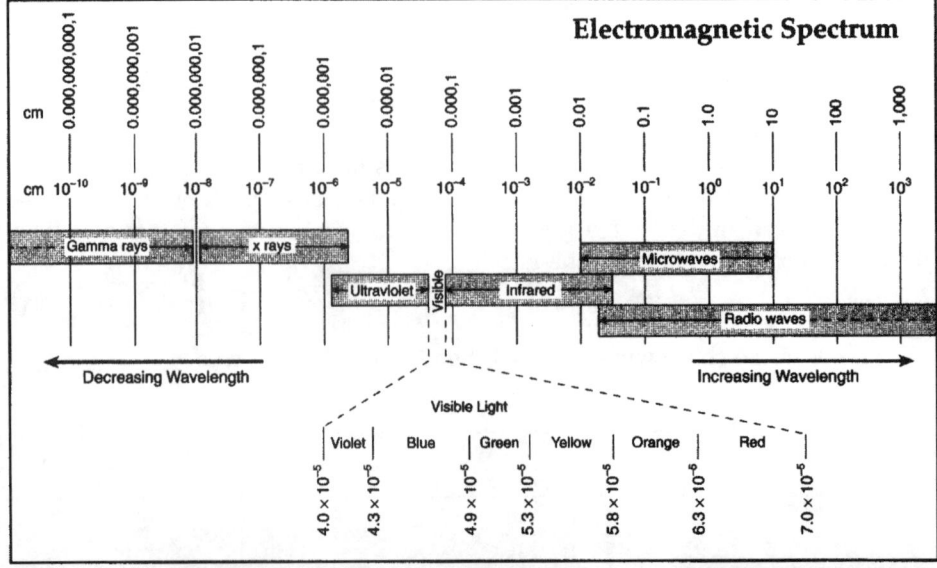

X-RAYS & GAMMA RAYS

X-rays and **gamma rays** are very high energy forms of radiant energy. Overexposure to these forms of radiant energy can be harmful to living organisms.

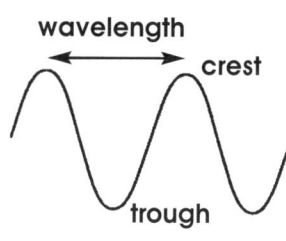

Electromagnetic energy travels in waves called **transverse waves**. These waves move perpendicularly to their direction of motion. The speed of light in a vacuum is a constant speed of 300,000 km/sec. The type of radiant energy depends on its wavelength. A **wavelength** is the distance from the top (crest) of one wave, through the bottom (trough), and back to the top (crest) and beginning of another wave. The wavelengths can be hundreds of kilometers apart such as in **radio waves**, while other waves such as x-rays are billionths of a meter apart. The smaller the wavelength, the more energy the light wave has. **Ultraviolet light** has more energy than **infrared light**.

40 What is the difference between infrared and ultraviolet light? _____

_____

THE SCIENCE BEHIND COLOR

When a light wave strikes matter, it can be **transmitted**, **absorbed**, or **reflected**. You can see through glass, because light waves are transmitted through glass. Matter that appears black to you absorbs all light from the visible spectrum. It also absorbs light you cannot see, such as infrared light. Infrared light waves are converted to heat energy when they strike matter. Therefore, wearing black clothing on a hot day can be very uncomfortable. Matter that appears white is reflecting waves from the spectrum. Wearing white on a hot day is a good choice of attire. The color that you see is the light being reflected off the matter. A red shirt appears red because the fabric is absorbing all colors of the spectrum – except the red – which it reflects. If there is no light, then there is no color.

41 Explain why you see color. _____

_____

SOUND (VIBRATIONAL ENERGY)

Sound is a form of vibrational energy that travels in waves. Unlike light waves, sound waves must be transmitted through a medium such as a gas, liquid, or solid. These waves are called **longitudinal** or **compressional waves**. As these waves travel through matter, particles in the matter compress together and then spread apart. This is similar to the motion of an accordion. The waves are also called longitudinal because they move parallel to the direction of its motion. The speed of sound is much slower than the speed of light and its speed is affected by temperature in addition to the density of the matter. Since particles of matter are moving faster at higher temperatures, sound travels faster. Sound also travels faster through mediums in which particles are packed closely together. This is why sound travels fastest through a solid. The average speed of sound is about 1200 km/hr through air, much slower than the speed of light.

Applications of Wave Energy

Disciplinary Core Idea:

PS4.C: Information Technologies and Instrumentation

Digitized signals (sent as wave pulses) are a more reliable way to encode and transmit information.

- *Lasers* (light amplification by stimulated emission of radiation) send information transmitted through fiber optics to devices such as the Internet and television. Many homes today receive Internet service this way. Fiber optics send information much faster than electrical wires because of the reduction in friction.

- *Satellite dishes* receive and transmit radio wave data to electronic devices such as cell phones and televisions.

- *GPS* (*Global Positioning Systems*) use radio waves to receive and transmit data from satellites.

- *Ultrasound* (Sound waves) can take pictures of a fetus because of the wave property of reflection. The energy from sound waves can also be used to boil water as some coffee pots use.

- *Photography*: Almost all pictures we take today is stored digitally. Because the byte size to store information keeps getting smaller, picture quality improves while a lot of information can be stored.

- *Music* is stored digitally. The early digital music on CDs did not have the quality of analog music on vinyl records, but today the byte size is so small that perhaps only a person with a very "musical" ear can tell the difference.

- *E-book*: Today, almost all information is stored digitally. The book you are reading right now is available as an e-book in addition to the analog on printed pages.

- *Security Issues*: Of course, that brings up the issue of security. Digital information can be hacked into. Many precautions are taken to prevent this which can sometimes be a pain to the user, but is very important. Hence we have passwords, codes, etc. before accessing some digital information.

- *Medical devices* use electromagnetic waves to take pictures of soft tissue in your body. For example, an MRI (magnetic resonance imaging) uses non-iodizing radiation (radio waves), a large magnet and computer to take pictures of your body. It works because, living things have a lot of hydrogen in them [we are about 60-70% water(H_2O)]. The hydrogen has a magnetic moment when it spins which is picked up by the magnet and translated into an image by a computer. Other medical devices use ionizing radiation to take pictures. X-rays are used to take pictures of bones. CT and CAT Scans also use X-rays.

- *Artificial Intelligence* (*AI*): Many homes already have a Siri or Alexa that can play your favorite tunes or time a boiling egg or tell you who is president of Belarus. This field is growing quickly and most engineers and scientists say it needs to be regulated. There is AI that can write your English paper and your teacher may have an AI device that can identify AI written papers. It will be your generation that determines the direction of this field.

A Different Wave – The Earthquake Wave

When an earthquake occurs, the motion of the tectonic plates releases tremendous amounts of energy. Some of the energy is transmitted through Earth as compressional waves (**primary wave**, *P-wave*) and some as transverse waves (**secondary wave**, *S-wave*). Knowledge about Earth's different layers is based on earthquake research. All waves travel in a straight line. However, scientists discovered that the direction of the waves changed as they moved through Earth. This change of direction could only occur if the wave is traveling through matter with differing densities. Since transverse waves cannot travel though a liquid, but compressional waves could, scientists were able to deduce that parts of the interior of Earth was liquid.

REFLECTION, REFRACTION, & DIFFRACTION

Reflection, **refraction**, and **diffraction** are properties of wave energy.

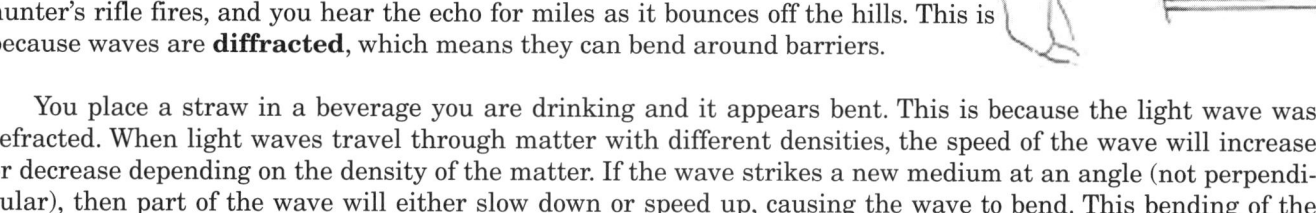

Since a wave can be reflected, you can see your image and hear an echo. The light waves bounce off a surface, called **reflection**. The smoother and shiner the surface, the less distortion in the reflection.

The Sun may set below the horizon, but you can still see light from the Sun. A hunter's rifle fires, and you hear the echo for miles as it bounces off the hills. This is because waves are **diffracted**, which means they can bend around barriers.

You place a straw in a beverage you are drinking and it appears bent. This is because the light wave was refracted. When light waves travel through matter with different densities, the speed of the wave will increase or decrease depending on the density of the matter. If the wave strikes a new medium at an angle (not perpendicular), then part of the wave will either slow down or speed up, causing the wave to bend. This bending of the wave is called **refraction**.

42 Describe the wave properties of reflection, refraction, and diffraction. _____

_____

_____

ELECTRICAL ENERGY: THIS ENERGY WILL GIVE YOU A CHARGE

Electrical energy is another type of electromagnetic energy. It differs from light energy in that it depends on the motion of charges.

STATIC ELECTRICITY

Have you ever received a shock when you touched a doorknob? What you have experienced is a type of electrical energy called **static electricity**. Static electricity energy occurs when there is a build up of charges on matter. The shock that you felt when you touched the door knob is the static electricity discharging (becoming neutral). Lightning is an example of static electricity discharging.

The charge may be positive or negative. For example, sometimes when you comb your hair with a plastic comb, the hair seems to "stick" to the comb. Your hair and the comb both contain atoms. The charge of an atom is neutral because the number of positive and negative charges are the same. As you comb your hair, the rubbing of the comb through your hair causes electrons from the comb to move to your hair. Your hair, with its negative charge, is attracted to the positively charged comb.

Static electricity is not very useful electrical energy. In order for electricity to do work, it needs to travel though a conductor. Since metals are excellent conductors, they are used to transmit electricity.

ELECTROSCOPE

Static electricity can also be detected with an apparatus called an **electroscope**. An electroscope is a device consisting of a metal rod and two pieces of metal foil leaves suspended in insulating glass. Note that insulating means a non-carrier of electricity or a non-conductor.

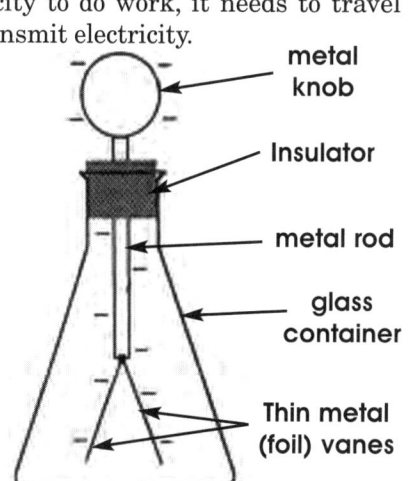

The metal foil leaves of an electroscope move apart or collapse together depending on whether they have a charge. If the knob comes in contact with negative charges, electrons travel down to the metal foil leaves giving them a negative charge. This causes the leaves to push away from one another because they have like charges.

LAW OF CHARGES

The *Law of Charges* states that like charges repel and unlike charges attract. The metal foil leaves push away from one another when the charges become alike. If the knob has contact with positive matter, electrons in the metal foil leaves travel up to the knob. Since the leaves lost electrons, they take on a positive charge. Since the charges are still alike, the leaves again move away from one another.

CIRCUITS

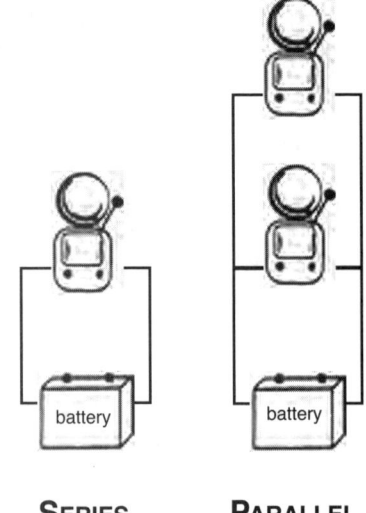

All electrical devices use current electricity (flow of electrical charges). Generators and batteries are sources of electrical energy. The electricity they produce must be conducted through a circuit. A **circuit** provides a complete path (closed) for electricity to follow from a negative to positive connection. A circuit contains a number of **loads** (devices that use electricity) such as a TV or radio and a switch that can open or close the circuit. Some circuits that carry a large amount of electricity are grounded with a third prong. There are two types of circuits:

- Some Christmas lights are placed on a **series circuit**. If one light goes out, all of the lights go out. This is because, electricity in a series circuit must travel through all of its loads before it travels back to its source of electrical energy. If one light goes out, it acts like a switch to open the circuit. This blocks the flow of electricity.

- Homes and apartments are wired in a **parallel circuit**. Each load in a parallel circuit has its own path back to the source of electricity. Therefore, if a lamp light goes out, it does not open the circuit for all of the other loads (electrical appliances).

SERIES CIRCUIT   **PARALLEL CIRCUIT**

43 What is the difference between a series and parallel circuit? _____

_____

MAGNETISM

Magnetism is a part of electromagnetic energy. It is a force of repulsion or **attraction** associated with matter. Magnetism was discovered in ancient times in a rock called magnetite or **lodestone**. When suspended from a string, one side of these rocks pointed to the North Star or leading star. The North Star has been used since ancient times to navigate. Lodestone (from the Greek word Lodestar, leading star) was used to make the early compasses.

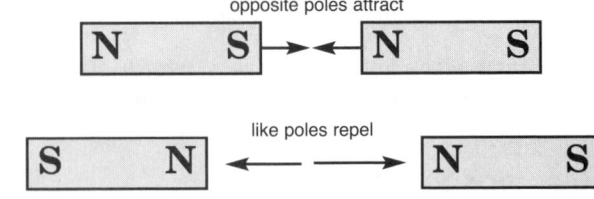

LAW OF POLES

Anyone who has ever played with magnets must have noticed that magnets can attract or **repel** one another. This is because magnets always have a north and south pole. The **Law of Poles** states that like poles repel and unlike poles attract.

44 How can a magnetic compass be used to navigate? _____

_____

_____

EXPLANATION OF MAGNETISM

As with electricity, magnetism is associated with the motion of electrons. As electrons orbit around the nucleus of an atom, they are also spinning on their axis. The spinning adds to the creation of a magnetic field around an electron. Regions in which magnetic fields are located are called **domains**. You can think of a domain as a miniature magnet. Magnetism only occurs when all of the domains are aligned in the same direction. This means that all of the north poles face in one direction and the south poles face in the opposite direction.

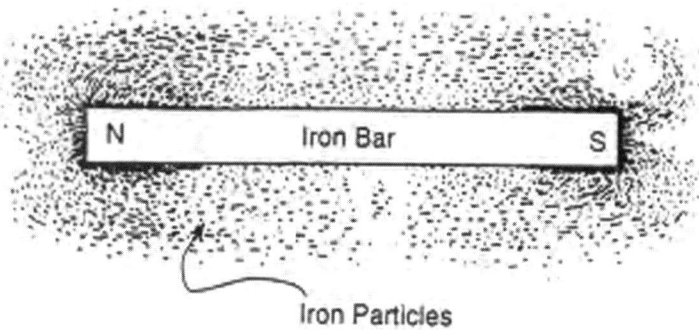

Iron Particles

ELECTROMAGNETS

If you place a compass by an electrical wire that has current flowing through it, you will notice that the compass needle deflects (moves). This phenomenon was discovered in 1820 by the Danish physicist, **Hans Christian Oersted**. When Oersted made this observation, he concluded correctly that current electricity gives rise to a magnetic field. Oersted also showed that when wire is coiled with a current flowing through it, it behaves like a magnet. Scientists later discovered that the strength of the magnetic coil can be improved by wrapping it around soft iron. These type of magnets are called electromagnets because the magnetic field is formed from the flow of current electricity.

ELECTROMAGNETIC FIELD

Electromagnetic energy can be detected over a distance. The area in which electromagnetic energy acts is called a **magnetic field**. A magnetic field can be visualized by placing iron filings around a magnet. The iron filings will align along the magnetic lines of force or **flux lines** coming out of the magnet. When like poles face one another, these flux lines bend away from one another. When unlike poles face one another, the flux lines bend toward one another. Magnetic fields are strongest by the poles.

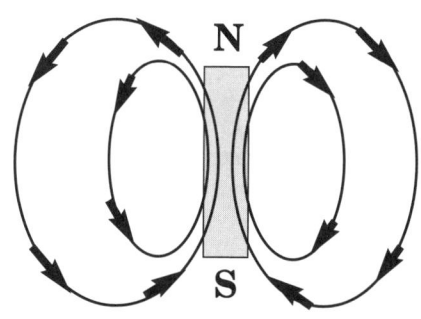

During an electrical storm, an electric field can be detected. If an electroscope was placed in the field, the metal foil leaves would separate verifying the presence of moving charges. A person may experience the hair on their arms beginning to stand up.

PERFORMANCE INDICATOR 4.5

DESCRIBE SITUATIONS THAT SUPPORT THE PRINCIPLE OF CONSERVATION OF ENERGY.

IT IS THE LAW: *LAW OF CONSERVATION OF ENERGY*

The principle known as the **Law of Conservation of Energy** states that energy cannot be created or destroyed. Energy is constantly being converted from one form to another. If you touch an electrical device such as a computer that has been running for awhile, it will feel warm. This is because some of the electrical energy is converted into heat energy. Incandescent light bulbs waste energy because a lot if its light energy is converted to heat energy. Led light bulbs don't do this as much so are a more efficient light bulb. A light bulb is a good example of the transformation of energy. Most energy in a home comes from a power plant. A **power plant** uses generators to create electricity. Suppose a power plant uses falling water to turn the turbines of a generator. The falling water can be considered mechanical energy. The mechanical energy is converted to electrical energy as it turns the turbines. Large power lines bring this electricity to your home. The friction of the electricity moving through the filament of a light bulb creates a tremendous amount of thermal agitation. As the filament gets hot, some of the heat energy is converted into light energy.

45   What does the Law of Conservation of Energy state? _____

_____

_____

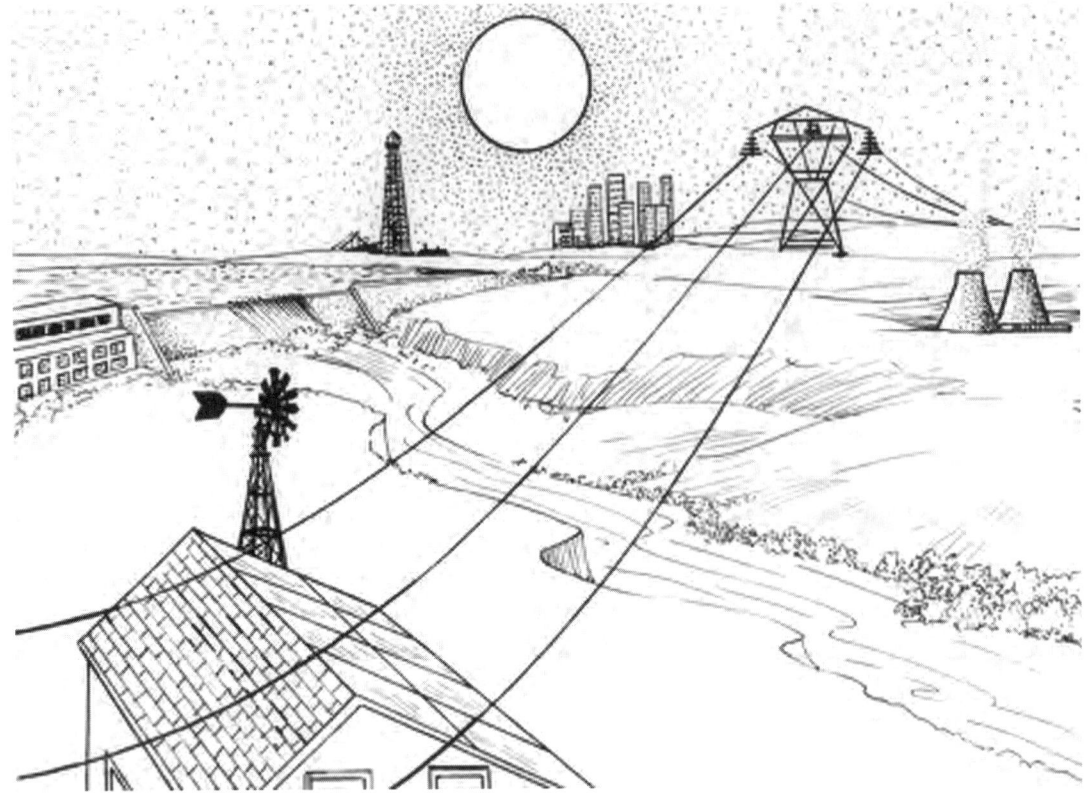

EXAMPLES OF ENERGY CONVERSIONS

The tremendous amount of heat energy created in an incandescent light bulb, makes it a very inefficient electrical device. In fact, any machine that produces a large amount of heat in addition to its major function is wasting electricity. Fluorescent light bulbs are much more energy efficient than incandescent light bulbs because they produce much less heat energy. Industry is always looking for better ways to make more efficient machines.

There are many other examples of energy conversions. When you talk on the telephone sound energy is being converted to electrical energy and then back to sound energy.

46 What type of energy conversion is involved with a hair dryer? _____

_____

_____

MULTIPLE CHOICE

1   What allows work to be done?
    (1) power   (3) mass
    (2) energy   (4) density

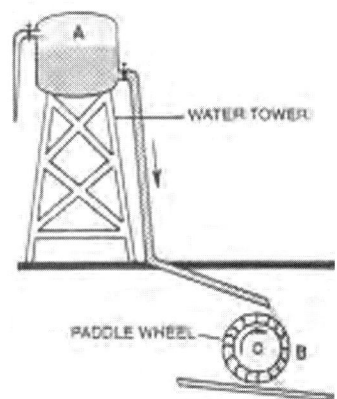

2   What form of energy is associated with the water
    tower in the diagram at the right?
    (1) heat
    (2) nuclear
    (3) mechanical
    (4) chemical

3   The energy obtained from the digestion of food is called
    (1) heat   (3) mechanical
    (2) nuclear   (4) chemical

4   Refer to the diagram of a thunderstorm at the right. What type of
    energy is being transmitted between the cloud and the ground?
    (1) electricity
    (2) sound
    (3) heat
    (4) chemical

5   Which statement below best reflects the energy of the rocks
    shown in the diagram at the right?
    (1) They have the same amount of potential energy.
    (2) Rock A has more potential energy than Rock B.
    (3) Rock B has more potential energy than Rock A.
    (4) Rocks A and B have only kinetic energy.

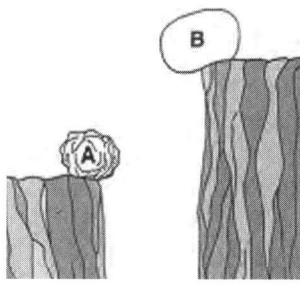

6   What type of energy conversion is taking place in the diagram
    at the right showing a green plant in sunlight during the process
    of photosynthesis?
    (1) chemical to light   (3) light to chemical
    (2) electric to light   (4) heat to chemical

7   The building in the photo at the right is using energy from
    the Sun. What type of energy is this considered?
    (1) solar
    (2) chemical
    (3) nuclear
    (4) heat

8   Which type of energy is considered nonrenewable?
    (1) coal
    (2) hydroelectric
    (3) solar
    (4) wind

Directions for questions 9 through 11: use the diagram below to answer the questions.

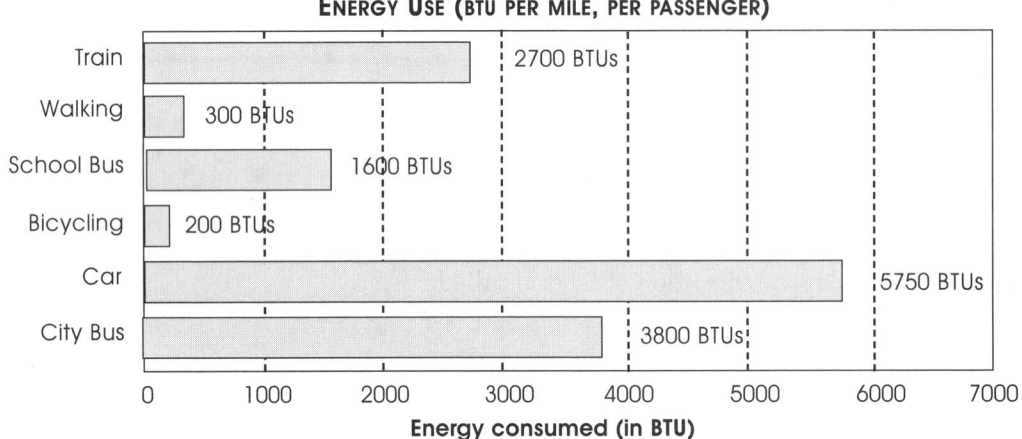

9   Which mode of transportation uses the most amount of fossil fuel per person?
    (1) train   (2) school bus   (3) city bus   (4) car

10 Which mode of travel uses the least amount of BTUs as compared to the forms of transportation given?
    (1) walking   (2) school bus   (3) bicycling   (4) car

11 How many BTUs does a car use after traveling 10 miles?

    (1) 575   (2) 57, 500   (3) 575,000   (4) 5,750,000

12 What type of heat transfer is illustrated in
    the diagram at the right?
    (1) radiation   (3) convection
    (2) conduction   (4) induction

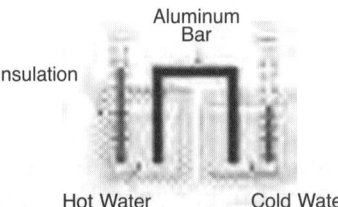

13 What type of energy conversion is illustrated in the diagram
    at the right. The diagram shows a hand crank generator.
    (1) electric to mechanical to light
    (2) mechanical to light to electric
    (3) mechanical to electric to light
    (4) light to mechanical to electric

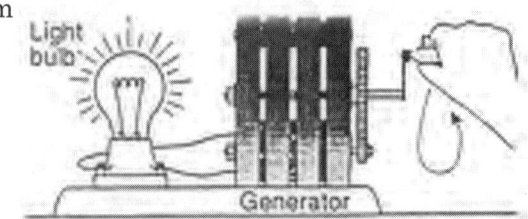

14 Which circuit below works?

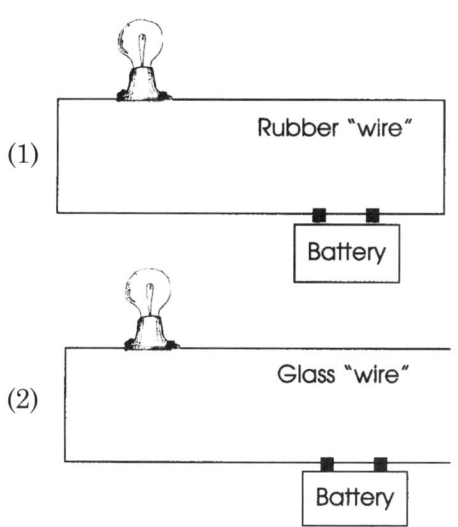

(1) Rubber "wire" / Battery

(2) Glass "wire" / Battery

(3) Metal "wire" / Battery

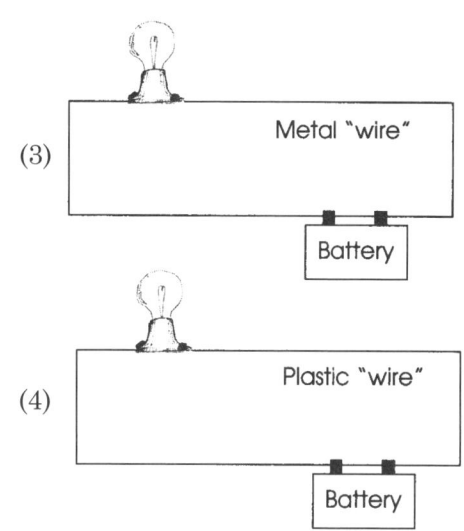

(4) Plastic "wire" / Battery

15 Using the information in the diagram at the right, determine which metal rod is the best conductor of heat.
   (1) iron
   (2) aluminum
   (3) copper
   (4) silver

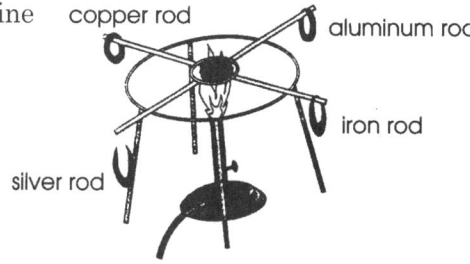

copper rod   aluminum rod   iron rod   silver rod

16 The diagram at the right shows an iron bar magnet surrounded by iron filings. At which point is the magnetic force the strongest?
   (1) A
   (2) B
   (3) C
   (4) D

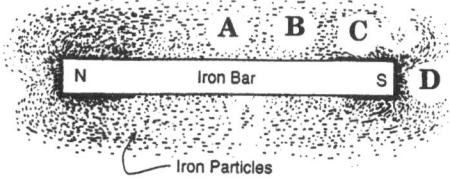

A B C
N   Iron Bar   S D
Iron Particles

Directions: use the diagram at the right of a classification scheme for the use of crude oil to answer questions 17 & 18.

17 According to the scheme shown in the diagram, what are the two major uses of crude oil?
   (1) diesel oil and petrochemicals
   (2) fuels and petrochemicals
   (3) fuel oil and plastics
   (4) gasoline and alcohol

Kerosene, Gasoline, Synthetic rubber, Plastics, Synthetic fibers, Dyes, Detergents, Fuel oil, Inks, Alcohol, Diesel oil, FUELS, PETROCHEMICALS, Fertilizers, Asphalt, Lubricating oil, Waxes, CRUDE OIL

18 An example of a petrochemical is
   (1) gasoline
   (2) fertilizer
   (3) diesel oil
   (4) wax

Directions: Use the pie chart at the right to answer questions 19 and 20.

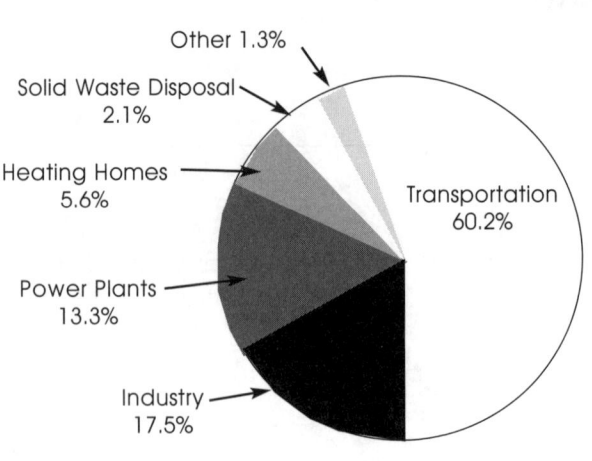

19 The pie chart is most likely illustrating the use of which natural resource?
  (1) water
  (2) oil
  (3) wood
  (4) coal

20 The smallest use of this natural resource is
  (1) solid waste disposal
  (2) power plants
  (3) industry
  (4) not identified

21 The photovoltaic cell in the diagram at the right converts
  (1) chemical to light energy
  (2) light energy to electric energy
  (3) electric energy to light energy
  (4) light energy to chemical energy

22 What is the purpose of prong 3 in the diagram of a plug at the right?
  (1) grounding
  (2) increase electric flow
  (3) decrease electric flow
  (4) create a tighter connection

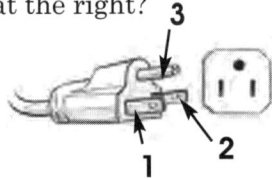

23 Which diagram would result in a working circuit once the switch is closed?

 (1)    (2)   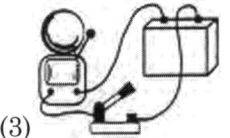 (3)    (4)

24 Chemical reactions involve heat exchanges. The diagram at the right gives evidence that the chemical reaction must have
  (1) lost heat
  (2) gained heat
  (3) had no heat exchange
  (4) had no change in temperature

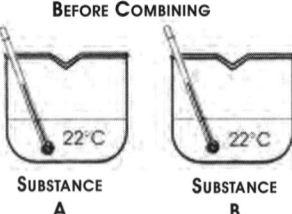

25 What properties of light are illustrated in the diagram at the right?
  (1) refraction and luster
  (2) reflection and color
  (3) refraction and reflection
  (4) color and luster

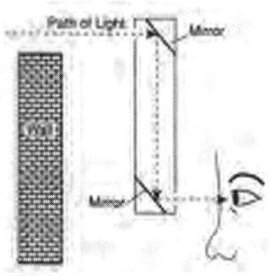

26 An electroscope is illustrated at the right. It is a device that detects static electricity. The leaves on the electroscope have separated because
(1) they have like charges
(2) they have unlike charges
(3) they are not charged
(4) one of the leaves is charged

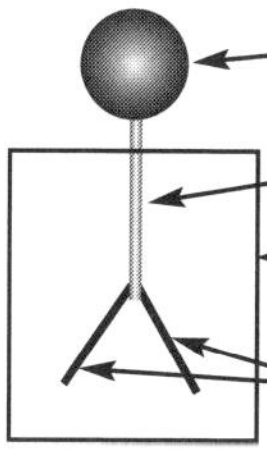

METAL KNOB
METAL BAR
INSULATING GLASS
METAL FOIL LEAVES

27 Which diagram illustrates the conversion of chemical energy to heat?

(1)   (2)   (3)   (4)

THE QUESTIONS THAT FOLLOW TEST YOUR UNDERSTANDING OF SCIENTIFIC PRACTICES, CORES IDEAS IN SCIENCE AND CROSS-CUTTING CONCEPTS SUCH AS PATTERNS AND CAUSE AND EFFECT RELATIONSHIPS.

SHORT ANSWER

1 Use the diagram of the spring car at the right to answer the questions below.

When would the car have the most potential energy? _____

_____

Explain your answer. (2 points) _____

_____

2 Use the diagram of the solar cells at the right to answer the questions that follow.

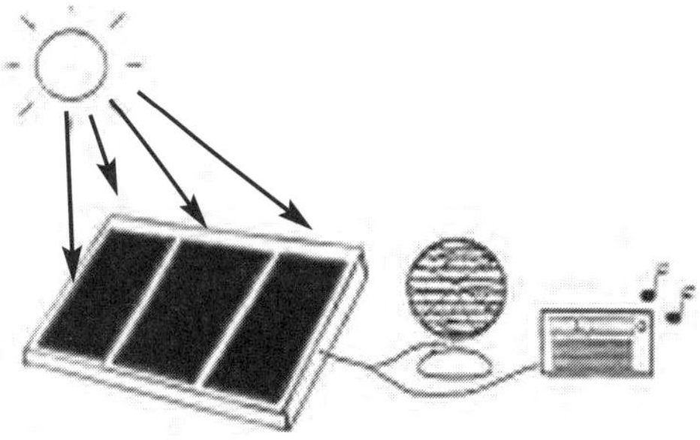

a Describe the type of energy conversion taking place when the solar cell is connected to the fan. (1 point)

_____

_____

_____

b Describe the type of energy conversion taking place when the solar cell is connected to the radio. (1 point)

_____

_____

c How does this diagram illustrate the *Law of Conservation of Energy*? (1 point)

_____

_____

3 The diagram at the right shows two types of soils being exposed to a heat lamp.

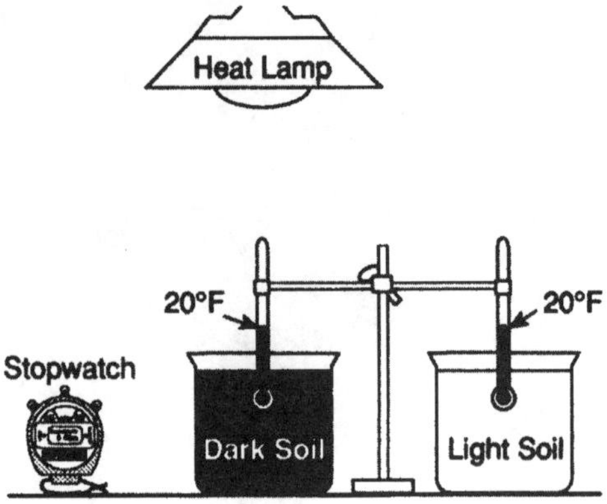

a Which soil's temperature will rise faster? Explain. (2 points)

_____

_____

_____

b What was the purpose of keeping the temperature of the soils the same at the start of the experiment? (2 points)

_____

_____

Short Essay

1 Design an experiment to determine if the height a ball is dropped affects the height of its bounce. Include the following information:

a Write your hypothesis. (1 point) _____

_____

_____

b Describe an experiment making sure to include the following:

(1) Identify the dependent and independent variables. (2 points) _____

_____

(2) List two factors that need to be controlled. (2 points) _____

_____

_____

2   Use the diagrams below of a metal ball being heated, to answer the questions that follow.

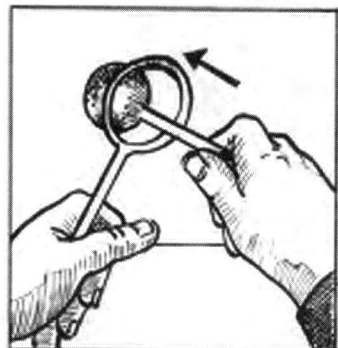 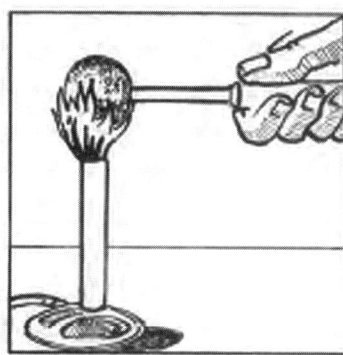 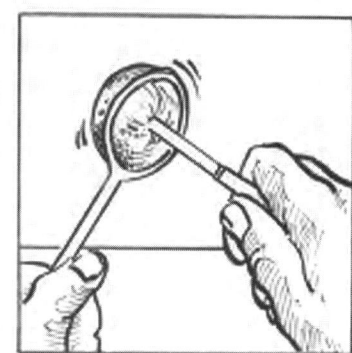

a   Why was the ball placed through the ring before it was heated? (2 points) _____

_____

_____

b   Explain why the ball would not fit through the ring. Include in your discussion how the relative motion
     of the particles that make up the metal ball have changed from before it was heated. (3 points)

_____

_____

_____

_____

_____

MOTION AND STABILITY

DISCIPLINARY CORE IDEAS:

PS2.A: FORCES AND MOTION

PS2.B: TYPES OF INTERACTIONS

DETERMINING MOTION – THE IMPORTANCE OF A FRAME OF REFERENCE

Motion occurs when an object changes its location in space over a period of time. Did you know that you are spinning about 1600 km/hr (1000 mi/hr) while you are sitting down reading this book? In fact, not only are you spinning, but you are cruising along at a speed of about 102.4 km/hr (64,000 mi/hr) through space around the Sun. Do you feel dizzy? Probably not. The reason you are not dizzy is because motion can only be detected relative to a frame of reference such as some other object or a point in space. The most common frame of reference is Earth.

Many drivers have experienced pulling into a parking space, braking, and then slamming on the brakes because they still think they are moving forward. A person pulling out of a parking space next to the driver will give that illusion. If you watch a movie from the 1940's, you may think you see a car cruising along a road. However, if you visited that movie set in which the scene was filmed, you would see that the car is stationery, and that the background is moving. Add some fans to create air motion and the audience's senses are tricked.

47   How is motion determined? _____

FORCES

Motion only takes place when a force has been applied to matter. A force may be a push or a pull. A four wheel drive automobile both pushes and pulls a car forward.

As the car moves forward on a road, the wheels rub against the road creating a force that opposes motion. This force is called friction. The motion of the car is affected by all the forces acting on it. The push or pull on the car and the friction on the wheels all affect its motion. The motion could also be affected if it was going uphill or downhill due to the pull of gravity. Of course, if the total force on the cars adds up to zero, its motion will not change. Cruise control on a car tries to keep the speed constant, but the gradient of the road and even its roughness can change its motion.

Did you know that your weight is a force? Weight is due to the pull of gravity on a given mass. The larger the mass, the greater its response to the pull of gravity. Since the mass of the Moon, and therefore its gravitational pull, is less than the mass of Earth, you weigh less on the Moon. The greater the mass of a person on Earth, the more that person will weigh. However, it must be understood that when an astronaut goes to the Moon, only his weight changes, his mass stays the same.

MOMENTUM

Some forces depend on both the mass of an object and its velocity. For example, if a bowling ball and a ping-pong ball are rolling at the same velocity on the ground, the bowling ball will exert a greater force because it has more mass. What characteristics does a football coach look for in a player who is targeted to be a tackle? The coach looks for a person with a lot of mass and who can also run very fast. These two traits will create a large amount of momentum. A quarterback on the opposing team will fall quickly when in contact with this type of tackle. The momentum of any object is always related to its velocity and mass.

Mathematically, momentum can be expressed as:

$$\text{Momentum} = \text{Mass} \times \text{Velocity}$$

SAMPLE PROBLEM

If the mass of an object is 50 g and its velocity is 5 km/hr, what is its momentum?

$$\text{Momentum} = 50\,g \times 5\,km/hr = 250\,g\,.km/hr$$

48 A small car and a large truck have a head-on collision. The small car is knocked backwards as the truck, although slowed, continues to move forward. Why did the small car move backwards?

_____

_____

49 What is momentum? _____

_____

MEASURING MOTION – HOW DO YOU MEASURE MOTION?

SPEED

What information do you need to determine the **speed** of a car between points **A** and **B**?

A ——————————————————————— B

First, you would need the distance (**d**) between points A and B. Second, you would have to time (**t**) how long it takes the car to travel between the two points. Speed (**s**) is simply the distance traveled per unit of time. This can be expressed mathematically with the following formula:

$$\text{Speed} = \frac{\text{Distance}}{\text{Time}} \qquad\qquad s = \frac{d}{t}$$

SAMPLE PROBLEM:

If the distance (**d**) is 100 km and it had taken the car two hours to travel the distance, what is the speed of the car?

$$s = \frac{d}{t} = \frac{100\,km}{2\,hrs} = 50\,Km/hr$$

VELOCITY

All motion has a direction. For example, a car traveling from Albany, NY to New York City would be going mostly south. The jet stream, (a moving "river" of air located in the upper atmosphere) travels from west to east across the country bringing with it new weather as it moves air masses. Speed with a direction is known as **velocity**.

ACCELERATION

Acceleration measures the rate of change in the velocity of an object. For example, a car may be accelerating 5 kilometers an hour every second. This means that every second the car is going 5 km/sec. faster. Therefore, the 1st second it is traveling 5 km/sec., the 2nd second, 10 km/hr, the 3rd second, 15 km/hr and so forth. To solve for acceleration, you must know the initial velocity (V_1) and final velocity (V_2) of a moving object and the time (t) between these two velocities.

SAMPLE PROBLEM

If a stopped car obtains a velocity of 60 km/hr in 10 seconds, what is its acceleration? In order to solve, first write the formula you are going to use.

$$\text{Acceleration (a)} = \frac{\text{Final Velocity } (V_2) - \text{Initial Velocity } (V_1)}{\text{Time (t)}}$$

Identify the two velocities and the time in seconds. V_1 = 0 km/hr; V_2 = 60 km/hr; t = 10 sec. Plug these values into the formula and solve.

$$a = \frac{V_2 - V_1}{t} = \frac{60 \text{ km/hr} - 0 \text{ km/hr}}{10 \text{ sec}} = \frac{60 \text{ km/hr}}{10 \text{ sec}} = 6 \text{ km/hr/sec}$$

This means that every second, the car is going 6 km/hr faster.

DECELERATION

Deceleration is simply *negative* acceleration. It measures the change in velocity of a moving object that is slowing down. When brakes are applied to a car, it is decelerating.

50   What is the difference between speed and velocity? _____

_____

51   What is the difference between acceleration and deceleration? _____

_____

GRAPHING MOTION

All motion continues in a straight line unless an outside force is exerted upon it. All motion can be expressed as a graphing relationship between distance and time. A graph for constant speed is always a straight line such as the one at the right measuring the speed of two cars.

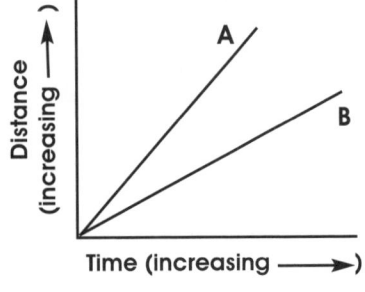

52   Which car is moving faster?_____ Explain. _____

_____

_____

53   Examine the graph at the right. Explain what happened at

section 2 of the graph. _____

_____

_____

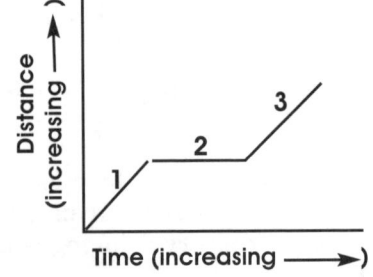

Directions (54 and 55): Use the data table below and the graph grid at the right. Graph the speed of a car that traveled 500 km in 5 hours. Make sure to correctly label the x-axis and y-axis.

DATA TABLE	
TIME (HOURS)	**DISTANCE (KM)**
1	100
2	200
3	0
4	400
5	500

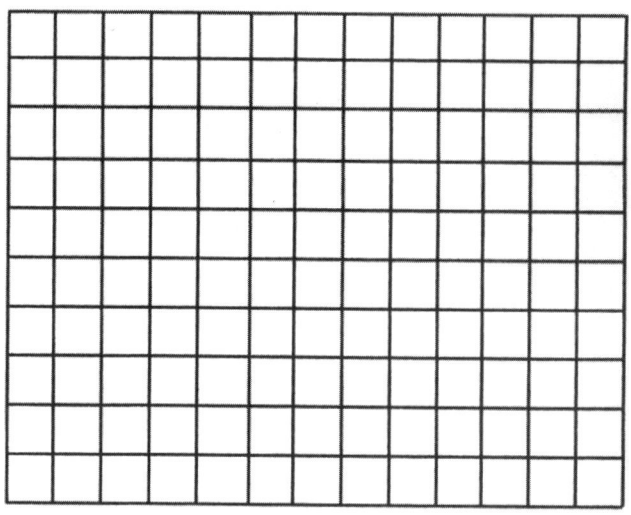

54 What was the car doing between the second and third hour?

_____

_____

55 *a* Did the car go a constant speed during its journey? _____

b Explain. _____

_____

56 *a* What is the shape of the line on the graph for acceleration?

_____

b What is the shape of the line on the graph for deceleration?

_____

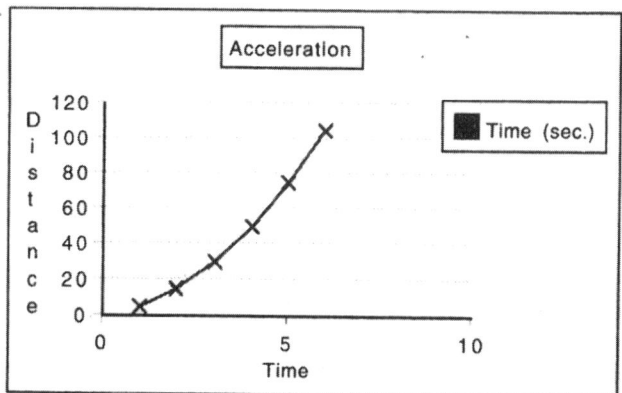

57 Using the deceleration graph, what happens to the distance traveled per unit of time?

_____

_____

_____

_____

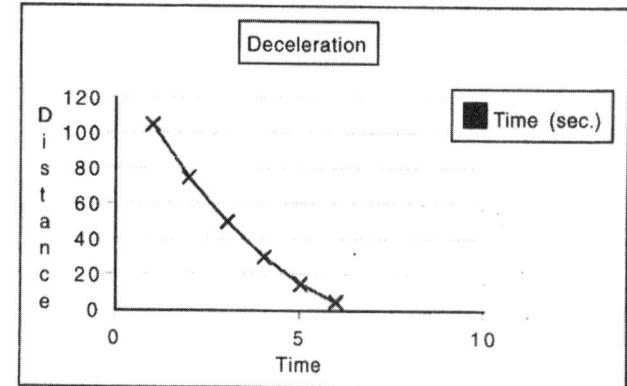

Newton's Three Laws Of Motion

It is important to understand that all motion is guided by a set of three laws developed by the English scientist, **Sir Isaac Newton** between 1665-1666. Whether you are rollerblading in your neighborhood or orbiting Earth in a space shuttle, Newton's three laws apply.

First Law Of Motion

Newton's first law of motion is called the ***Principle of Inertia***. This law states that an object at rest will stay at rest while an object in motion will stay in motion unless acted upon by an outside force. Have you ever seen a magician pull a tablecloth away from a set table without disturbing the settings? This is not magic, but simply Newton's first law. The table setting is staying at rest.

Second Law Of Motion

Newton's second law of motion shows the relationship between **force (f)**, **mass (m)**, and **acceleration (a)**. This law shows how the amount of force is directly related to mass and acceleration. This relationship can be expressed mathematically in the formula:

$$\text{Force (f)} = \text{Mass (m)} \times \text{Acceleration (a)} = m/s/s = \text{Newtons (N)}$$

Sample Problem

What is the force needed to accelerate a car 3 m/s/s, if its mass is 850 kg?

$$\text{Force} = 850 \, kg \times 3 \, m/s/s = 2{,}550 \, N$$

If you were an engineer and were asked to design a very fast race car, what criteria (standards) would you use? First, you would want a big engine. The bigger the engine, the greater the push or pull to accelerate the car. Second, you would want a lightweight (low mass) car which would be a lot easier to push or pull. Using Newton's second law of motion to solve for acceleration, this becomes clearer.

If you solve for acceleration, it can be written as:
$$\text{Acceleration (a)} = \frac{\text{Force (f)}}{\text{Mass (m)}}$$

When the formula is written this way, it can be seen that acceleration is directly related to the size of the force and indirectly related to the size of the mass. Since mass is in the denominator of the formula, it can be seen that if this number increases, the acceleration will decrease. On the other hand, if the force, which is in the numerator of the formula is increased, then the acceleration becomes greater. Think about how much force is needed to place the space shuttle in orbit around Earth!

Third Law Of Motion: Action/Reaction

Newton's third law of motion is called Action / Reaction. This law states that for every action, there is an equal and opposite reaction. Push your hand against your desk or a table. What do you feel? You should feel the desk pushing against your hand with the same force you applied to the desk or table. Push harder. Now you should feel greater pressure against your hand, because the force of the desk's reaction will equal the force of your action.

58 Refer to the launching of the shuttle at the right. What is the action and reaction in the launching of the rocket?

_____

_____

_____

_____

NEWTON'S *LAW OF UNIVERSAL GRAVITATION*

One of the forces associated with matter is the **force of gravity**. All objects exert a gravitational force on other objects. The amount of force is dependent on the amount of matter. Students exert a gravitational force on their teacher and vice versa. However, since the mass of a person is relatively small, you do not notice it. The Earth's force on you is significant because Earth has so much more mass than you. Newton determined that the strength of the gravitational force is not only dependent on the amount of mass, but on the square of the distance between the masses. He expressed this mathematically in the following formula:

m_1 = **Mass** of object 1

m_2 = **Mass** of object 2

d_2 = Square of the **distance** between the two masses

g = Force of **gravity**

$$\text{Gravity (g)} = \frac{\text{Mass } (m_1) \times \text{Mass } (m_2)}{\text{distance } (d^2)}$$

Gravity causes all objects to fall to Earth at the same rate. For example, if you drop a bowling ball and a ping-pong ball from the same height at the same time, they both will land on the ground at the same moment. This is because Earth's gravitational pull causes all objects to accelerate at the same rate which is 9.8 m/sec/sec (9.8 m/s^2). This means that for every second, a falling object is going 9.8 m/sec faster. Therefore, an object falling for three seconds would have a velocity of 29.4 m/sec (3 x 9.8 m/sec).

59 What is the velocity of a rock that falls for 5 seconds? _____
Show the formula and your calculations below.

ORBITAL MOTION

The **Moon** orbits Earth, Earth orbits the Sun and numerous artificial satellites orbit Earth. What is orbital motion? Remember that all objects move in a straight line unless an outside force acts upon them. When a satellite is launched, it is traveling in a straight pathway. However, Earth's gravity is pulling on the satellite at the same time as it is traveling in a straight pathway. When a balance has been reached between the forward motion (inertia) of the object and the pull of gravity, orbital motion results.

If you tie a ball on a string and begin spinning the cord, the ball would go straight, but the pull of the string places the ball into an orbital motion.

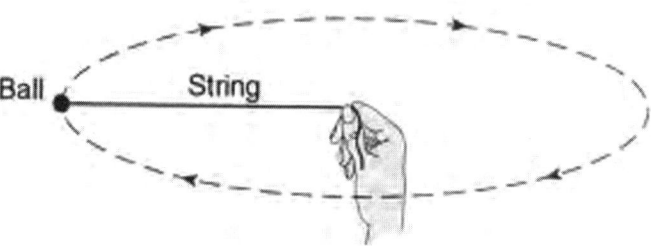

60 What causes orbital motion?

_____

_____

Gravity is a force that works across a distance. As noted before, our solar system is held together by the sun's gravitational pull. Electricity and magnetism also work across a distance. However, they not only attract matter but can also repel or push it away. (See Law of Charges and Law of Poles on Page 196).

PROJECTILE MOTION

Have you ever watched the path of a baseball? Does it move in a straight line or is it curved? It always moves in a curved path because all objects moving through air or any fluid are affected not only by the **horizontal force** applied to them, but the **vertical force** of gravity. Therefore, whether you are throwing a baseball for a strike, shooting arrows to hit the bulls eye, your throw must compensate for these two forces. Projectile motion is always a curved pathway.

61 Describe the path of an arrow from the bow to the target.

MICROGRAVITY ON THE SPACE SHUTTLE

When objects **orbit** Earth, such as the space shuttle, they are also in a "free fall" toward Earth. Though the space shuttle is moving forward, Earth's gravitational pull is also causing it to fall toward Earth. Millions of adventurous people experience "free fall" in Disney's Tower of Terror. When the elevator drops, the passengers are experiencing **weightlessness** or **microgravity**. In fact, every time you jump off something, you are experiencing microgravity while you are falling. Although there is gravity pulling on the space shuttle, since it is in a free fall toward Earth, the astronauts experience microgravity.

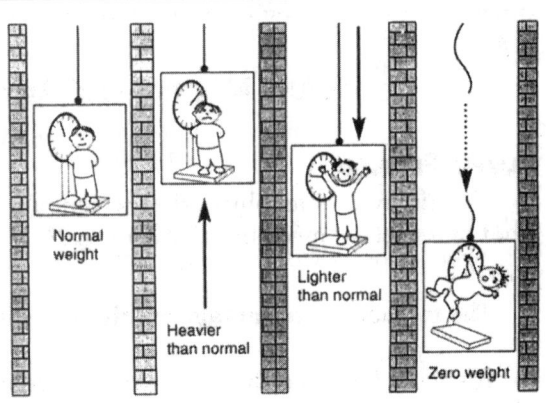

62 What causes microgravity on the space shuttle? _____

WORK

Work is accomplished when a force is applied through a distance. Every time you climb a flight of stairs you are doing some work because a force (your weight) is being applied over a distance (height of stairs). Mathematically, work can be expressed as

$$\text{Work} = \text{Force} \times \text{Distance.}$$

SAMPLE PROBLEM:

How much work do you do climbing a staircase that is 2.5 meters high? Suppose your mass is 110 Kg. This means you weigh (110 x 9.8) 1,078 Newtons*.

$$\text{Work (W)} = \text{Force (f)} \times \text{Distance (d)}$$

$$w = 1{,}078\,N \times 2.5\,m = 2{,}695 \text{ Newton-meters (N-m)}$$

*To convert mass to Newtons, multiply your mass in Kg by 9.8 (constant for Earth's pull of gravity)

POWER

Power is work done per unit of time. Mathematically it is expressed as:

$$\text{Power (P)} = \frac{\text{work done (W)}}{\text{Time (t)}}$$

SAMPLE PROBLEM:

If a machine does 550 N-m of work in 5 seconds, what is the power?

$$P = \frac{W}{t} = \frac{550\,N\text{-}m}{5\,\text{sec}} = 11\,N\text{-}m/\text{sec} = 11 \text{ watts*}$$

* A watt (w) is a unit of power.

MS-ETS1 Engineering Design

This sections deals with how engineering helped to develop tools to make work easier. Simple machines can be made into compound machines and with today's technology artificial intelligence can run many compound machines which are still often based on simple machines. This is an optional unit for the science teacher or to work with a math and technology teacher. Many of the disciplinary core ideas for engineering can be addressed.

MACHINES

Machines are devices that make work easier. Machines can make work easier by:

- **multiplying your effort force** (makes objects that need to be moved seem lighter than they are)
- **changing the direction** of an applied force
- **reducing the time** needed to do the work
- **reducing friction**

These advances in Engineering and Technology provide a mechanical advantage to make work a lot easier by reducing the force a human needs to apply to get work done. The mechanical advantage of simple machines can be expressed mathematically. Levers, pulleys and inclined planes are all simple machines. When combined together, they form a complex machine.

MECHANICAL ADVANTAGE

The mechanical advantage of a machine is the number of times it multiplies the **effort force**. The **effort force (F_e)** is the force applied when a machine is being used to move an object. The weight of the object is the **resistant force (F_r)**. **The actual mechanical advantage (A.M.A.)** of a machine can be expressed mathematically as:

$$\text{Actual Mechanical Advantage (A.M.A.)} = \frac{\text{Resistant Force } (F_r)}{\text{Effort Force } (F_e)}$$

SAMPLE PROBLEM

If a pulley lifts an object that weighs 50 N with a force of 10 N, what is its actual mechanical advantage?

$$\text{A.M.A.} = \frac{F_r}{F_e} = \frac{50 \text{ N}}{10 \text{ N}} = 5$$

The mechanical advantage of this pulley is 5. This means it multiplies the effort force 5 times (5 times easier).

It should be noted that the actual mechanical advantage of a machine also takes into consideration the force of friction. Friction is a force that opposes motion. The **ideal mechanical advantage** of a machine measures how much a machine could multiply the effort force if there was no friction. The ideal mechanical advantage of a machine is always greater than the actual mechanical advantage. The **Ideal Mechanical Advantage (I.M.A.)** compares the **effort distance (d_e)** to the **resistance distance (d_r)**. The effort distance is the distance the object travels using a machine while the resistance distance is the actual distance the machine has to travel. The greater the effort distance, the greater the mechanical advantage. For example, when a chain has to be cut, very long handle cutting pliers are used to give a greater mechanical advantage.

$$\text{Ideal Mechanical Advantage (A.M.A.)} = \frac{\text{effort distance } (d_e)}{\text{resistance distance } (d_r)}$$

SAMPLE PROBLEM

If a ramp that is 15 meters long is used to move a car onto a car carrier that is 1.5 meters above the ground, what is the ideal mechanical advantage of the ramp?

$$\text{I.M.A.} = \frac{d_e}{d_r} = \frac{15 \text{ N}}{1.5 \text{ N}} = 10$$

15 m (d_e)

1.5 m (d_r)

The mechanical advantage of this ramp is 10. This means it multiplies the effort force 10 times.

TYPES OF SIMPLE MACHINES

A **pulley** is a chain or rope wrapped around a wheel. A flag is not very heavy to lift, but it would be difficult to climb a flagpole everyday to "raise the flag." Fortunately, a simple machine called a pulley was invented. You pull down and the flag goes up. This is the same way in which blinds on a window work. A movable pulley not only changes the direction of the applied force, but makes an object appear to weigh less. Movable pulleys are common at construction sites. If a motor is attached to a pulley system, the work can be done at a faster rate.

A rake, golf club, can opener, a seesaw, and a wheel barrel are examples of **levers**. A lever consists of a bar that is free to rotate around a fixed point called the **fulcrum**. Levers come in three classes.

Class I – The fulcrum is located between the points of effort and resistance. Examples: seesaw and crowbar.

effort ←— d_e —→ ←— d_r —→ resistance
△ fulcrum

force (effort)

resistance

fulcrum

Class II – The resistance (or load) is between the point of effort and the fulcrum. Examples: bottle opener and wheel barrow.

effort   resistance
△ fulcrum

Class III – The effort is located between the point of resistance and the fulcrum. Examples: the human forearm and tweezers.

resistance   effort
△ fulcrum

A **wheel and axle** is a lever that rotates in a circle. It consists of a large wheel and a smaller wheel. A door knob, ferris wheel and the wheels on a bike or car are examples.

An **inclined plane** is a slanted surface. Any ramp is an inclined plane. When you climb a flight of stairs that can be considered an inclined plane.

Stadium →

The **wedge** and **screw** are also examples of types of inclined planes. A wedge is a moving inclined plane. A screw is a spiral inclined plane.

63 Refer back to the Ideal Mechanical Advantage problem on the previous page. What type of machine

was used with the car carrier? _____

COMPLEX (COMPOUND) MACHINE

A machine combining more than one **simple machine** is called a compound machine. A bicycle is an example of a complex machine.

64 Examine the drawing of the wheelchairs at the right. List the names of the simple machines used on the wheelchairs.

_____

65 Identify and give the advantage of using some other simple machines. _____

_____

EFFICIENCY OF A MACHINE

The **efficiency** of a machine is closely associated with reducing friction. When parts rub, some of the energy of the machine is converted into wasteful heat energy. How can you reduce the friction between the moving parts of a machine? The most common method is to apply lubricants. Bearings also help reduce friction. Automobiles with balanced wheels cut down on unnecessary motion that increases friction. Automobiles with unbalanced wheels use more petroleum making them less energy efficient (they get less mileage to a gallon).

66 Explain how to improve the efficiency of a machine. _____

_____

MULTIPLE CHOICE

1 All motion is relative to a
   (1) moving object   (2) frame of reference   (3) stationery object   (4) distant star

2 The force(s) applied by a car with front wheel drive is (are)
   (1) a push   (2) a pull   (3) a push and pull   (4) without friction

3 The response of an object to the force of gravity is called
   (1) volume   (2) mass   (3) weight   (4) density

4 As the mass of an object increases, its response to the pull of gravity
   (1) increases   (2) decreases   (3) doubles   (4) remains the same

5 Which of the following balls would have the most momentum when falling from the same height?

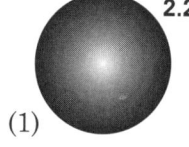 2.2 Kg    1.5 Kg   1.0 Kg   0.5 Kg

(1)   (2)   (3)   (4)

6 A car travels 120 km in 3 hours. What is its speed?
   (1) 123 km/hr   (2) 60 km/hr   (3) 40 km/hr   (4) 20 km/hr

7 What is the acceleration of a car that travels from 0 km/hr to 88 km/hr in 10 seconds?
   (1) 98 km/hr/sec   (2) 88 km/hr/sec   (3) 9.8 km/hr/sec   (4) 8.8 km/hr/sec

8 What information is required in order to measure speed?
   (1) time and distance   (2) velocity and time   (3) distance and velocity   (4) momentum and time

9 Which of the following graphs best shows acceleration?

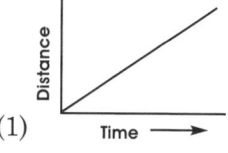   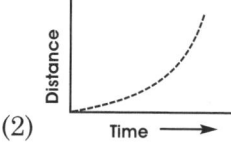   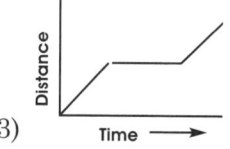   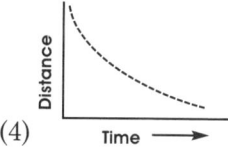

(1)   (2)   (3)   (4)

10 When a car is accelerating, the distance traveled per unit of time
   (1) decreases   (2) increases   (3) remains the same   (4) doubles

11 You wear seat belts in a car because of
   (1) gravity   (2) inertia   (3) force, mass, and acceleration   (4) action/reaction

12 A ball bounces because of
   (1) gravity   (2) inertia   (3) force, mass, and acceleration   (4) action/reaction

13 A force is needed to move a stationery car because of
   (1) gravity   (2) inertia   (3) force, mass and acceleration   (4) action/reaction

14 As the distance between two masses increases, the gravitational pull
   (1) increases   (2) decreases   (3) remains the same   (4) doubles

15 Which diagram best displays the motion of a baseball being thrown to a catcher?

(1) 　　　　(2)　　　　(3) 　　　　(4)

16 The magnets shown at the right are placed near one another. It is most likely they will
(1) move closer　　(3) not move
(2) move apart　　(4) change poles

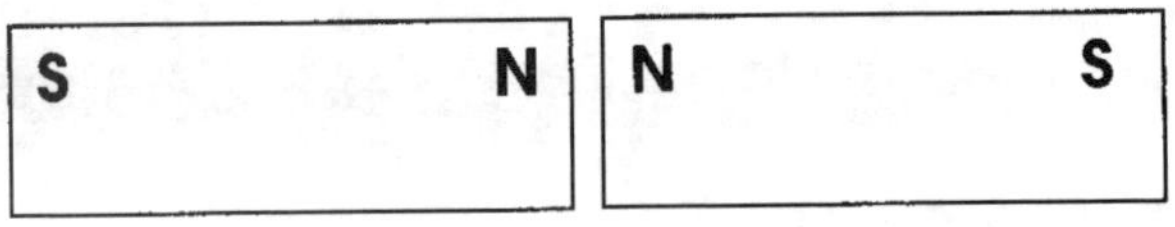

S　　N　　　N　　　S

17 What type of simple machine is shown at the right?
(1) pulley
(2) wheel and axle
(3) screw
(4) wedge

18 Which of the following machines is an

(1) 　　　　(2) 　　　　(3) 　　　　(4)

19 What are two simple machines found on a handtruck as shown at the right?
(1) wedge and pulley
(2) inclined plane and wheel and axle
(3) lever and wheel and axle
(4) screw and pulley

THE QUESTIONS THAT FOLLOW TEST YOUR UNDERSTANDING OF SCIENTIFIC PRACTICES, CORES IDEAS IN SCIENCE AND CROSS-CUTTING CONCEPTS SUCH AS PATTERNS AND CAUSE AND EFFECT RELATIONSHIPS.

SHORT ANSWER

1　Carefully examine diagrams A-D which show a person moving a box up a ramp.

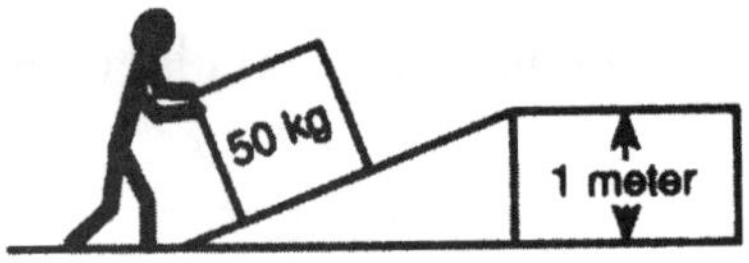

a　In which diagram would the person need the least effort to move the box up the ramp? _____ Explain. (2 points)

_____

_____

_____

50 kg　1 meter
Diagram A

50 kg　1 meter
Diagram C

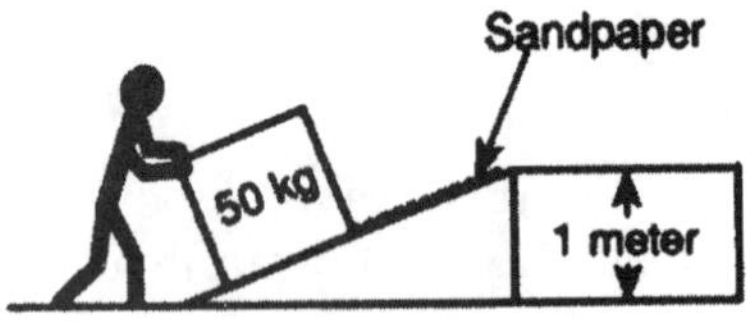

Sandpaper
50 kg　1 meter
Diagram B

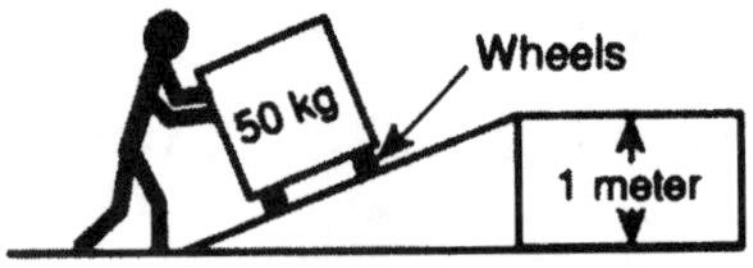

Wheels
50 kg　1 meter
Diagram D

b　What type of simple machine is a ramp? (1 point) _____

2　The air from a balloon is let out, and the balloon sails across a classroom. Which of Newton's laws of

Motion control the motion of the balloon? _____ Explain. (2 points) _____

_____

_____

3   Refer to the picture of a bicycle at the right.

   a   What type of type machine is a bicycle? (1 point) _____

   b   Identify one simple machine on a bicycle and describe how it works.

   (2 points) _____

   _____

   _____

SHORT ESSAY

1   Construct a graph showing the motion of a
   turtle displayed in the data table. Make sure
   to label the axes correctly and title the graph.
   After the graph is complete, answer the following questions. (3 points)

Distance (m)	Time (minutes)
0	0
3	2
6	4
9	6
9	8
9	10
12	12

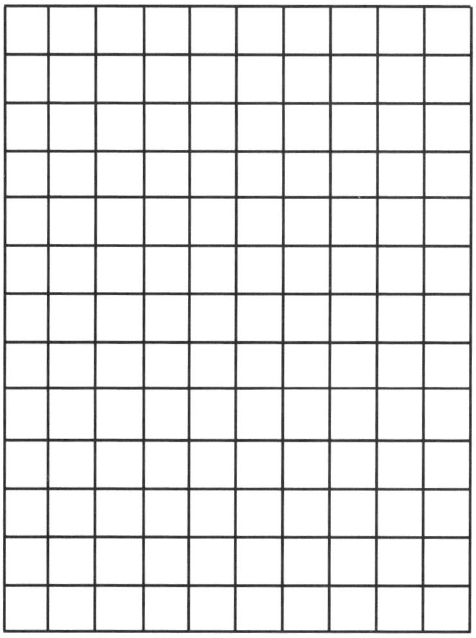

   a   What is happening after the turtle traveled 9 meters? (1 point)

   _____

   b   What is the average speed of the turtle? (1 point) _____

2   Cars have a considerable number of safety features built into them.

   a   How do seat belts protect a passenger. (2 points) _____

   _____

   _____

   b   Why do signs warn drivers to *slow down* when traveling around a sharp curve in the road. (2 points)

   _____

   _____

   _____

Section Five

GENERAL SCIENCE SKILLS
(CROSSCUTTING AND PRACTICES)

SKILL AREA 1
SAFETY IN THE SCIENCE CLASSROOM

Working in a science classroom can be great fun, but there are many safety rules (procedures) you must follow.

FOLLOWING DIRECTIONS
Make sure you read all rules for laboratories before beginning any science investigation.

CHEMICAL SAFETY
- Treat all chemicals as harmful.
- Avoid direct contact with your hands.
- Wear safety glasses at all times.
- Dispose of chemicals as outlined by the teacher.
- If a chemical needs to be smelled, use your hand in a wafting motion toward your nose. Do not inhale the fumes of the chemical all at once. This could irritate your lungs.
- Ask the teacher what to do if there is a chemical spill.
- You should not eat any food (gum included) in a lab.
- Do not mix chemicals together unless specified by the teacher. When mixed together, certain chemicals can release toxic fumes. Some chemicals might be so reactive that they can explode, causing harm to you and others.

GLASS SAFETY
- Do not place heated glass on a cold surface. The sudden change in temperature may cause the glass to crack.
- Lubricate glass rods before placing through a rubber cork. Use a gentle twisting motion to move the tube through the cork.
- Heat test tubes so that the open end faces away from you and others.

OPEN FLAMES SAFETY
- Never reach over an open flame. If not using the open flame, turn it off and relight it only when needed.

EYE SAFETY
- Always wear safety glasses when required.

CLOTHING SAFETY
- Avoid loose clothing.
- Pull long hair back in order to avoid contact with chemicals or an open flame.
- Wear aprons if provided by your school.

ELECTRICAL SAFETY
- Use dry hands when plugging wires into sockets.

WORK AREA SAFETY
- Keep your work area uncluttered, clean, neat, and organized.
- Have only the equipment you need for the investigation in your work area.

1 Why is it important to follow all directions when working in a laboratory? _____

_____

_____

2 Describe two safety procedures to follow when handling chemicals. _____

_____

_____

3 Why is it always important to wear safety glasses in a laboratory when it is required? _____

_____

SKILL AREA 2
MEASUREMENT IN THE SCIENCE CLASSROOM

Scientists use the **metric system** instead of the English system. In fact, the metric system is used by most countries of the world. It is based on the **decimal system**. Prefixes (see table at bottom of the page) are used with the standard measurement to make measures.

BASIC UNITS OF METRIC MEASUREMENT

MEASUREMENT	BASIC UNIT
Length (distance from point to point)	meter (m)
Mass (amount of matter)	kilogram (kg)
Volume (space matter occupies)	liter (L)
Temperature (average kinetic energy in matter)	Celsius or Kelvin(C or K)
Weight/Force (push or pull)	Newton (N)
Work (force applied through a distance)	Joule (j) or Newton-meter (N-m)
Power (work per unit of time)	watt (W)
Speed (distance per unit of time)	meter per second (m/s)
Acceleration (speed per unit of time)	meter per second squared (m/s^2)

BASIC UNITS OF METRIC MEASUREMENT

Some basic units and what they measure are summarized in the table above and some common prefixes are given at the right.

4 What is the basic unit for: *a* mass? _____

  b distance? _____ *c* volume? _____

5 One gram is equal to _____ kg

COMMON PREFIXES

PREFIX	VALUE OF BASIC UNIT
nano-	0.000 000 001
milli-	0.001
centi-	0.01
giga-	1 000 000 000
mega-	1 000 000
kilo-	1 000

EQUATING METRIC & ENGLISH UNITS

METER
- A **meter** (m) is a little longer than a yard. It is equal to 39.37 inches.
- One inch is is equal in length to 2.54 **centimeters** (cm) .
- A **kilometer** (km) is about 0.6 miles or think a little longer than half a mile.

LITER
- 1 **liter** (1,000 milliliters, mL) is equal to 1.1 quarts – a little more than a quart.
- 1 ml = 1 cm³ This means that a cube with a 1 cm side can hold exactly 1 mL of volume.
- 250 ml is equal to about 1 cup.

CELSIUS
The best way to remember the **Celsius** temperature scale (metric) is to use reference points to **Fahrenheit** (English). Water boils at 100°C (212°F) and freezes at 0°C (32°F). Body temperature is 37°C (98.6°F). A comfortable room temperature is about 22°C (72°F).

You can convert Celsius to Fahrenheit with the following formula:

$$°C = {}^5/_9 \times (°F - 32)$$

SAMPLE PROBLEM
If the temperature is 212°F, what is the Celsius temperature?

$$°C = {}^5/_9 \times (212°F - 32) = {}^5/_9 \times (180°F) = 100°C$$

The **Kelvin** temperature scale is also a metric scale. It begins at **absolute zero** (the temperature at which no heat can be transferred out of matter). Water freezes at 273°K and boils at 373°K.

GRAM
- A **gram** (g) has a mass of about one small aluminum paper clip or one M&M candy. It is about 0.04 ounces.
- A **kilogram** (kg) is equal to 2.2 pounds. Think of your weight in terms of a little more than one-half your weight in pounds and you have your mass in kilograms.

NEWTON
The **newton** (N) is used to measure force. It is the amount of force required to accelerate a mass of one kilogram one meter per second .

- 9.8 Newtons = 1000 grams of force
- 4.5 Newtons = 1 pound of force

6 What would happen to the size of a football field, if it measured 100 meters instead of 100 yards? _____

_____

7 *a* Would you be running a fever if your temperature was 38°C? _____ *b* Explain. _____

_____

8 *a* Will a liter of milk fit nicely into a cup? _____ *b* Explain. _____

_____

_____

9 Mass (weight) is measured in _____ (metric).

TRIPLE BEAM BALANCE

A beam balance is used to measure mass.

For example, the mass of the item in the diagram is determined by adding together the mass of each slider scale.

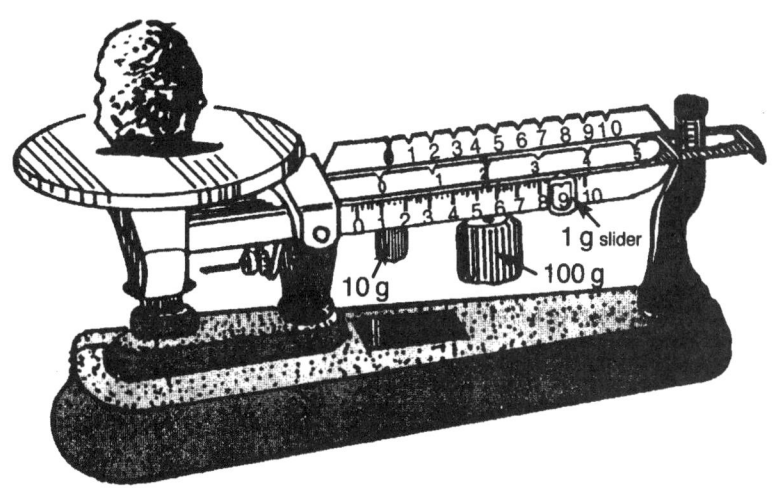

10 What is the mass of the object pictured on the triple beam balance?

_____

11 What measurement is made with a

triple beam balance? _____

12 Explain how to use a triple beam balance. _____

_____

GRADUATED CYLINDER

A **graduated cylinder** is used to measure volume. Each calibration in the graduated cylinder shown in the diagram is equal to 1 mL. The value of the liquid is read at the **meniscus** (the lowest level of the liquid).

13 What is the volume of the liquid in this cylinder? _____

14 Explain how to read the measurement on a graduated cylinder. _____

_____

15 Explain how to determine the volume of an irregularly shaped object.

_____

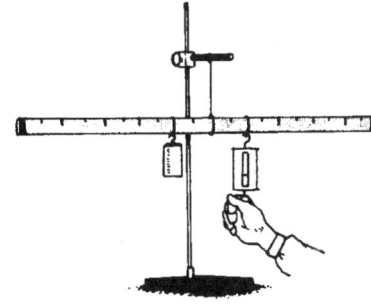

A graduated cylinder can be used to determine the volume of an irregularly shaped solid such as a rock. The volume of the object is equal to the amount of water it displaces. For example, if the starting volume of the water is 20 mL and the final volume is 25 mL, then the volume of the object is 5 mL.

SPRING SCALE

A **spring scale** measures force (at right) or weight (at left). A kilogram or 1000 g is equal to 9.8 Newtons or about 10 N.

VOLTMETER

A **voltmeter** measures voltage or potential difference which is measuring the "push" from the flow of electricity. Voltmeters are adjusted for whether the voltage is direct current ("DC" – voltage from a a battery) or alternating current ("AC" – current from a generator). Your teacher should show you how to use a voltmeter.

METRIC RULER

A **metric ruler** measures length. Each small unit is equal to a millimeter. There are 10 mm in each segment shown.

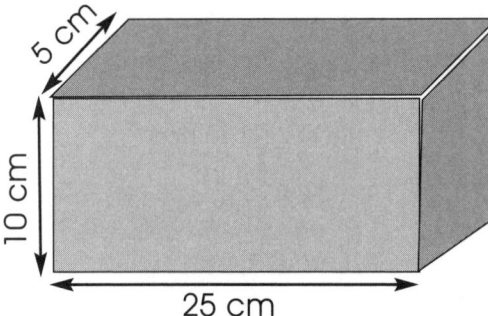

16 What is the distance shown on the metric ruler? _____

The volume of some regularly shaped objects can be determined by using a metric ruler. For example, the volume of all rectangular blocks is determined by the formula:

Volume = Length x Width x Height.   v = l x w x h

17 *a* What is the volume of the block shown below? _____
    b Show work in space provided below and be sure to show
        the measuring units as cm³.

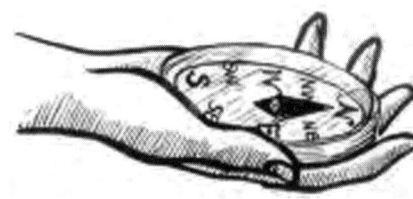

MAGNETIC COMPASS

A **magnetic compass** shows cardinal directions N, S, E, W. The magnetic needle points to the magnetic north pole of Earth. This compass is showing a north (N) orientation.

STOP WATCH

A **stop watch** can be used to make precise time observations and measurements.

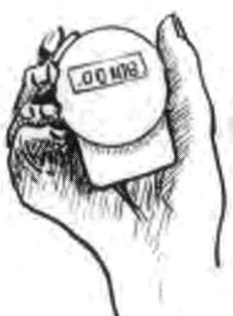

THERMOMETER

Scientists use metric **thermometers**. Metric temperature can be measured in Celsius or Kelvin. The temperature chart at the right compares the three temperature scales: °Fahrenheit (English system) to °Celsius to Kelvin.

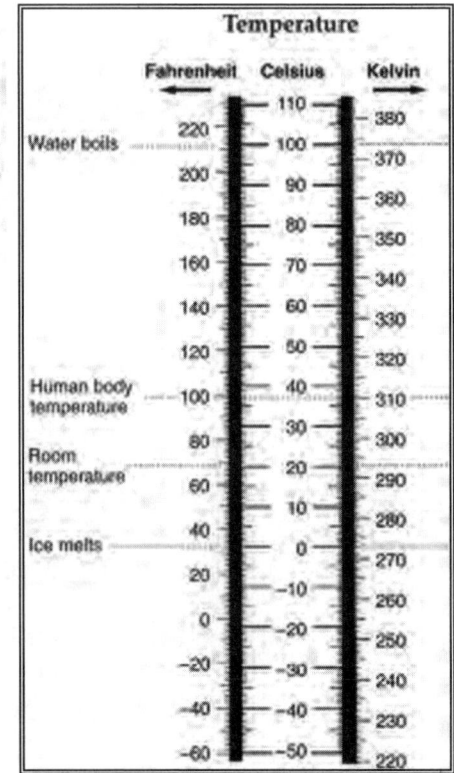

18 At what temperature does water freeze? _____°F _____°C _____K

19 At what temperature does water boil? _____°F _____°C _____K

20 What is the human body's normal temperature? _____°F _____°C

21 What is 140° F in Celsius degrees? _____°C Kelvin? _____K

INDICATORS

Scientists use **indicators** to determine the presence of a certain substance or to measure the acidity or basicity of a substance.

The table at the right summarizes the use of some common indicators.

TESTING INDICATORS

INDICATOR	POSITIVE TEST
Red Litmus Paper	turns blue in presence of a base
Blue Litmus paper	turns red in presence of an acid
Hydrion Paper	pH less than 7 indicates acid; pH greater than 7 indicates base
Limewater	turns cloudy in presence of carbon dioxide
Bromthymol Blue	turns yellow in presence of carbon dioxide
Iodine	turns blue black in presence of starch or cellulose
Universal Indicator	pH less than 7 indicates acid; pH greater than 7 indicates base
Benedict's Solution	turns light green to brick red in presence of sugar
Methylene Blue	stains DNA
Phenolphthalein	turns pink in presence of base

pH SCALE

The **pH scale** is used to measure the strength of an **acid** or a **base**. The scale goes from 0-14. A solution with a pH of 7 is considered **neutral**. A pH less than 7 is an acid, while a pH greater than 7 is a base. The lower the pH, the stronger the acid, while the higher the pH, the stronger the base.

22 Give an example of a solution with a high pH. _____

23 Give an example of a solution with a low pH. _____

NEUTRALIZATION REACTION

When an acid and base of equal concentration are combined, a neutral substance with a pH of 7 is formed. The product of a **neutralization reaction** (acid plus base) is a salt plus water. Example:

$$HCl + NaOH \rightarrow NaCl + H_2O$$

Hydrochloric Acid + Sodium Hydroxide (base) → Sodium Chloride (salt) + water

SCIENTIFIC NOTATION

Scientists find it useful to express very large numbers as powers of ten. This is called **scientific notation**. For example, the distance to the Sun is 148,000,000 km. This can also be expressed using scientific notation as 14.8×10^7 ($10 \times 10 \times 10 \times 10 \times 10 \times 10 \times 10$).

24 Express the number 158,000,000,000 in scientific notation.

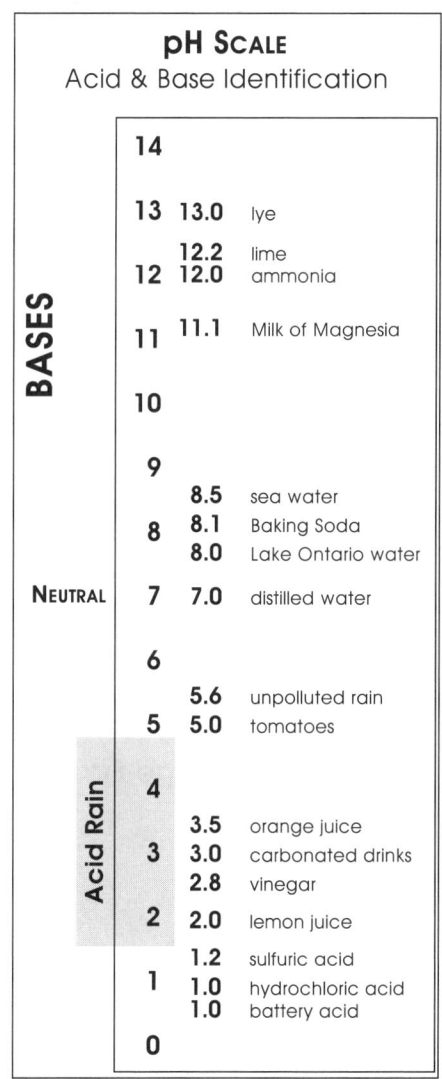

pH SCALE
Acid & Base Identification

BASES

14		
13	13.0	lye
12	12.2	lime
	12.0	ammonia
11	11.1	Milk of Magnesia
10		
9		
	8.5	sea water
8	8.1	Baking Soda
	8.0	Lake Ontario water
7 NEUTRAL	7.0	distilled water
6		
	5.6	unpolluted rain
5	5.0	tomatoes
4		
	3.5	orange juice
3	3.0	carbonated drinks
	2.8	vinegar
2	2.0	lemon juice
	1.2	sulfuric acid
1	1.0	hydrochloric acid
	1.0	battery acid
0		

Acid Rain

SKILL AREA 4
MATH IN THE SCIENCE CLASSROOM

SOLVING LINEAR EQUATIONS

Scientists often use **mathematics** to describe order in the natural world. The following steps are useful when applying a mathematical equation.

1   Decide which equation is appropriate.
2   Solve for the **unknown** by plugging in the known values.
3   Use the correct units when expressing the answer.

SAMPLE PROBLEM

What is the density of a block of wood that has a mass of 50 g and a volume of 100 cm³?

1   **Write the formula**:   Density = Mass/Volume
2   **Plug in the values**:   Density = 50 g/100 cm³
3   **Solve**:   Density = 0.5 g/cm³

The formula for speed is $\textbf{Speed (s)} = \dfrac{\textbf{distance (d)}}{\textbf{time (t)}}$

25 *a* What is the speed of a car that travels 400 km in 3 hours? _____
    b Show all work in the space provided below.

STATISTICS

Scientists use statistics to help determine the validity of data.

PERCENTAGE OF ERROR

The **percentage of error** compares data (observed value) to a standard value. For example, you took an exam and received a grade of 90%. All scientists are interested in the size of their error. If you received a grade of 90% on the exam, it means you had a 10% error. Mathematically, percent of error can be expressed as

$$\textbf{Percentage Error} = \frac{\textbf{difference between your value and a standard value}}{\textbf{standard value}} \times \textbf{100}$$

SAMPLE PROBLEM

What is the percentage of error, if you measure the density of aluminum and find it to be 2.9g/cm³ while the standard value is 2.7g/cm³? Plug in the given information and solve. Remember, always subtract the lower value from the higher value. You cannot have a negative value for percent of error.

$$\textbf{Percentage Error} = \frac{\textbf{0.2 g/cm}^3 \text{ or } (\textbf{ 2.9g/cm}^3 - \textbf{2.7g/cm}^3)}{\textbf{2.7g/cm}^3} \times \textbf{100} = \textbf{7.4 \%}$$

26 What is the percentage of error if the length of a room was estimated to be 15.5 m and when measured,

the length turned out to be 14 m? _____ %

MEAN

The **mean** is the value obtained by dividing the sum of a set of quantities by the number of quantities in the set, or in other words, the average. Mathematically, the mean is expressed with the following formula:

$$\text{Mean (Average)} = \frac{\text{Sum of the Trials (samples)}}{\text{Total Number of Trials}}$$

Suppose a student was studying the effects of fertilizers on the growth of bean plants. The length of the primary stalks was measured and recorded in the data table at the right for both groups of bean plants (with and without fertilizer).

EFFECTS OF FERTILIZER ON BEAN PLANT GROWTH

PLANT WITH FERTILIZER	HEIGHT (CM)	PLANT WITHOUT FERTILIZER	HEIGHT (CM)
A	9	A	6
B	7	B	5
C	8	C	7
D	7	D	8
E	7	E	6
F	9	F	6
G	10	G	7

What is the mean growth of each group of bean plants?

$$\text{Mean (Average) Plants with Fertilizer} = \frac{9 + 7 + 8 + 7 + 7 + 9 + 10}{7} = \frac{57}{7} = 8.14 \text{ cm}$$

$$\text{Mean (Average) Plants No Fertilizer} = \frac{6 + 5 + 7 + 8 + 6 + 6 + 7}{7} = \frac{45}{7} = 6.43 \text{ cm}$$

MODE:

The mode is the sample size that occurs most frequently. In the case of the plants that received fertilizer, the mode was 7, while the mode for the group that did not receive fertilizer was 6.

MEDIAN:

This is the score that occurs in the middle of a sample of data that is ranked from lowest to highest.

After ranking the fertilizer group:
7 7 7 8 9 9 10, it can be seen that 8 is the median.

After ranking the non-fertilizer group:
5 6 6 6 7 7 8, it can be seen that 6 is the median.

27 What are the mean, mode, and median of the speeds of a toy car recorded in the data table at the right?

a mean _____

b mode _____

c median _____

SPEEDS OF TOY CAR

TRIAL	SPEED (CM/S)
1	10
2	12
3	12
4	11
5	9
6	12
7	8

DEVELOP & USE A DICHOTOMOUS KEY

CLASSIFY THINGS ACCORDING TO A STUDENT GENERATED SCHEME AND AN ESTABLISHED SCHEME

Scientists often organize the natural world by **classifying** or grouping living and nonliving things. Classifying looks more closely at how things are similar (placed in the same group) or different (placed into separate groups).

One way to organize things is through the use of a **dichotomous key**. In a dichotomous key, things are organized by arranging them based on their traits or characteristics. A dichotomous key is a series of couplets that describe two different sets of characteristics. Choosing the set of characteristics that fits the object you are examining leads you to further couplets and further choices. This continues until the object you are examining is identified. Each couplet has two choices, that is why it is called a dichotomous key (di, meaning two).

For example, you are trying to organize all the vertebrates (animals with back supporting structures) into groups. How would you classify humans? Use the following dichotomous key to determine the answer.

Directions: Read couplet 1 and choose *A* or *B*. Follow the directions until you find how humans are classified.

A KEY TO THE CLASSES OF VERTEBRATES

	Couplet	Description
1	A	Legs and arms as fins; one or more pairs of gill slits .go to 2
	B	Legs and arms not as fins; legs usually present . . . . .go to 4
2	A	Fins not in pairs; no scales; round mouth . . . . .Class Agnatha
	B	Some paired fins; scales; mouth with jaw . . . . .go to 3
3	A	Skeleton of cartilage . . . . .Class Chondrichthyes
	B	Skeleton of bone . . . . .Class Osteichthyes
4	A	Body smooth; no scales; no claws . . . . .Class Amphibia
	B	Body covered with scales, feathers, hair or fur . . . . .go to 5
5	A	Body covered with scales or in shell; no feathers, hair or fur . . . . . . .Class Reptilia
	B	Feathers, hair or fur present . . . . .go to 6
6	A	Body covered with feathers . . . . .Class Aves
	B	Hair or fur present on body . . . . .Class Mammalia

CLASSIFY HUMANS

Read the first couplet. Which of the descriptions, A or B, fits humans? . . ._____
  Since humans have legs and arms you should have chosen 1B.
1B sends you to couplet 4.

Read couplet 4. Which of the descriptions, A or B, fits humans? . . . . ._____
  Since humans have hair you should have chosen 4B.
4B sends you to couplet 5.

Read couplet 5. Which of the descriptions, A or B, fits humans? . . . . ._____
  Since humans have hair you should have chosen 5B.
5B sends you to couplet 6.

Read couplet 6. Which of the descriptions, A or B, fits humans? . . . . ._____
  Since humans have hair you should have chosen 6B.
6B leads you to our answer. Humans are in the Class Mammalia, mammals.

28 The diagram below represents ten types of fish that are found in the northeastern United States. On the next page, create a dichotomous key so that each type of fish can be distinguished. To start, determine which traits you will use to distinguish the different fish. Look at the labeled diagram of a fish for some hints

TEN TYPES OF FISH FOUND IN NORTHEASTERN UNITED STATES

Fish Key: **General Characteristics**

(1) Pectoral fin

(2) Pelvic (ventral) fin

(3) Anal fin

(4) Lateral Line

(5) Caudal fin

(6) Adipose fin

(7) Dorsal fin

(8) Barbel

(9) Operculum

(10) Maxillary

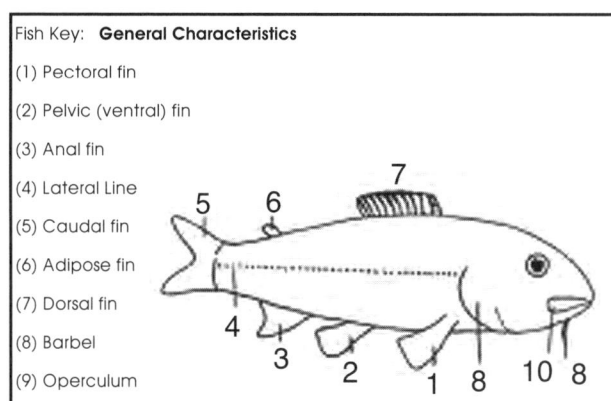

Family Centrarchidae
Micropterus salmoides
- Largemouth Bass

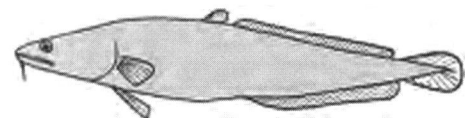

Family Gadidae
Lota lota - The Burbot

Family Cyprinodontidae
Fundulus diaphanus - Barred killfish

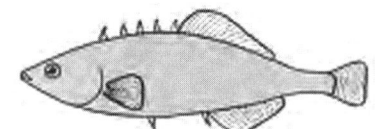

Family Gasterosteidae
Eucalia inconstans - Brook Stickleback

Family Poeciliidae
Gambusia affinis - Mosquitofish

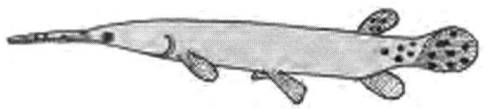

Family Lepisosteidae
Lepisosteus osseus - Longnose gar

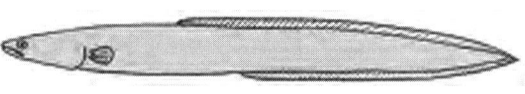

Family Anguillidae
Anguilla rostrata - American eel

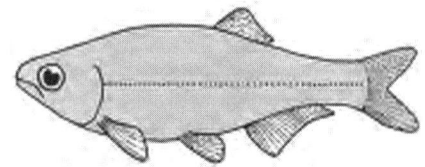

Family Hiodontidae
Hiodon tergisus - Mooneye

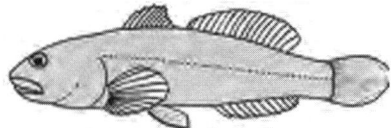

Family Cottidae
Cottus bairdi - Mottled sculpin

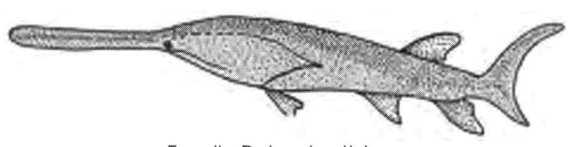

Family Polyodontidae
Polyodon spathula- Paddlefish

Here is the beginning of a key to help you get started.

- 1A _____

  B _____

- 2A _____

  B _____

- 3A _____

  B _____

- 4A _____

  B _____

- 5A _____

  B _____

- 6A _____

  B _____

- 7A _____

  B _____

- 8A _____

  B _____

- 9A _____

  B _____

- 10A_____

  B _____

INTERMEDIATE SCIENCE- 3D LEARNING & ASSESSMENTS Topical Review Book Company ©

THE USE OF THE COMPOUND MICROSCOPE

Observing is the first step in learning about your world. Scientists have invented many instruments as tools to help you observe the world more closely. One of the most useful tools is the microscope. The **microscope** allows you to observe things that are too small for you to see. Microscopes open up the microscopic world to observation and study. Steps for study:

Steps:

A) Examine the diagram of the compound microscope. Refer back to it as you review the use of the microscope.

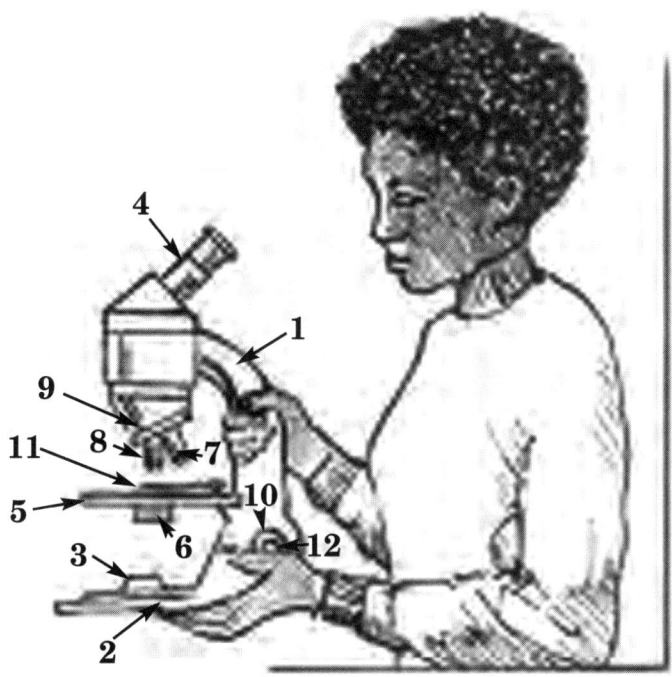

THE MICROSCOPE

MY MICROSCOPE
by Joan Wagner
I use my trusty microscope
to view the worlds around me
of tiny little specimens
with features quite unsightly!
They dash about my viewing field
swimming freely to and fro;
and even though they are so small
they reproduce and grow.
My microscope has many parts
to help me study these creatures
and even special lenses
that magnify the smallest features:
The coarse and fine adjustments,
and numerous objectives; the
diaphragm and special light
that focus these tiny "subjectives."
And so I highly recommend
this scientific tool
to better understand the world
and pass science in my school.

	PART	FUNCTION
1	**Arm**	connects body tube and base; supports stage; used to carry microscope
2	**Base**	supports microscope; may hold lamp or mirror
3	**Mirror** or **Lamp**	provide light needed to view specimens (light source)
4	**Eyepiece**	contains one of the lenses
5	**Stage**	where slide is placed
6	**Diaphragm**	regulates the amount of light reaching the specimen
7	**Low Power Objective**	contains one of the lenses; less magnification than high power objective
8	**High Power Objective**	contains one of the lenses; greater magnification than low power objective
9	**Nosepiece**	holds the objectives; can be rotated so that objectives can be switched
10	**Course Adjustment**	moves stage and objectives closer or farther apart quickly to allow for large focusing changes
11	**Stage Clip**	holds slide in position on stage
12	**Fine Adjustment**	moves stage and objectives closer or farther apart slowly to allow for fine focusing

B) Notice how the student is holding the microscope. Always carry the microscope with two hands, one holding the arm and one under the base. It is safest to carry the microscope in an upright position.

C) In order for a compound microscope to work, light must travel up through the specimen to your eye. Directly above the base of the microscope will be the mirror or lamp.

If you have a mirror, while looking through the eyepiece, move the mirror until it reflects light from a source in the room. You should see a bright circle of light. NEVER use direct sunlight as a light source. Direct sunlight can seriously damage your eye.

If you have a lamp, turn on the lamp and look through the eyepiece. You should see a bright circle of light.

Whenever looking through the eyepiece keep both eyes open. This will reduce eye strain. If you have trouble focusing with the one eye, keep your other eye open but cover it with your hand.

D) Attached to the stage of the microscope is the diaphragm. The diaphragm is used to regulate the amount of light. Turn the diaphragm while looking through the eyepiece. You should be able to make your field brighter and darker. It is usually best to start with a brightly lit field.

E) The reason that your microscope is called a 'compound' microscope is that it has two lenses. One is in the eyepiece and one is in the objective. Each lens focuses the light so that the object being viewed is magnified. Using two lenses compounds this effect. In other words, if an eyepiece magnifies 10 times (10x) and the objective magnifies 10 times (10x) together in a compound microscope they magnify the object 100 times (100x).

Your microscope may have more than one objective. Each one is labeled according to its magnification power. The low power objective is usually 10x. The high power objective is usually around 40x.

29 What is the magnification of your low power objective?_____

30 What is the magnification of your high power objective?_____

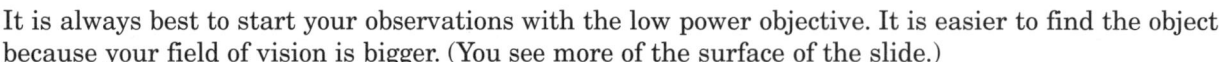

Turning the nosepiece allows you to switch objectives.

Switch from the low power to the high power objective.

You should hear a click when the objective is in place

Switch back to low power.

It is always best to start your observations with the low power objective. It is easier to find the object because your field of vision is bigger. (You see more of the surface of the slide.)

Before using your microscope, it is always best to clean the lenses using lens paper.

F) Look for the large knob on the side of the microscope. This is the coarse adjustment wheel. This wheel is used to adjust the focus of the lenses. While looking from the side, turn the coarse adjustment wheel. Notice that it changes the distance between the stage and the objective.

31 As you move the coarse adjustment clockwise, the objective and stage move _____.

32 As you move the coarse adjustment counter-clockwise, the objective and stage move _____.

NEVER move the coarse adjustment clockwise while looking through the eyepiece. When looking at slides, this may cause the objective to hit the slide and damage the slide and the microscope.

G) Place the slide you are going to use on the stage of the microscope. Place the slide so that the object you wish to view is directly over the opening in the stage. Use the clips to secure the slide.

H) Looking from the side, turn the coarse adjustment clockwise until the low power objective is as close to the slide as you can get it without touching.

I) While looking through the eyepiece, turn the coarse adjustment wheel in a counter-clockwise direction slowly until the object comes into focus. If you turn the wheel too quickly and pass the object, look from the side again and bring the objective and slide closer together. Repeat this step until you have the object in focus.

J) Use the fine adjustment wheel to bring your object (specimen) into a finer (more clear) focus. Fine means slight or small. Make a detailed, labeled diagram of what you observe. Label the diagram low power and include the magnification you are using.

It is important to note how things are positioned under a compound microscope. While looking through the eyepiece, move the slide to the right.

33 Which direction does the object appear to move? _____

Move the slide in a downward direction.

34 Which direction does the object appear to move?_____

Objects under a compound microscope appear upside down and backwards. Remember this when moving the slide around to find objects.

K) Now you will switch to the high power objective.

Before switching, it is important to understand what will happen to your field of vision. The high power objective takes the center of the low power objective field and magnifies it. Therefore, you will see less of what is on your slide. It is important to center what you want to observe before switching to high power.

 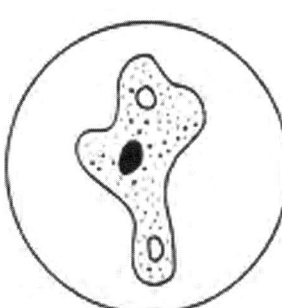

Low Power   **High Power**

35 Low power objective = more area and _____ magnification.

36 High power objective = _____ area and _____ magnification.

Turn the nosepiece until the high power objective clicks into place. If your object (specimen) is focused under low power, it will usually be focused when you switch to high power. Only use your fine adjustment wheel when observing under high power. If you use the coarse adjustment, you may drive the objective into the slide, damaging both. While using high power, you may need more light. Adjust the diaphragm if needed.

L) Focusing the microscope actually changes the depth or level of the specimen you are observing. You may be observing things from above the coverslip to down within the slide itself. Think of what you are observing as a ham and cheese sandwich. You can examine the upper slice of bread, the ham, the cheese, or the lower slice of bread depending on how you focus. This is especially evident when you are using high power objectives.

While looking at your specimen under high power, turn the fine adjustment wheel slowly back and forth. Observe the different layers of your specimen.

37 Make a detailed, labeled diagram of what you observe. Label the diagram high power and include the magnification you are using. Magnification. _____

DETERMINING THE SIZE OF A MICROSCOPIC OBJECT USING A COMPOUND MICROSCOPE

How large are the objects (specimens) viewed under the microscope? An easy way to determine this is by using a transparent plastic ruler.

Steps:

A) Place a transparent metric ruler on the stage of the microscope with its marked edge centered in the opening of the stage.

B) Using the low power objective, focus in on the marked edge. Measure the diameter of your field of vision (the white circle you see) in millimeters. Estimate to the nearest 0.001 of a millimeter (micron, μ or micrometer, mμ).

38 The diameter of my field under low power is _____mm.

Convert the diameter into microns. 1 mm = 1,000 μ

39 The diameter of my field under low power is _____ μ.

C) Switch to your high power objective. Measure the diameter of your field of vision. Estimate to the nearest tenth of a millimeter.

40 The diameter of my field under high power is _____mm.

41 The diameter of my field under high power is _____μ.

D) Once you have estimated the diameter of your field of vision you can estimate the size of objects (specimens) in the field. For example, a microscopic organism is observed to be as long as one half of the low power field. The low power field is estimated to have a size of 1500 μ. The length of the microorganism would be estimated to be 750 μ.

42 Estimate the size of the following organisms. First, estimate what fraction of the field the specimen covers. If it covers 1/5 of the field, then: 1/5 x 1500 μ = 300 μ (size - low power). If this was done under high power, then the object takes up 4/5 of the field of vision. Its size is then 4/5 x 375 μ = 300 μ (same as low power).

Low power	High Power
Field diameter 1500 μ	Field diameter 375 μ

43 Organism A is .2 under Low Power. _____   Organism A is .8 under High power. _____

PREPARING A WET MOUNT SLIDE

Objects are placed on a glass slide before being observed under a microscope. In order to make the object (specimen) easier to observe, it is placed in water between the slide and a coverslip. Make a **wet mount** in the follow way:

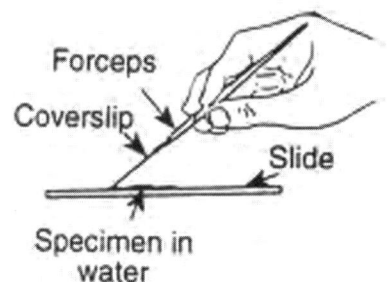

Steps:
A) Place a drop of water on the center of the slide.

B) Place the object (specimen) to be viewed in the water. Add another drop of water onto the object.

C) Holding the edges of the coverslip, tilt the coverslip so that it touches the water on the slide only along its lower edge. Gently drop the coverslip onto the object. (This helps reduce the making of air bubbles which cause problems with observations.)

44 Why is it necessary to place a coverslip onto the object on the slide? _____

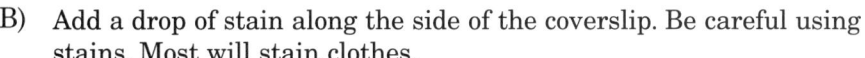

STAINING TECHNIQUES

In order to make objects (specimens) easier to see when using a compound microscope, they may be stained. **Staining** makes an object (specimen) darker by adding a dark chemical to it.

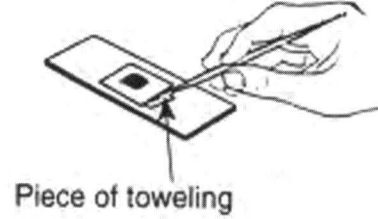

Steps:
A) Prepare a wet mount of the object (specimen) to be observed.

B) Add a drop of stain along the side of the coverslip. Be careful using stains. Most will stain clothes.

C) Place a piece of tissue paper at the opposite end of the coverslip from where the stain was added. Be sure it touches the water of the wet mount. The stain will be drawn across the slide into the object.

D) Be sure to follow your teacher's directions on how to clean up the extra stain.

45 What is the reason you use stain on some specimens? _____

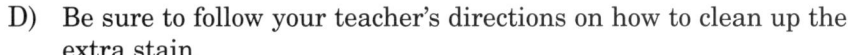

46 Place a drop of stain at one edge of the slide and coverslip. Next, place a piece of tissue paper at

the opposite side of the coverslip. Explain what happens with the stain and the specimen. _____

_____

_____

FIELD MAPS

Field maps describe the physical characteristics of a region by assigning values to all points in the region. When points of equal field value are connected, **isolines** are created. If the field value measures temperature, the lines are called **isotherms**. If the values connect points of equal atmospheric pressure, the lines are called **isobars**. **Topographic maps** are created by connecting points of equal elevation on a map. These lines are called **contour lines**.

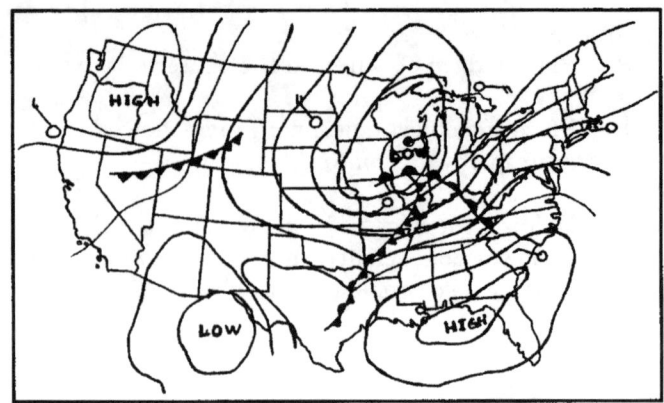

The rate of change between field values is called its gradient. The **gradient** on any field map can be determined with the following formula:

$$\text{gradient} = \frac{\textbf{Change in the Field Value}}{\textbf{Change in the Distance}}$$

In order to determine the gradient between point C and D, the distance between them must be determined. Using the scale, the distance is equal to 4 km.

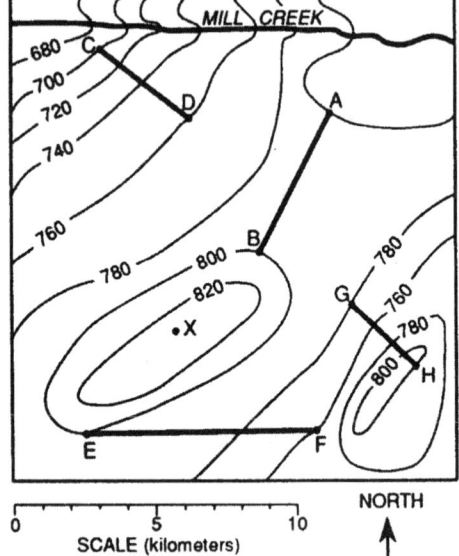

47 Now use the formula above to determine the gradient between those two points. Show your work in the space provided below.

Gradient = _____

MEASURING ALTITUDE

The angular elevation of an object measured from Earth is called its **altitude**. It is equal to a celestial body's angular distance above the **horizon**. An object on the horizon has an altitude of 0°, while an object directly above an observer (at the zenith) has an altitude of 90°. Using instruments that measure angles, altitude can be measured such as the altitude of the airplane, a hot air balloon, or the Moon.

A type of protractor called an **astrolabe** can be used to measure the altitude of an object. A weight is suspended from the protractor. One edge of the protractor is aligned to the object. The angle through which the weight passes is equal to the altitude of the object.

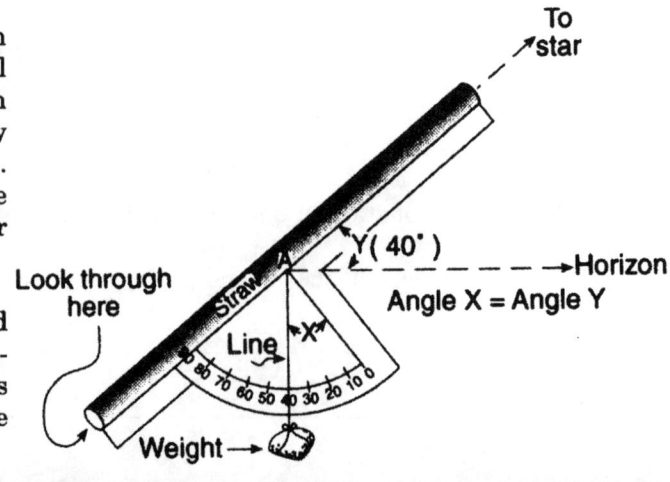

48 On a bright Moonlit night, you observe a Full Moon. It appears to be about halfway between the

horizon and directly over your head (**zenith**). What is the Moon's approximate altitude? _____

THE COORDINATE SYSTEM OF LATITUDE & LONGITUDE

LONGITUDE

Lines that run north / south on the globe and merge at the poles measure longitude. The 0° meridian passes through Greenwich, England and is called the **Prime Meridian**, while the 180° meridian is the approximate location of the **International Date Line**. Since Earth is basically round, the higher the latitude the shorter the distance between two longitudes. In order to determine your longitude, you must know two noon times, local noon and the time of local noon in Greenwich, England. Since Earth rotates west to east at the rate of 15° every hour, the longitude of an area is 15 x the difference in time between Greenwich noon time and local noon time. Therefore, since New York has a five hour difference from England, 5 x 15° places New York at the 75° longitude West.

LATITUDE

Lines that run east/west and are parallel to the equator are called lines of **latitude**. They form circles also called **parallels** around Earth. The **equator** has a latitude of 0° and forms the a **great circle** around Earth. The degrees of altitude of Polaris is the same as the latitude of the observer in the Northern Hemisphere.

USING THE COORDINATE SYSTEM

Any point on Earth can be determined by using the **coordinate system** of latitude and longitude. For example, Washington, DC is located at latitude 39° N and longitude 77°W.

49 Using the coordinate system on the globe at the right, explain how the longitude of a place can be determined using noon times.

_____

_____

_____

_____

_____

_____

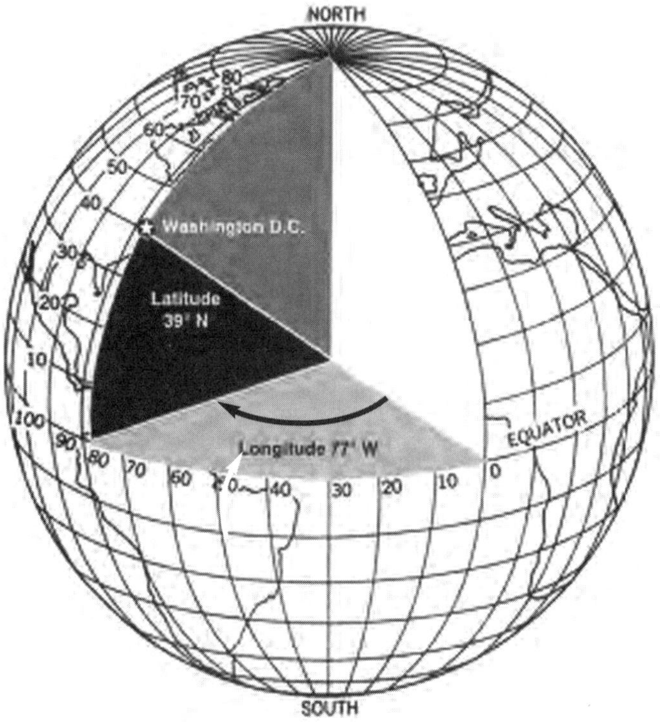

Directions: Circle the best answer in each of the following questions.

1 From what direction does the light used to view a specimen with a compound microscope come?
(1) above the specimen
(2) below the specimen
(3) the side of the specimen
(4) the stage

2 What procedure should NOT be used while viewing a specimen with a compound microscope?
(1) raising the coarse adjustment
(2) lowering the coarse adjustment
(3) raising the fine adjustment
(4) lowering the fine adjustment

3 What is the magnification if the eyepiece is 10x and the objective is 40x?
(1) 10   (2) 40   (3) 50   (4) 400

Use the field of vision diagram at the right to answer questions 4 and 5.

4 What is the diameter of the field of vision?
(1) 1.5 mm   (3) 1.5 cm
(2) 150 mm   (4) 150 cm

5 What is the length of the organism in the diagram?
(1) 0.5 mm
(2) 0.5 cm
(3) 5.0 mm
(4) 5.0 cm

0 mm   1mm

6 When making a wet mount slide, why do you touch the coverslip to the water before dropping it?
(1) to trap the specimen
(2) so that it does not fall off
(3) to increase the air bubbles
(4) to reduce the air bubbles

7 When staining a specimen, why is a piece of paper towel placed at the edge of the coverslip?
(1) it draws the stain across
(2) it cleans up the slide
(3) it removes the extra stain
(4) to position the coverslip

Use the diagram and the key to answer questions 8 and 9.

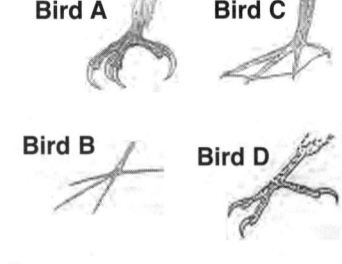

Bird A   Bird C

Bird B   Bird D

A Key To Identifying Birds		
Couplet	**Description**	
1	A	Toes webbed 2
1	B	Toes not webbed 3
2	A	All 4 toes webbed together . . . . .Cormorant
2	B	Three toes webbed together .Duck
3	A	Claws curved 4
3	B	Claws not curved . . . . .Jacana
4	A	Claws large and heavy . . . . .Eagle
4	B	Claws small . . . . .Kingfisher

8 Bird A is a
(1) cormorant   (2) duck   (3) eagle   (4) jacana

9 The characteristic used to tell eagles and jacanas apart is
(1) curve of the claw
(2) number of toes webbed
(3) size of the claw
(4) webbing

10 What type of blood vessel is used to count pulse rates?
   (1) arteries   (2) capillaries   (3) ducts   (4) veins

11 A student took her pulse for one minute four times and got the following results; first trial 80, second trial 80, third trial 79, and fourth trial 85. What is the student's average pulse rate per minute?
   (1) 80   (2) 324   (3) 81   (4) 20.25

12 How many grams make up a kilogram?
   (1) 10   (2) 100   (3) 1000   (4) 10,000

13 The temperature at which water freezes on the Celsius scale is
   (1) 0°C   (2) 32°C   (3) 100°C   (4) 212°C

14 A metric unit of volume is a
   (1) gram   (2) liter   (3) meter   (4) cubic foot

15 A millimeter is equal to
   (1) 0.1m   (2) 0.01 m   (3) 0.001 m   (4) 0.0001 m

16 What is the average speed of a car that travels 1500 km in 12 hours. The formula for speed is: Speed = d/t.
   (1) 105 km/hr   (2) 110 km/hr   (3) 125 km/hr   (4) 130 km/hr

17 Using the information shown on the thermometers at the right, what temperature would represent an average fall day in New York State?
   (1) –5°C
   (2) 0°C
   (3) 15°C
   (4) 30°C

Directions for question 18. The formula for converting Fahrenheit temperature to Celsius is:

$$°C = 5/9 \times (°F - 32)$$

18 If the temperature is 68°F, what would the temperature be in Celsius?
   (1) 5°C
   (2) 10°C
   (3) 15°C
   (4) 20°C

19 Water boils at 212° F. Using the conversion temperature scale at the right, what is the boiling point of water in degrees Celsius?
   (1) 32°C
   (2) 100°C
   (3) 212°C
   (4) 372°C

20 What is the volume of the water in the graduated cylinder at the right?
   (1) 5 mL
   (2) 7 mL
   (3) 8mL
   (4) 10 mL

21 How many cm long is the snail in the diagram at the right?
   (1) 2.5 cm
   (2) 3.0cm
   (3) 20 cm
   (4) 25 cm

22 What is the volume of the pebbles in the diagram at the right?
   (1) 15 cm³
   (2) 14 cm³
   (3) 5 cm³
   (4) 4 cm³

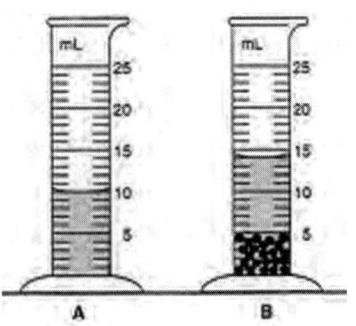

23 What is the volume of the wooden block shown below?
   (1) 35 cm³
   (2) 205 cm³
   (3) 1000 cm³
   (4) 2050 cm³

24 According to the triple beam balance at the right, what is the mass of the object?
   (1) 209 g
   (2) 219 g
   (3) 409 g
   (4) 419 g

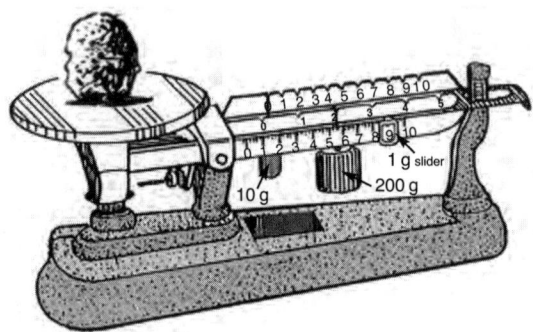

25 What direction is represented by the compass?
   (1) NW
   (2) N
   (3) W
   (4) SE

26 What is the spring balance shown at the right used to measure?
   (1) temperature   (3) density
   (2) force   (4) direction

27 What is the approximate height of the highest point of the plant the diagram at right?

(1) 7 cm
(2) 6.5 cm
(3) 5.6 cm
(4) 4.5 cm

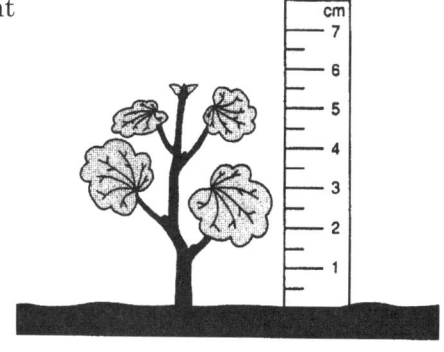

28 The diagram at the right shows a student using a pencil to submerge a beaker in water. The reason the student is pushing the beaker down is because the beaker

(1) has the same volume as the water
(2) is denser than the water
(3) is less dense than the water
(4) has no volume unless submerged in water

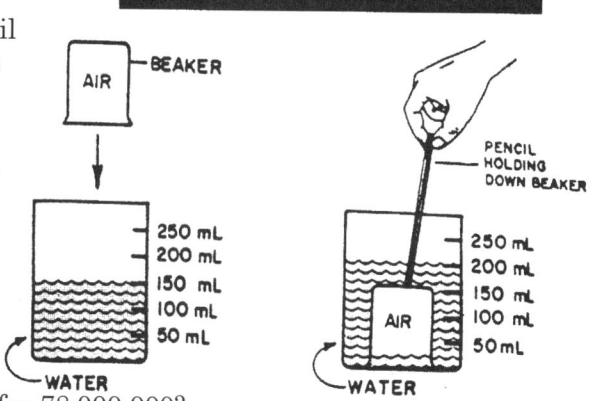

29 Which of the following would be the scientific notation for 78,000,000?

(1) 7.8×10^6 (3) 7.8×10^5
(2) 78×10^6 (4) 78×10^5

30 The topographic map at the right shows contour lines. Which letter represents the highest elevation?

(1) F
(2) B
(3) C
(4) D

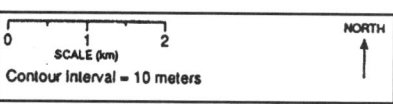

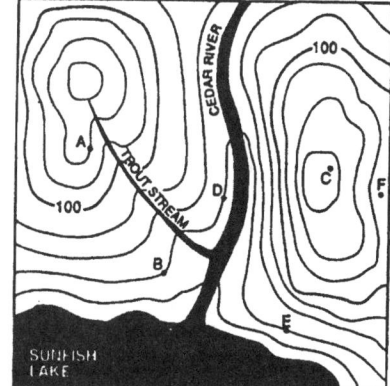

31 Using the map of the world at the right, what letter represents a location that is at latitude 60° N and longitude 100°E?

(1) A
(2) B
(3) C
(4) D

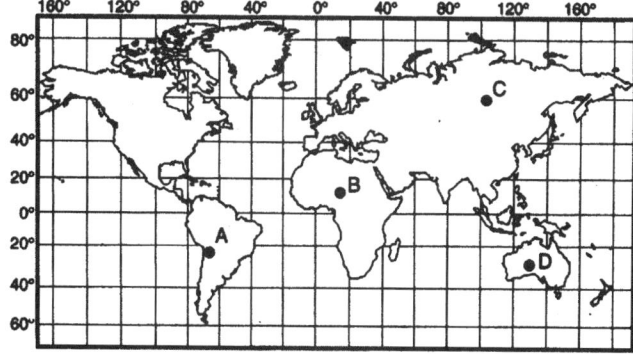

32 The illustration at the right is part of a procedure used with a microscope and object study. The piece of toweling is being used to

(1) support the cover slip
(2) keep the cover slip away from the object
(3) absorb extra fluids from the object
(4) pull the stain across the object

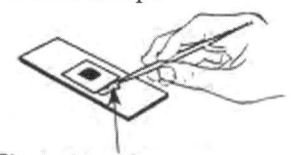

Piece of toweling

THE QUESTIONS THAT FOLLOW TEST YOUR UNDERSTANDING OF SCIENTIFIC PRACTICES, CORES IDEAS IN SCIENCE AND CROSS-CUTTING CONCEPTS SUCH AS PATTERNS AND CAUSE AND EFFECT RELATIONSHIPS.

SHORT ANSWER

CHART OF THE PLANETS

	DISTANCE FROM SUN (MILLIONS OF KM)	PERIOD OF ROTATION	PERIOD OF REVOLUTION	NUMBER OF SATELLITES
Mercury	58	59 days	88 days	0
Venus	108	243 days	225 days	0
Earth	150	23 hours 56 minutes 4 seconds	365 1/4 days	1
Mars	228	24 hours 37 minutes 23 seconds	687 days	2
Jupiter	778	9 hours 50 minutes 30 seconds	11.9 years	11 (4 rings)
Saturn	1,427	10 hours 14 minutes	29 1/2 years	15 (1000 rings)
Uranus	2,869	10 hours 49 minutes	84 years	5
Neptune	4,496	16 hours	165 years	2

1 Use the Chart of The Planets above to answer the following questions.

a State a general relationship between a planet's period of revolution and distance from the Sun? (1 point)

_____

b Which planet's revolution breaks this rule? (2 points) _____

Give a possible reason.

_____

_____

INTERMEDIATE SCIENCE- 3D LEARNING & ASSESSMENTS Topical Review Book Company ©

2 Based on how they obtain energy, develop a dichotomous key to distinguish the following groups. (2 points)
   Carnivores, Herbivores, Omnivores, Producers (The first few couplets are provided. Use more if needed.)

   1A _____

   1B _____

   2A _____

   2B _____

   3A _____

   3B _____

3 Use the diagram of the pH scale at the right
   to answer the questions that follow.

   a Which substance(s) would be considered a strong base? Explain. (1 point) _____

   _____

   _____

   b Phenolphthalein turns pink in the presence of a base. Which substance(s) will turn pink? Explain. (1)

   _____

   _____

4 Use the diagram of the weather map
   of the United States to answer the
   following questions.

   Schenectady Gazette, January 15, 2000

   a What type of front is moving
      out of Canada? (1)

      _____

   b In which direction is the jet
      stream moving? (1)

      _____

   c How would you characterize the temperature of air the jet stream is carrying and describe one factor
      that affects the temperature of the air carried by the jet stream? (2 points)

   _____

   _____

   _____

INTERMEDIATE-LEVEL SCIENCE TEST

PART A: PRACTICE TEST ONE - QUESTIONS 1-44

Directions (1-44): Each question is followed by four choices. Decide which choice is the best answer.

1 What is the basic unit of structure and function in plants and animals?
   (1) cell   (2) tissue   (3) organ   (4) system

2 Photosynthesis is the process by which
   (1) chemical energy is converted into light energy   (3) light energy is converted into nuclear energy
   (2) light energy is converted into chemical energy   (4) nuclear energy is converted into light energy

3 When most foods are digested completely they can
   (1) easily pass through cell membranes   (3) be considered to be gasses
   (2) contain long chains of fats   (4) be found in the stomach

4 Which system transports hormones and nutrients?
   (1) circulatory   (2) digestive   (3) immune   (4) respiratory

5 What is the process of producing genetically identical plants from the cells of a single plant called?
   (1) cloning   (2) fission   (3) graphing   (4) sexual reproduction

6 Which cell process is illustrated at the right?
   (1) sexual reproduction in a plant cell
   (2) asexual reproduction in a plant cell
   (3) sexual reproduction in an animal cell
   (4) asexual reproduction in an animal cell

7 In the process of evolution, the effect of the environment is to
   (1) prevent mutations   (3) provide conditions so that fossils form
   (2) select for variations   (4) provide stable conditions

8 Skeletal similarities between two different types of animals are probably due to the fact that both
   (1) live in the same environment   (3) are related to a common ancestor
   (2) perform the same function   (4) have survived until the present time

9 Asexual reproduction differs from sexual reproduction in that, in asexual reproduction
   (1) new organisms are usually genetically identical to the parent
   (2) reproduction includes the formation of sex cells
   (3) two cells join together
   (4) offspring show great genetic differences

10 A plant with 12 chromosomes undergoes normal cell division. What is the total number of chromosomes in each of the resulting daughter cells?
   (1) 4   (2) 8   (3) 12   (4) 24

11 What type of organism is not shown in the following food chain?

Grass ⟶ Mouse ⟶ Snake ⟶ Hawk

   (1) producer   (2) decomposer   (3) carnivore   (4) herbivore

Base your answers to questions 12 through 14 on the the diagram diagram on the right.

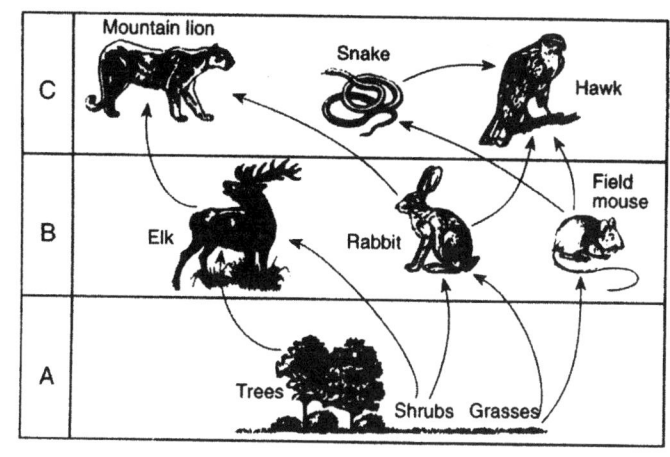

12 Which two organisms are classified as producers?
(1) elk and rabbit
(2) shrubs and grasses
(3) grass and field mouse
(4) rabbit and shrubs

13 Carnivores are represented by the
(1) mountain lion and hawk
(2) snake and rabbit
(3) rabbit and field mouse
(4) trees and grasses

14 Consumers are represented by the
(1) trees and snake   (2) trees and rabbit   (3) mountain lion and elk   (4) shrubs and trees

15 In a particular area, the living organisms and nonliving environment function together as
(1) a population   (2) a community   (3) an ecosystem   (4) a species

16 The diagram at the right represents a wet mount of a specimen seen under a microscope. If the specimen swims toward the left side of the slide, which diagram represents the image that will be observed through the microscope?

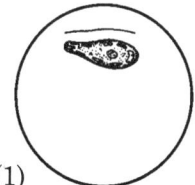
(1)

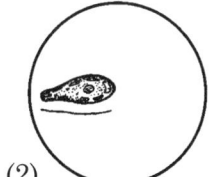
(2)

(3)

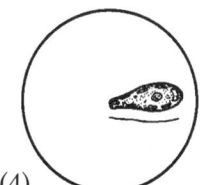
(4)

17 Which part of the microscope should be used with the low-power objective, but not the high power objective?
(1) coarse adjustment   (2) fine adjustment   (3) diaphragm   (4) mirror

18 According to the diagram at the right, the if a student was looking through the high power lens of this microscope with an eyepiece that was 10x, the total magnification would be
(1) 10x
(2) 40x
(3) 100x
(4) 400x

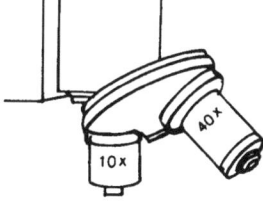

19 A microscope has four lenses labeled 4x, 10x, 43x, 97x. Which lens would provide for the largest field of vision?
(1) 4x   (2) 10x   (3) 43x   (4) 97x

20 In pea plants, the trait for smooth seeds is dominant over the trait for wrinkled seeds. When two hybrids are crossed, which results are most probable?
(1) 75% smooth and 25% wrinkled seeds   (3) 50% smooth and 50% wrinkled seeds
(2) 100% smooth seeds   (4) 100% wrinkled seeds

21 Which of the following graduated cylinders shows a volume of 16 mL ?

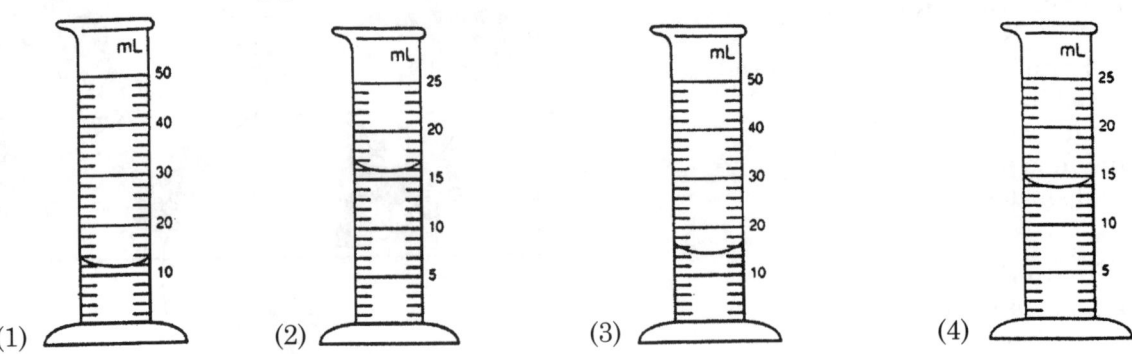

(1)   (2)   (3)   (4)

22 What is the length of the nail in the diagram shown at the right?
   (1) 24cm
   (2) 24mm
   (3) 34cm
   (4) 34mm

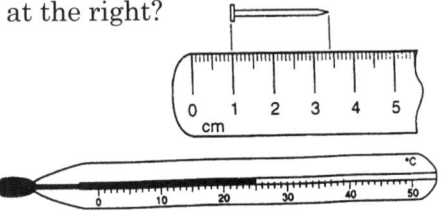

23 What is the temperature of the thermometer shown?
   (1) 20.5°C   (3) 26°C
   (2) 25°C   (4) 26.5°C

24 Using the diagram below, in which picture will the sugar dissolve the fastest?

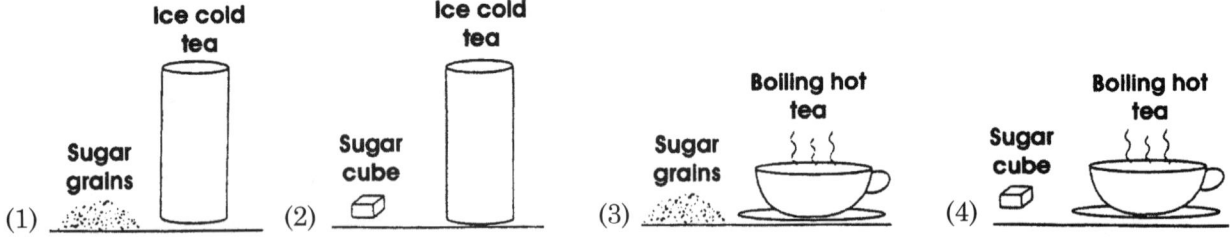

(1)   (2)   (3)   (4)

25 Examine the diagram of the ball and ring. Which of the following statements best describes what is happening.
   (1) The ring expanded.
   (2) The ball expanded.
   (3) The ring contracted.
   (4) The ball contracted.

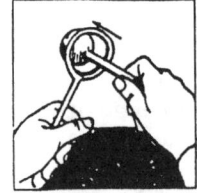  

Before heating metal ball   Heating metal ball   After heating metal ball

26 What information should be placed above the arrow in the diagram to the right?
   (1) condensation Freezing
   (2) evaporation
   (3) heating
   (4) melting

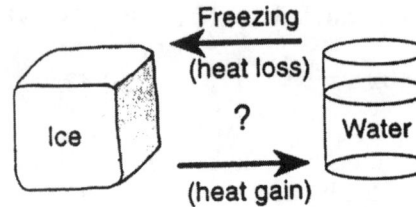

27 Examine the diagram at the right.
Which statement is true about objects A and B?
(1) Object A has a greater mass than object B.
(2) Object B has a greater mass than object A.
(3) Object A and B have the same mass.
(4) Object A and B have no mass.

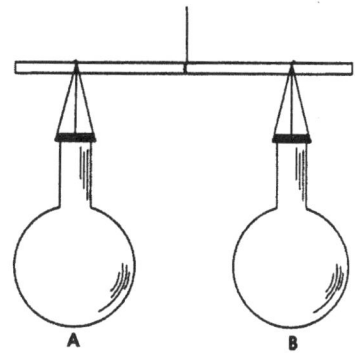

28 The diagram at the right shows a picture of a lever. Which
statement best describes the advantage of using this lever?

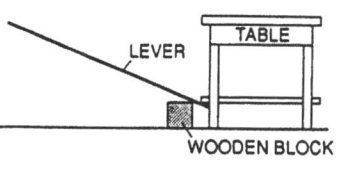

(1) The lever multiplies the effort force, but does not change
the direction of the applied force.
(2) The lever multiplies the effort force and changes the
direction of the applied force.
(3) The lever reduces the effort force, but does not change the direction of the applied force.
(4) The lever reduces the effort forces and changes the direction of the applied force.

29 Examine the experiment with antibiotics and
bacteria. What would be the dependent and
independent variables in this experiment?
(1) The dependent variable is the amount of
antibiotic, while the independent variable
is the bacterial growth.
(2) The dependent variable is the bacterial
growth, while the independent variable
is the amount of antibiotic used.
(3) The dependent variable is the cotton plug
while the independent variable is the
amount of antibiotic.
(4) The dependent variable is the amount of time it takes to kill the bacteria, while the independent
variable is the cotton plug.

30 Which of the following circuits will light the bulb?

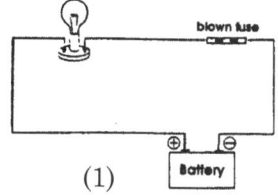   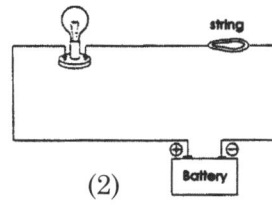   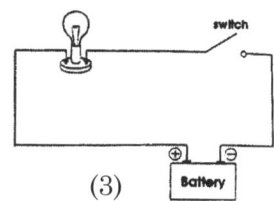   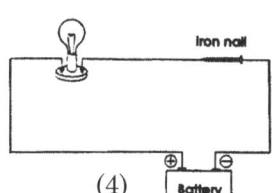

(1)   (2)   (3)   (4)

31 What type of energy conversion is represented in the diagram at the right?
(1) light to electric
(2) heat to light
(3) chemical to light
(4) light to chemical

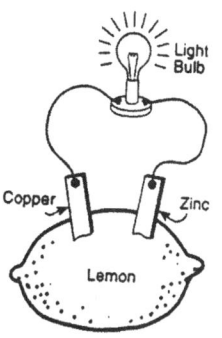

32 Which of the diagrams below best illustrates the transfer of heat in the enclosed box at the right?

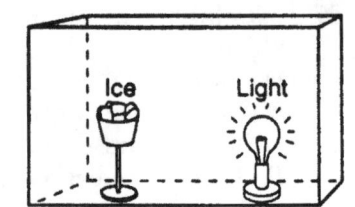

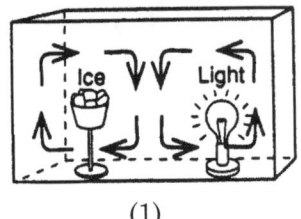

(1)

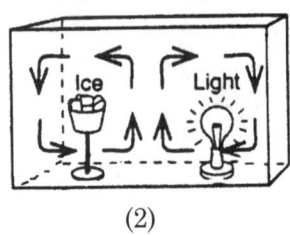

(2)

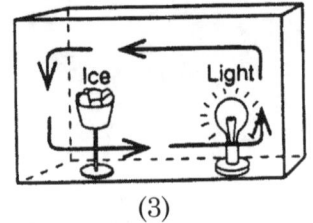

(3)

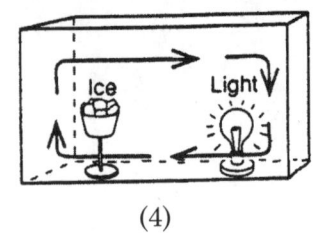

(4)

33 The diagram at the right shows a type of energy called
(1) solar
(2) hydroelectric
(3) geothermal
(4) nuclear

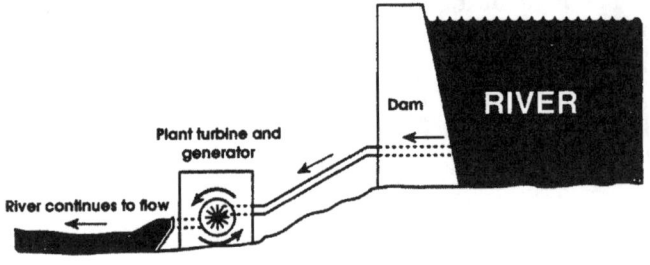

34 Examine the diagram of the basketball and hoop. Which letter represents the greatest potential energy for the basketball?
(1) A
(2) B
(3) C
(4) D

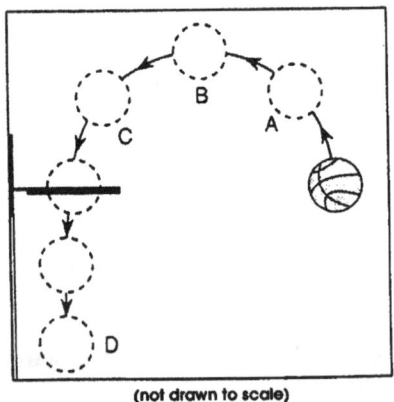

(not drawn to scale)

35 The diagrams below represent three phases of matter. Energy transfer by convection takes place best in phase(s).
(1) A and B
(2) A and C
(3) B and C
(4) A only

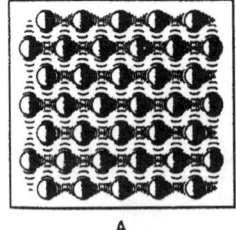

A

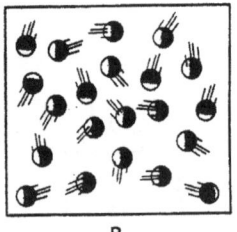

B

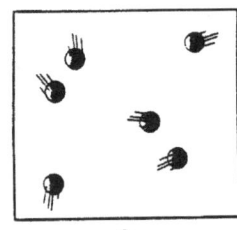

C

36 The formation of the San Andreas fault shown in the diagram at the right was due to
(1) moving water
(2) weathering rocks
(3) moving plates
(4) volcanic activity

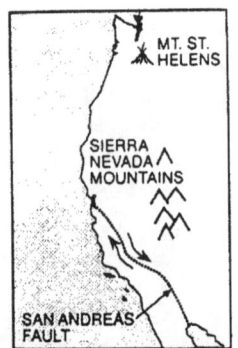

37 According to the map at the right showing the incidence of confirmed rabies cases in New York State, which animal is most responsible for the reported cases?
   (1) gray fox
   (2) raccoon
   (3) cat
   (4) bat

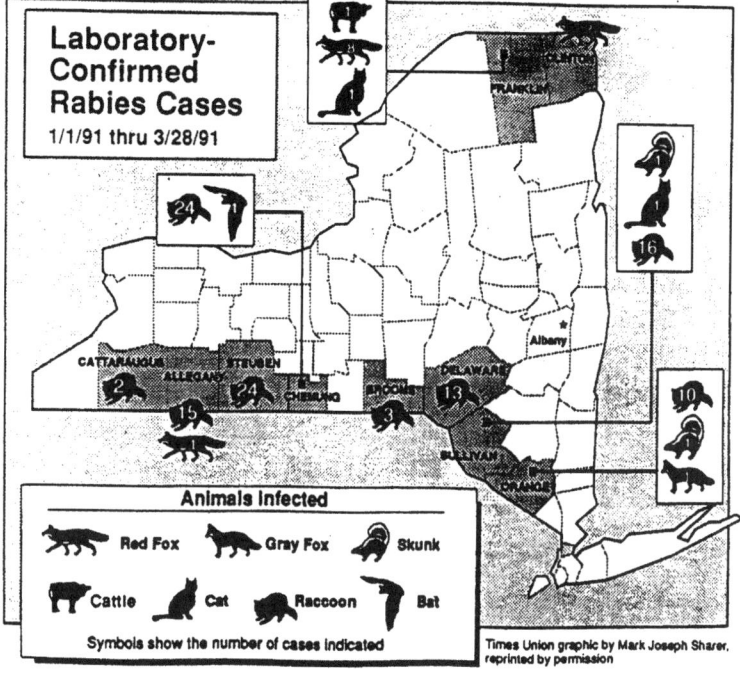

Laboratory-Confirmed Rabies Cases
1/1/91 thru 3/28/91

Animals Infected

Red Fox   Gray Fox   Skunk
Cattle   Cat   Raccoon   Bat

Symbols show the number of cases indicated

Times Union graphic by Mark Joseph Sharer, reprinted by permission

38 Examine the isobars in the map on the right. What is the barometric pressure at point *B*?
   (1) 1008 mb
   (2) 1012 mb
   (3) 1016 mb
   (4) 1020 mb

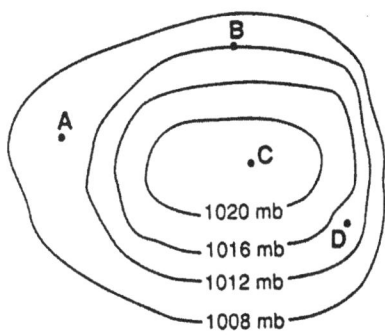

1020 mb
1016 mb
1012 mb
1008 mb

39 In the diagram below, which point is located on a weather front?
   (1) *A*
   (2) *B*
   (3) *C*
   (4) *D*

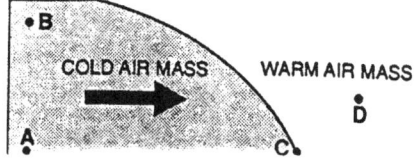

COLD AIR MASS   WARM AIR MASS

40 According to the weather map on the right, what type of weather is approaching Atlanta?
   (1) cold and dry
   (2) cold and moist
   (3) warm and dry
   (4) warm and moist

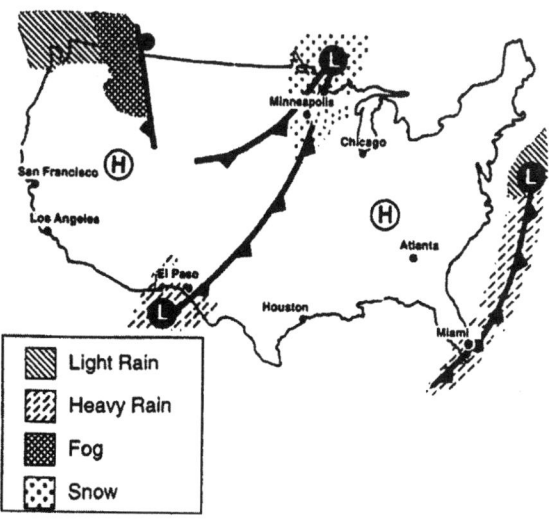

Minneapolis
Chicago
San Francisco
Los Angeles
El Paso
Houston
Atlanta
Miami

Light Rain
Heavy Rain
Fog
Snow

41 According to the map at the right, which air mass is characterized by dry, cold air?
(1) *A*
(2) *B*
(3) *C*
(4) *D*

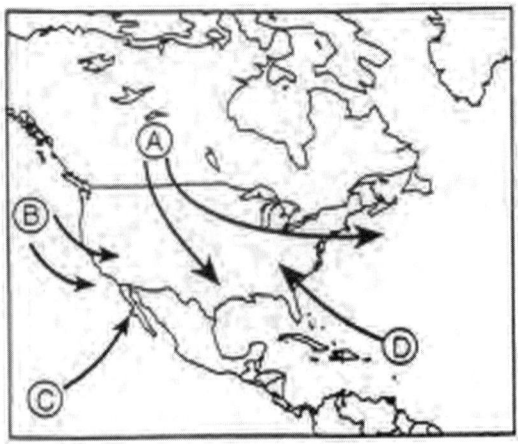

42 Based on their location in the sedimentary rock diagrammed at the right, which organisms are the oldest form of life shown in the rock?
(1) early horses
(2) dinosaurs
(3) armored fish
(4) trilobites

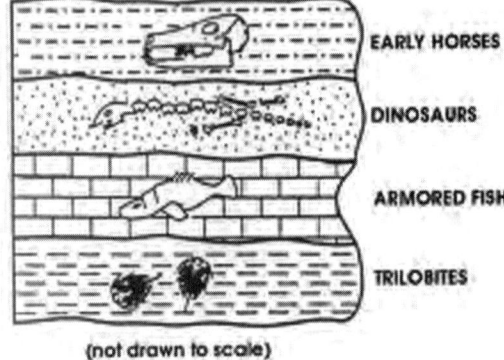

(not drawn to scale)

43 According to the diagram at the right, the northern hemisphere is experiencing
(1) summer
(2) fall
(3) winter
(4) spring

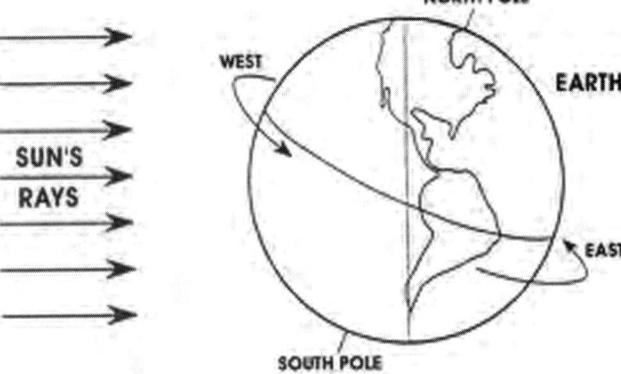

44 In the diagram at the right of Earth and the Moon orbiting it, in which position would there be a full Moon?
(1) *A*
(2) *B*
(3) *C*
(4) *D*

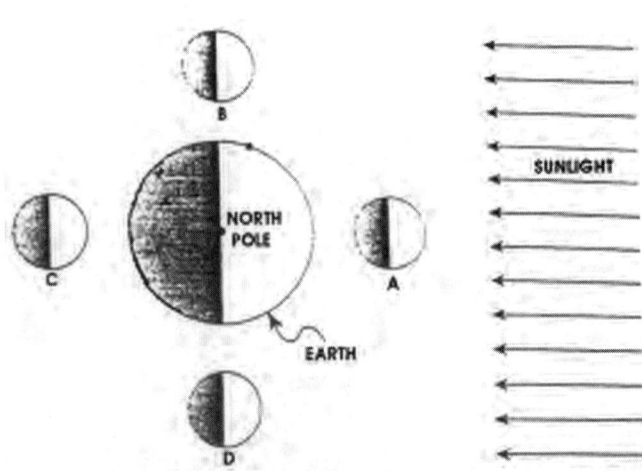

PART B - SHORT ANSWER

Answer each question as thoroughly as you can.

1   Base your answers on the diagram at the right that illustrates the transport of oxygen and carbon dioxide.

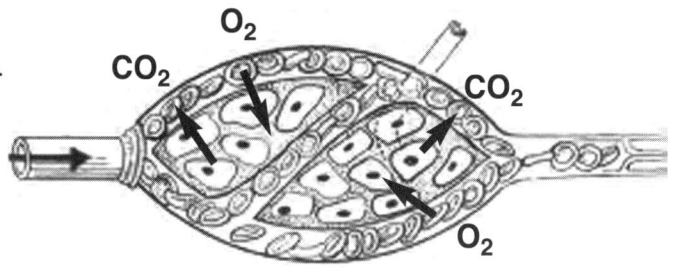

 a   Why must this structure be located in the

lungs? [1] _____

_____

_____

 b   Why must the capillaries be so thin for this process to occur? [1] _____

_____

2   Base your answers to the following questions on the data table at the right.

 a   What interpretations about the deer population can be made from the

general pattern of the entire 40 years? [1] _____

_____

 c   What reason can you give that could have caused both the increase and

the decrease in deer population over the years? [1] _____

_____

_____

Data Table

Year	Deer Population (thousands)
1900	3.0
1910	9.5
1920	65.0
1924	100.0
1926	40.0
1930	25.0
1940	10.0

3   The diagram below shows an experiment with a water plant.

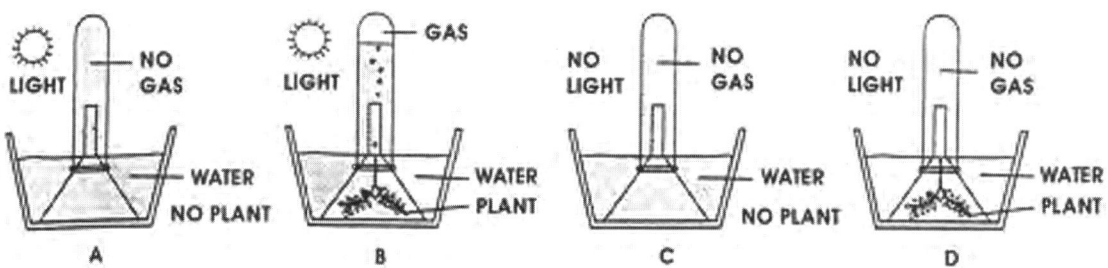

 a   What is the purpose of this experiment? [1] _____

_____

 b   What is the purpose of set up A & B? [1] _____

_____

 c   What is the independent variable in this experiment? [1] _____

_____

4   The diagram at the right, shows internal and Ce) external forces that have shaped Earth.

Choose any two forces and explain how they have shaped the landscape of Earth. [2]

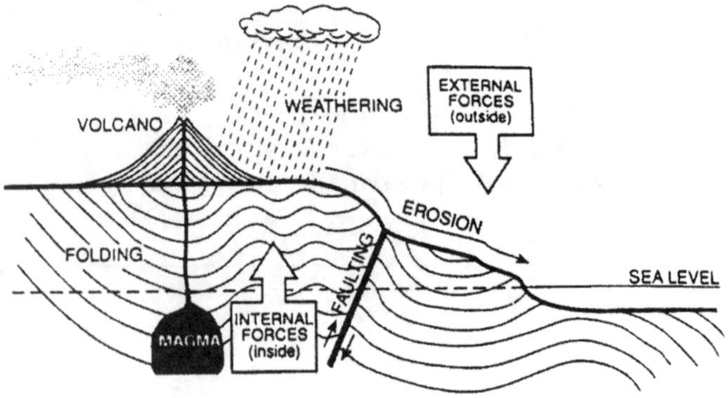

_____

_____

_____

_____

_____

5   The diagram at the right shows two circus performers.

a   What is the name of the simple machine on to which one circus performer jumped? [1]

_____

_____

b   Apply your knowledge of the Laws of Motion to explain why the second performer was able to land on the elephant. [2]

_____

_____

6   Examine the diagram at the right of a tree growing on a hill. Sections of the roots have been labeled A and B.

What is the factor that caused some of the roots to be longer? Explain. [2]

_____

_____

_____

_____

7   Examine the diagrams at the right of a spring scale pulling against a nail which has been hammered into a block of wood.

What happened to the effort force after pencils were placed under the wood? Explain. [2]

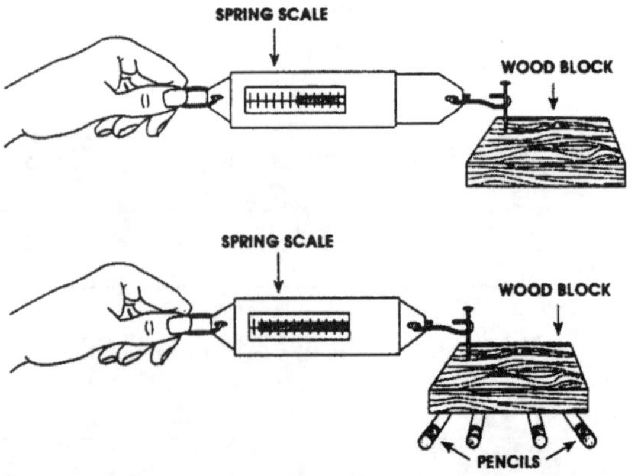

_____

_____

_____

_____

PART C - SHORT ESSAY

Read each question carefully and the answer the questions that follow.

1   An investigation was designed to determine the effect of temperature on the respiration of germinating seeds. The data table at the right shows the results of the investigation.

EFFECTS OF TEMPERATURE ON RESPIRATION OF GERMINATING SEEDS		
	TOTAL OXYGEN CONSUMPTION (ML)	
TEMPERATURE °C	BEADS	PEAS
5	0	9
10	0	12
15	0	15
20	0	18
25	0	20
30	0	23

a   Label each axis. [1]

b   Mark an appropriate scale on each axis. [1]

c   Plot the oxygen consumption for the peas. [1]

d   Describe the effect of temperature on respiration in germinating peas. [1]

_____

_____

_____

_____

e   What is a possible control for this experiment? [1]

_____

_____

_____

_____

2   Dylan and Hunter performed the experiment illustrated below for their teacher Mr. Norman.

Amount of Fertilizer Added Daily

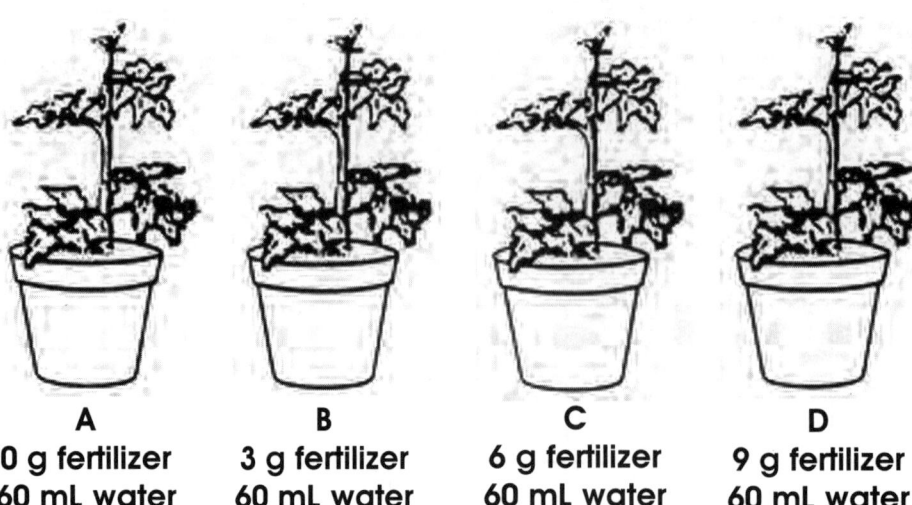

A	B	C	D
0 g fertilizer	**3 g fertilizer**	**6 g fertilizer**	**9 g fertilizer**
60 mL water	**60 mL water**	**60 mL water**	**60 mL water**

a   What is the purpose of their experiment? [1] _____

_____

_____

b   Describe two types of data Dylan and Hunter would need to collect before they can come to any

conclusion about their experiment. [2] _____

_____

_____

c   Describe two sources of error that can possibly affect the outcome of their experiment. [2] _____

_____

_____

_____

3 Examine the diagram of the falling ball.

   a Describe the ball's change in potential and kinetic energy as it falls. [1]

_____

_____

   b Describe how its potential energy changed on its second bounce, noting the percentage of change. [2]

_____

_____

_____

   c Explain why the ball's potential energy changed after its second bounce. [1] _____

_____

_____

4 Sarina, Alana, and Caitriona wanted to prove to their parents, Alan and Kathleen, that if soft rock music is played while taking a test, it has no effect on how a student will perform on the test. They asked their teacher if they could use their science class for their experiment.

   a Describe an experiment that they might do. Make sure to identify the control group and experimental group. [3]

_____

_____

_____

_____

_____

   b Design a data table for collecting their data. Make sure the data table identifies the dependent and independent variables. [2]

_____

_____

PERFORMANCE EXPECTATIONS CLUSTERS

LIFE SCIENCE

1) SYSTEMS AND SYSTEM MODELS: LS1-1

When you examine organisms you may notice that almost every part of it helps it to survive in some way. Choose one organism and fill in the chart of how 5 of its structures help it to survive.

Organism: _____

Structure	How it helps in survival

2) STRUCTURE AND FUNCTION: LS1-1

Engineers often take information from science to help them build things that are useful to humans. Think about how plants and animals protect themselves. Pick 3 traits of plants or animals and describe how we could copy their trait for us as humans for protection.

Living Thing	Trait	Human Copy
Example: Roses	thorns	Barbed wire

3) PATTERNS: LS3-1

Create a timeline for the life cycles of a marigold plant and a butterfly. They both have a lifespan of about one year. Include on the timeline: birth, adult, producing eggs, producing seeds, seeds, death, young plant, larva, and pupa.

Marigold

Time (0)_____(1 year)

a Examine the two life cycles. What do both the marigolds and butterfly have in common?

_____

_____

_____

b Examine the two life cycles. What are the differences in the life cycles of marigolds and butterflies?

_____

_____

_____

4) PATTERNS: LS1-2

Human babies cry when they need something from their parents. How does this type of behavior happen in other living things that care for their young. Pick 3 other types of living things and describe how their babies signal their parents that they need help.

5) ENERGY AND MATTER: LS1-1

The scientist, Van Helmont did an experiment on plant growth. He planted a tree in a pot and watered it regularly for 5 years. The following chart shows his results.

Starting Mass (kg)	Mass after 5 years (kg)
Plant = 2.3	Plant = 76.8
Soil = 90.9	Soil = 90.8

Besides the water, explain how the plant mass increased?

_____

_____

6) PATTERNS: LS1-1

Plants and animals are both living things. They have some needs in common. Fill in the chart below listing things they both need and things only one may need.

Things both plants and animals need	Things only plants need	Things only animals need

7) SYSTEMS AND SYSTEM MODELS: LS1-2

Living things interact with the world through their senses. Some information brought in by the senses travels through nerves to the brain. The brain responds to the information. Draw a diagram using everyday objects to represent the following: A sense taking in the signal, nerves, brain, and a way to respond. Use arrows to show the flow of the information.

8) STRUCTURE AND FUNCTION:
CONNECTIONS TO ENGINEERING, TECHNOLOGY, AND APPLICATIONS TO SCIENCE: LS1-1

Bats and dolphins can sense where objects are located by sending out bursts of sound waves which echo or bounce back to them. This sense is called echolocation. The ocean floor has been mapped by using echolocation. For example, underwater mountains and trenches have been found. Explain how echolocation can distinguish between a mountain and a trench.

_____

_____

_____

9) Systems and System Models: LS2-1
Structure and Function: LS2-2

Construct a food web diagram based on the plants and animals that live in your area. Include and label at least two of each of the following: plants, animals, and decomposers.

10) Cause and Effect: LS2-1

Use the chart below, showing how a plant grew with different amounts of light and water, to answer the following questions.

Week	Amount of Water (ml)	Amount of Light (hours)	Growth (cm)
1	5	2	10
2	5	4	20
3	10	2	20
4	10	4	40
5	20	2	20
6	20	4	40

a Explain how the amount of water affected the plant's growth. Include data from the chart to support your answer.

_____

_____

_____

b Explain how the amount of light affected the plant's growth. Include data from the chart to support your answer.

_____

_____

_____

11) SYSTEMS AND SYSTEM MODELS: LS2-1

Energy pyramids show that there is more usable energy at the beginning of a food chain. About 10% of the energy from one level is usable by the next level. Draw a model of a pyramid with four levels showing these amounts in proper scale. Place a tree, eagle, mouse and snake in each level. Explain your choices.

_____

_____

_____

12) SYSTEMS AND SYSTEM MODELS:
CONNECTIONS TO ENGINEERING, TECHNOLOGY, AND APPLICATIONS TO SCIENCE: LS4-4

A pesticide was sprayed in an area killing almost all of the grasshoppers.

_a_ Diagram a food web that includes the grasshoppers. Using your web describe which living things would increase and which living things would decrease due to the pesticide. Explain your answers.

_____

_____

_____

_b_ Farmers said that the area was sprayed because grasshoppers were eating the crops. Describe a different solution to the problem. (Hint: Use the food web)

_____

_____

13) Cause and Effect: LS2-1

a   What advantages does living in a group have for animals? Choose one of the following and list the advantages provided by living in a group.

a. Herd of cattle   **b. Swarm of bees**   **c. Flock of geese**

_____

_____

_____

b   What advantages does living in a society have for humans? Think about the community in which you live. How do humans help each other?

_____

_____

_____

14) Patterns: LS3-1

Each parent passes traits to their children. In some cases each parent gives one of the two codes needed for a trait. If T is the code for tall and t is the code for short how might each of these children of the following parents look.

Dad code T t	Mom code T t	
	Mom gives T	Mom gives t
Dad gives T	Child 1 ___ ___	Child 3 ___ ___
Dad gives t	Child 2 ___ ___	Child 4 ___ ___
	Offspring	Offspring

What percentage of the children are tall? _____%

15) Cause and Effect: LS3-1

People are living longer and longer in our modern world. What factors may be helping us to live longer?

_____

_____

_____

16) PATTERNS: LS3-1

Two sisters were talking. One said," You look like Mom." The other said, "You look like Dad, but we both look like Grandma." Explain how this might occur based on how traits are inherited.

_____

_____

_____

_____

17) SCALE, PROPORTION, AND QUANTITY: LS4-1

Examine the diagram of fossils of animals found in sedimentary rock. Describe the changes in environment over the years that might have favored the changes in the organisms found. Where are the oldest fossils found?

Top Layer 1	Mammals, reptiles
Layer 2	Reptiles
Layer 3	Fish, Amphibians
Bottom Layer 4	Fish

_____

_____

_____

_____

18) CAUSE AND EFFECT: LS4-2

In England there was a moth that came in two colors, white and brown. The moths often landed on white trees where the white moth was hard to see. Many more white moths compared to brown moths survived. During the Industrial Revolution in England many trees were darkened by the burning of coal.

Hypothesize what might happen to the two types of moths. Explain your hypothesis.

_____

_____

_____

_____

19) CAUSE AND EFFECT: LS4-3

The warming of the arctic has been attributed to climate change by over 97% of scientists. The arctic is a region made up mostly of ocean and sea ice. How do you think this will affect the organisms living there now?

a The polar bear lives in the arctic and uses sea ice to hunt, how will climate change impact this animal?

_____

_____

_____

b Read about others animals that live in the arctic and describes some adaptations they would have to make as the sea warms and there is less ice.

_____

_____

_____

20) INTERDEPENDENCE OF SCIENCE, ENGINEERING, AND TECHNOLOGY: SYSTEMS AND SYSTEM MODELS: LS4-4

The Albany Pine Bush in New York State is a fire-adapted community. This means that native plants and animals that live there have ways to reduce the impact of fire on their lives. For more that 50 years people have been putting out the natural fires in the Pine Bush. Because of this, the native animals are being driven out by the change and other species are moving in.

To help the native species, the Albany Pine Bush Preserve Commission conducts controlled burns to try to restore the original environment.

a Write an argument for why this might be a good solution to the problem.

_____

_____

_____

b Write an argument for why this might be a poor solution to the problem.

_____

_____

_____

PHYSICAL SCIENCE

1) SCALE, PROPORTION, AND QUANTITY: PS1-2

Alana heated 84 grams of substance A. As it was heated it separated into two substances B and C. Below are her results.

Before	Mass (g)	After	Mass (g)
Substance A	84	Substance B	61
		Substance C	23

Construct a bar graph that shows that the total mass of matter was conserved.

2) CAUSE AND EFFECT: PS1-1, PS1-3, PS1

You are given a sample of an unknown substance. It may be substance *A*, *B*, *C* or *D* as noted in the chart below.

Substance	Magnetic	Conducts electricity	Density g/cm³	Soluble in water
A	No	Yes	25	Yes
B	Yes	Yes	50	No
C	No	No	25	Yes
D	No	No	50	Yes

Based on the information provided in the chart above, describe a procedure you would follow to identify the unknown substance.

_____

_____

_____

_____

_____

3) SCALE, PROPORTION, AND QUANTITY: PS1-1, PS1-3

Atoms and molecules are too small to be seen with the unaided eye. Scientists develop models of them that explain their observations. Draw a diagram (model) of water molecules as a solid, liquid and gas. Explain your model.

_____

_____

_____

4) ENERGY AND MATTER: PS1-3, PS1A

To the right are different shapes that are attached to each other. Use this as a model of substance _A_.

a What do the different shapes represent?

_____

_____

b Is substance _A_, a compound or element? Explain. _____

_____

c Suppose substance _A_ is heated and two new substances _B_ and _C_ are formed. Draw a model to show this.

d Will the mass of substances _B_ and _C_ be equal to the number of shapes in substance _A_?

_____

5) Scientific knowledge assumes an order and consistency in natural systems: Scale, Proportions, and Quantity: PS1-4

You are asked to determine which of the following two experiments results in the formation of a new substance.

Experiment #1: Mix 100 mL of water with 20 g of sugar

Experiment #2: Mix 100 mL of vinegar with 20 g of baking soda

a Describe observations you would expect to see if a new substance had formed.

_____

_____

b Support your observations supporting no new substance(s) was formed.

_____

_____

c If you were to mass the products of each experiment after the substances were placed together, how would its total mass compare to the mass of the starting substances? Explain your answer.

_____

_____

6) Cause and Effect: PS1-4, PS1B

When matter is heated or cooled it changes in some way. These changes may be physical or chemical.

a Describe a type of matter that does not lose its chemical properties when heated to 100°C or cooled to 0°C. Explain how you would know.

_____

_____

b What happens to the particles of matter when heated and cooled?

_____

_____

c The density of most substances increases when it is cooled and decreases when it is heated. Ice floats on water. What happens to the density of water when it is cooled?

_____

_____

7) CAUSE AND EFFECT: PS2-1

Suppose you have three toy cars. Two cars pull with 5N of force. The third car pulls with 10N of force. Design an experiment to prove each of the following:

a When forces are balanced motion does not change.

b When forces are unbalanced motion changes.

8) PATTERNS: PS2-2

Marco dropped a ball from several heights and recorded how high up the ball bounced. Based on the chart of his results below, predict how high the ball will bounce if he drops it from 30 cm. Explain.

Trial	Height of drop (cm)	Height of bounce (cm)
1	50	10
2	20	4
3	10	2
4	70	14
5	40	8

9) CAUSE AND EFFECT: PS2-1

a Lincoln said, "Objects always fall down and heavier objects will hit the ground at the same time as lighter objects." Use your knowledge about gravity to write an argument supporting his idea. Include a diagram that shows Earth and the objects that are falling. Use arrows to represent the force of gravity and other forces operating on the objects.

b Lincoln's friend Anna repeated Lincoln's experiment but dropped a ball and a piece of paper. Use a model to explain what will happen. Include arrows to show the forces affecting the experiment.

c The international Space Station is falling to Earth. How come it never hits Earth but orbits it instead?

_____

_____

_____

10) CAUSE AND EFFECT: PS2-3

Many scientific investigations begin with a question. What questions could we investigate about how magnets affect each other using a dowel and four donut shaped magnets as shown in this picture.

a List at least 3 questions.

_____

_____

_____

b What can you conclude about magnetic forces?

_____

_____

11) INTERDEPENDENCE OF SCIENCE, ENGINEERING, AND TECHNOLOGY: PS2-4

Magnets are used in many ways in everyday life. Design a new simple invention that uses magnets to solve an everyday problem. Use a diagram to explain how your invention works.

12) CAUSE AND EFFECT: PS3-1

Examine the following graph. Describe how speed relates to probability of injury. Explain in terms of forces.

_____

_____

_____

_____

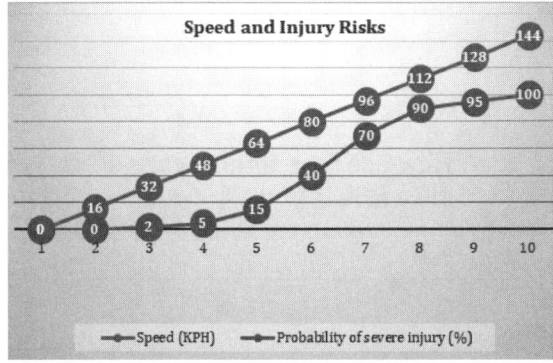

13) CAUSE AND EFFECT: PS3-1

Sloan wanted to see how long a ball will bounce after being dropped.

a   Draw a model to show how the gravitational potential energy and kinetic energy of the ball changes as it falls to the ground.

b   Eventually, the ball stops bouncing. Explain.

_____

_____

_____

14) CAUSE AND EFFECT: PS2-3

The chart below shows the effects of a collision between Object A and Object B. The effect of friction is not shown in this chart.

Trial	Object A Initial speed (m/sec)	Object A Speed after collision (m/sec)	Object B Initial speed (m/sec)	Object B Speed after collision (m/sec)
1	60	50	40	50
2	40	30	25	35
3	25	15	20	30
4	15	5	10	20

a Compare the energy lost by Object A and the energy gained by Object B. What pattern is shown?

_____

_____

_____

b How does the total energy of the system compare after each collision? Provide evidence in your answer.

_____

_____

_____

c Predict the speeds at the end of a collision where Object A is traveling 50 m/sec and Object B is traveling 40 m/sec. Explain how you got your answer.

_____

_____

_____

15) Energy and Matter:
Connections to Engineering, Technology, and Applications to Science:
Science as a human endeavor: PS3,4

An engineer is asked to design a new battery for an electric car so that it can travel further between charges. Describe at least two challenges the engineer would face and how the engineer might overcome them.

_____

_____

_____

_____

16) Cause and Effect: PS3-1

On a sunny summer day, the inside of your cars feels very hot.

a Explain how the sun causes this problem.

_____

_____

b Design an invention that would help solve this problem.

_____

_____

17) Interdependence of Science, Engineering, and Technology:
Science is a human activity that affects everyday life:
Energy and Matter: PS3-2

Energy is transformed from one form to another many times in our everyday lives. Fill in the chart below with three energy transfers that occur in your everyday life.

Activity	Energy Transfer
Example: Making toast	Electrical to heat

18) *a* CAUSE AND EFFECT: PS3-1

Emilia released a small toy car (Car *A*) from different heights on a ramp. Each time the car collided with another car (Car *B*) at the base of the ramp. The chart below shows the results.

Height of Car A when released (cm)	Distance Car B moves away from ramp when hit (cm)
100	24
50	12
75	18
25	6
10	2.4

Based on Emilia's results what is the relationship between the speed of the first car and the energy it contains. Use data from the chart to support your answer.

_____

_____

_____

b ENERGY AND MATTER: PS3-2

After each collision the two cars soon stop moving. Cayden wondered, "What happened to the energy?" List all the ways the energy from the first car was transferred to other things.

_____

_____

_____

c CAUSE AND EFFECT: ENERGY AND MATTER: PS2-1, PS3-3

Emilia then wondered what would happen if Car *B* was moving towards Car *A* when they collided? List a few questions you might ask about the results of this collision. Make a predication about what would happen when the cars hit.

_____

_____

_____

19) ENERGY AND MATTER: PS1

Draw a diagram that shows how energy from the sun ends up as energy for animals. Include and label: Carbon dioxide, water, oxygen, sugar, sunlight, plant, and animal. Also label where photosynthesis is occurring.

20) ENERGY AND MATTER: PS3-4

When you press down on the brakes to slow a car a great deal of energy is transferred as heat and not used in any way. Think of a device we could invent in the future that could take this energy and make it useful. Describe how your device would work. Include a sketch of the device and the car. Use arrow to show what is happening.

21) WAVES AND THEIR APPLICATIONS IN TECHNOLOGIES FOR INFORMATION TRANSFER PATTERNS: PS4-1

Tarun noticed that when a fire truck was moving away from him, the siren became lower pitched and when it moved closer it had a higher pitch. Use a model of sound waves to explain the pattern he notices.

22) PATTERNS: PS4-1

a Sound waves can be reflected, absorbed or transmitted:
   (1) What type of material would you need to sound proof a room? Explain.

   _____

   _____

   (2) What wave characteristic causes an echo? Explain.

   _____

   _____

   (3) What wave characteristic allows you to hear? Explain.

   _____

   _____

b How does a guitar produce sound? Can a guitar make a sound in a vacuum? Explain.

   _____

   _____

c Many mothers have ultrasound pictures of their babies before they are born. Use your knowledge of the characteristic of sound waves to explain why this is possible and not harmful to the fetus.

   _____

   _____

23) CAUSE AND EFFECT: PS4-2

Why must there be a source of light to see? Draw a diagram that shows how light is needed to see. In the diagram include: light source, object, and eye. Use arrows to show the path of the light.

24) CAUSE AND EFFECT: PS4-2

a What causes a shadow to form?

_____

_____

b Using diagrams show how a person may cast no shadow, a slight shadow or a full shadow.

25) PATTERNS:
CONNECTIONS TO ENGINEERING, TECHNOLOGY, AND APPLICATIONS TO SCIENCE: PS4-3, PS4-4

a You are asked to design a new communication system that sends messages across long distances using flashes of light. Write out the code you would use to send these two messages.

1. Send help now _____

2. No help needed _____

b You are asked to convert the code you created so it could be used with sound. Describe what changes, if any, you would have to make.

_____

_____

26) PATTERNS: PS4-1

Caitlyn is having trouble telling the difference between wavelength and amplitude. What could you do to demonstrate what each is for light and for sound? You may use diagrams, models, or acting.

27) PATTERNS: PS4-2

a Can sound waves cause an object to move? Give an example of how you could use sound waves to make something move.

_____

_____

b How does a vibrating object produce sound? Draw a diagram to show how sound travels through air.

_____

_____

c Sound needs a medium through which to travel. Explain why sound travels faster in water than in air and fastest in a solid?

_____

_____

28) CAUSE AND EFFECT/MODELING: MS-PS2-2

The table below is for describing the forces on and the velocity of a rocket aimed in a horizontal direction: before ignition, after ignition, after the burn stops, and when the rocket comes to rest. Use arrows and word symbols to describe the forces acting on the rocket under each condition and the velocity at each point. Gravity (G), Thrust (T), (Drag (D), Lift (L) and your choice.

	Attached to Launch Pad; Not Ignited	Attached to launch Pad; Burning	Burning Time = 1 second	Burning Time = 2 seconds	Burning Time = 3 seconds	Burning Time = 4 seconds	Burning Time = 5 seconds	Burn stops Time = 6 seconds	Time = 7 seconds	Rocket rests Time = 8 seconds
Vertical Forces										
Horizontal Forces										
Velocity (vertical)										
Velocity (horizontal)										
Velocity (net)										

1) PATTERNS: ESS1-2

Construct a graph from the data on the chart.

Amount of Daylight during the Year in Albany, New York

Date (Month/Date)	Daylight in hours
1/1	9.07
2/1	9.59
3/1	11.30
4/1	12.43
5/1	14.05
6/1	15.06
7/1	15.16
8/1	14.29
9/1	13.10
10/1	11.45
11/1	10.19
12/1	9.17

Describe the pattern shown in your graph. How does the amount of daylight relate to the seasons? Be sure to include data from the graph/chart in your answer.

_____

_____

2) PATTERNS: ESS1-1

Roberto noticed that constellations (patterns of stars) appear to move in a predictable pattern over the course of a year. What motion(s) of Earth would explain this? Create a diagram or model to illustrate this.

_____

_____

_____

3) PATTERNS:
SCIENTIFIC KNOWLEDGE ASSUMES AN ORDER
AND CONSISTENCY IN NATURAL SYSTEMS: ESS1-1

How do each of the diagrams show evidence of change in the landscape over time?

White Sands

Bryce National Park

_____

_____

_____

4) STABILITY AND CHANGE: ESS1-1

Sarina saw this lake when her family was visiting Rimrock, Arizona. She wondered how the lake formed. What would you tell Sarina?

_____

_____

5) PATTERNS:
CAUSE AND EFFECT: ESS2-1

Amelia noticed that the rain created a gully in the soil in her backyard. She recorded how deep the gully became after each rain. What cause and effect relationship do you see between the amount of rain and the depth of the gully? Explain how the gully got deeper after each rainfall.

Date	Amount of Rain (cm)	Depth of gully (cm)
3/7	1	3
3/12	3	5
4/2	2	7
4/5	5	12
4/10	4	14

_____

_____

_____

6) SYSTEMS AND SYSTEM MODELS: ESS2-1

The geosphere, hydrosphere, atmosphere, and biosphere interact with each other. For each of the following construct a diagram that shows an interaction. Be sure to label how the "spheres" affect one another.

a The ocean (hydrosphere) and the beach (geosphere).

b The wind (atmosphere) on landform (geosphere).

c The ocean (hydrosphere) on climate (atmosphere).

d The climate (atmosphere) and living things (biosphere).

7) SCALE, PROPORTION, AND QUANTITY: ESS2-2

a Create bar graphs for each of the charts below that quantifies Earth's total water and its fresh water.

Chart 1: Earth's Water

Earth's Water	Percent (%)
Salt Water	97
Fresh	3

Chart 2: Freshwater

Fresh Water	Percent (%)
Icecap and glaciers	68.7
Groundwater	30.1
Other	.9
Surface	.3

Chart 3: Fresh Surface Water (Liquid)

Fresh Surface Water (Liquid)	Percent (%)
Lakes	87
Swamps	11
Rivers	2

b How does the amount of freshwater in all the lakes, swamps, and rivers compare to all of the water on Earth?

_____

_____

8) PATTERNS: ESS2-2

The following chart shows the average high and low temperatures in New York City for one year. Construct a line graph comparing the months and low temperatures. Plot the high temperatures on the same graph.

Temperatures in New York City over a One Year Period

Month	Low C°	High C°
Jan	0	5
Feb	0	6
Mar	1	13
Apr	10	17
May	13	21
June	16	24
July	21	28
Aug	22	29
Sept	18	26
Oct	10	17
Nov	3	15
Dec	2	13

a What pattern do you see as you compare the temperatures to the seasons?

_____

_____

b If you had to estimate the temperature in the middle of July next year what would you predict? Support your answer with information from the graph.

_____

9) PATTERNS: ESS2-2

Examine the chart of January weather below. What type of weather would you predict for the first days in February? Use evidence from the chart in your answer. Why might your prediction turn out wrong?

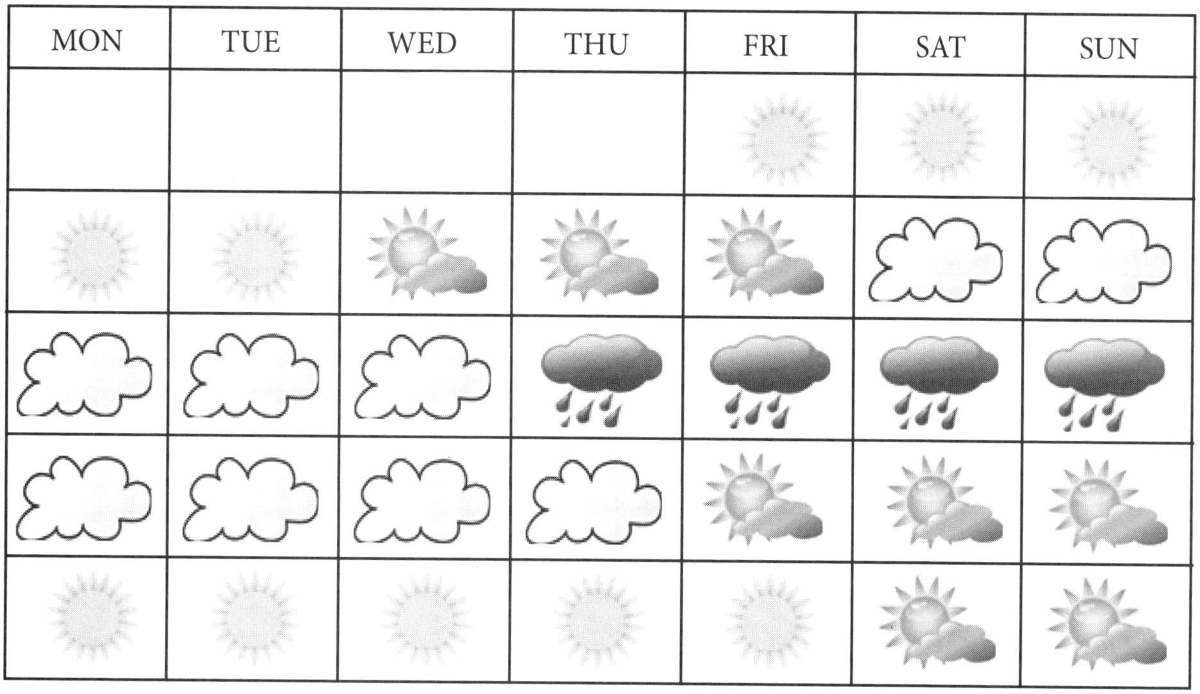

MON	TUE	WED	THU	FRI	SAT	SUN
				☀	☀	☀
☀	☀	⛅	⛅	⛅	☁	☁
☁	☁	☁	🌧	🌧	🌧	🌧
☁	☁	☁	☁	⛅	⛅	⛅
☀	☀	☀	☀	☀	⛅	⛅

_____

_____

_____

10) CAUSE AND EFFECT: ESS2-3

Does the water cycle have an effect on weather? Plan an experiment where one part of the water cycle is changed. Describe the results (changes in the weather) you might expect to see.

a What change will you make to the water cycle?

_____

_____

b How do you think the weather will change because of the change in the water cycle?

_____

_____

c What data will you collect to measure the change in the weather?

_____

11) <u>Patterns</u>: <u>Cause</u> and <u>Effect</u>: ESS2-2

Use the Internet to research TWO of the following climates.

Desert　　**Tropical Rain Forest**　　**Grasslands**　　**Deciduous Forest**　　**Coniferous Forest**

a Determine and compare the average temperature and rainfall for each.

_____

_____

_____

b What are some factors that affect climate?

_____

_____

12) <u>Systems</u> and <u>System</u> <u>Models</u>: ESS2-1

Living Things and the Carbon Cycle

Describe how both plants and animals, including humans, play a role in the carbon cycle.
Explain ways these roles can change in order to reduce greenhouse gases.

_____

_____

13) CAUSE AND EFFECT:

SCIENCE IS A HUMAN ENDEAVOR: INFLUENCE OF ENGINEERING, TECHNOLOGY, AND SCIENCE ON SOCIETY AND THE NATURAL WORLD: ESS2-1

Read the following descriptions of several ways to prevent the problems caused by flooding. Choose one of the methods affecting the environment and create an argument for why it is best. Include evidence in your answer.

Reducing the impact of Flooding

1. *Better Flood Warning Systems*: Improving flood warning systems will give people more time to prepare for a flood. This will help prevent loss of life and reduce damage to property.

2. *Better Construction of Buildings*: Ways to reduce the effect of floods on building include: the use of concrete floors, the replacement of the use of plasterboard for walls, and moving electric sockets higher up on walls.

3. *Location of Buildings*: New building should be constructed above the flood level for that area. Newer construction methods that lower the risk of flood damage should be used.

4. *Reduce the Effect of Climate Change*: Governments, businesses, and individuals should make greater efforts to reduce the increase in global temperatures.

5. *Adapting the Environment*: Changing the physical environment can reduce the effects of flooding. Creating and preserving wetlands will provide areas that can take up more of the water. Planting trees can help slow down waters when rivers overflow and hold soil in place.

6. *Returning Rivers and Streams to Their Natural Paths*: The paths of many rivers have been changed for many reasons. By adding more bends the length of the river and the speed of its flow may be reduced. This would reduce the effect of the flood downstream.

7. *Improving Soil Drainage*: Soil that is compacted, by machines and people, cannot absorb water. Instead of holding the water it may run off quickly. Well-drained soil can hold a larger amount of water. This will keep the water from entering rivers and increasing flooding.

8. *Building More Flood Barriers*: Raised embankments can reduce the effect of flooding in areas where floods are common. Temporary barriers may also be used in areas during times when floods are common.

14) <u>Systems and System Models</u>: ESS3-1

Most living things cannot live everywhere on Earth. Pick one living organism, except humans, and draw a diagram showing what it needs to live in a particular environment. Label the living thing and what it needs.

15) <u>Interdependence of Science, Engineering, and Technology Research Project</u>: ESS3-1

Use the Internet to complete the following chart.

Energy/Fuel	How it is obtained	Effect on the environment
Wind		
Water behind a dam		
Sunlight		
Oil		
Nuclear		

16) CAUSE AND EFFECT:

CONNECTIONS TO ENGINEERING, TECHNOLOGY, AND APPLICATIONS TO SCIENCE: ESS3-1

Anshu has an assignment to prepare for a meeting with a meteorologist (someone who studies weather). She is supposed to ask the meteorologists questions about how to prepare for and how to respond to severe weather. Prepare a list of five questions Anshu could ask to complete her assignment. Try to be specific about the type of severe weather you are asking about.

1) _____

_____

2) _____

_____

3) _____

_____

4) _____

_____

5) _____

_____

17) CAUSE AND EFFECT:

INFLUENCE OF ENGINEERING, TECHNOLOGY, AND SCIENCE ON SOCIETY AND THE NATURAL WORLD) HOW CAN WE REDUCE THE IMPACT OF NATURAL PROCESSES ON HUMANS?: ESS3-2

Use the Internet to search for multiple ways to reduce the effect of each of the following:

1. Earthquakes   **2. Floods**   **3. Wildfires**   **4. Hurricanes**

Choose one of the natural processes you researched and explain why one method that reduces the impact on humans may be better than another.

_____

_____

_____

_____

_____

_____

18) SCIENCE ADDRESSES QUESTIONS ABOUT THE NATURAL AND MATERIAL WORLD: ESS3-1

How has your local community/state used science, engineering and technology to protect Earth's resources and environments and reduce the impact caused by humans? Choose one living (biotic) or nonliving factor (abiotic) and describe how its negative impacts are being reversed.

_____

_____

_____

_____

_____

_____

_____

19) PATTERNS: ESS1-1

The Moon is a natural satellite of Earth.

a Why does it glow?

_____

_____

b From Earth, it looks like the Moon is changing its shape or going through phases we call: Full, waning gibbous, third quarter, waning crescent, new, waxing crescent, waxing first quarter, waxing gibbous, and full again. Use a model to show why the moon appears to change shape. Draw the Moon phases and the location of the Sun.

c Explain why Earth can only see one side of the Moon.

_____

_____

-A-

absolute zero: [pg 220] 0 K (see Kelvin) temperature at which no heat can be transferred from matter absolute zero; substances possess no thermal energy, equal to -273.15°C, or -459.67°F.

acceleration: [pgs 208, 210] measures the rate of change in velocity

acid: [pg 223] any of a large class of sour-tasting substances whose aqueous solutions are capable of turning blue litmus indicators red, of reacting with and dissolving certain metals to form salts, and of reacting with bases or alkalis to form salts. It must have a pH level below 7.

acid rain: [pgs 116, 141] formed when pollutants such as sulfur and nitrogen oxides, often from industrial emissions, combine with water in the atmosphere

Action/Reaction: [pg 210] Newton's third law of motion, states for every action, there is an equal and opposite reaction

actual mechanical advantage: [pg 213] benefit realized by using a machine; the measured amount that a machine multiplies force; includes the force of friction

adaptation: [pg 76] a change that helps an organism better fit the environment

aerate: [pg 145] to expose to the circulation of air for purification

agents of erosion: [pg 142] something that causes erosion. Erosion is the deterioration of something. It is most common to talk about soil erosion. Common agents are moving water, wind, and glaciers.

air: [pg 112] a mixture with varying amounts of moisture and particulate matter, enveloping Earth; the atmosphere.

air current: [pg 155] stream of air having similar air mass characteristics (see Jet Stream)

air mass: [pg 157] a large body of air that has the same characteristics of moisture and temperature as the surface over which it formed; examples, continental: forms over land; maritime: forms over water; polar: forms over higher latitudes; tropical: forms over lower latitudes

air pressure: [pg 139] pressure exerted by Earth's atmosphere

algae: [pg 101] any of various chiefly aquatic organisms, ranging in size from single-celled for ms to the giant kelp

alloy: [pg 167] homogeneous (similar) mixture or solid solution of two or more metals

altitude: [pgs 155, 234] distance above Earth's surface.

alveoli: [pg 51] tiny sacs in the lungs where gases are exchanged

anaphase: [pg 64] stage of mitosis and meiosis in which the chromosomes move to opposite ends of the nuclear spindle

Aristotle: [pgs 48, 74] 384-322 B.C. Greek philosopher,author of works on logic, metaphysics, ethics, natural sciences, politics, and poetics, he profoundly influenced Western thought

antibodies: [pg 57] substances produced by the body to fight disease

arteries: [pg 52] a branching system of muscular, elastic tubes that carry blood away from the heart to the cells, tissues, and organs of the body

asexual reproduction: [pgs 63, 83] form of reproduction requiring only one parent

asteroids: [pg 129] small planet-like bodies formed in the universe; not big enough to become a planet

astrolabe: [pg 234] an instrument, now replaced by the sextant, that can be used to determine the altitude of the Sun or other celestial bodies

atmosphere: [pg 137] layers of gases that surround the Earth

atom: [pgs 146, 173] smallest particle of an element that still has the properties of the element

atomic mass unit: [pg 173] (a.m.u.s) the mass of an electron atomic number

atomic number: [pg 174] number of protons in an element

attraction: [pg 196] unlike charges and unlike poles move toward one another

Autumnal Equinox: [pg 126] marks the beginning of fall, about September 22; point at which the Sun passes the celestial equator (on its way south) causing equal amounts of daylight and darkness; see vernal equinox

axis: [pg 124] imaginary line, extending from the north pole to the south pole, through the center of the Earth about which Earth rotates

-B-

bacteria: [pg 57] microorganisms lacking a nucleus; many cause disease in plants and animals; examples, bacillus (rod shaped), coccus (spherical), and spirillum (spiral)

banding: [pg 148] strip or stripe that contrasts with something else in color, texture, or material

bar graph: [pgs 20, 27] visual representation of a relationship between two factors; used when the relationship is not continuous

base: [pg 223] any of a large class of compounds, having a bitter taste, a slippery solution, the ability to turn litmus blue, and the ability to react with acids to form salts. It must have a pH level above 7.

Bichat, Marie Francois Xavier: [pg 45] 1771-1802 French anatomist who pioneered the histological study of organs

Big Bang Theory: [pg 129] a theory holding that the universe originated approximately 20 billion years ago from the violent explosion of a very small agglomeration of matter of extremely high density and temperature

biomass: [pg 188] total mass of living matter within a given unit of environmental area, including plant material, vegetation, or agricultural waste used as a fuel or energy source

binomial nomenclature: [pg 48] scientific naming of species whereby each species receives a Latin or Latinized name of two parts, the first indicating the genus and the second being the specific epithet

biodiversity: [pg 117] Any environmental factor not caused by living organisms

biome: [pg 155] major regional or global biotic community, such as a grassland or desert, characterized chiefly by the dominant forms of plant life and the prevailing climate

blending: [pg 67] when the effects of two different genes for the same trait are both expressed

blizzards: [pg 160] storms with high winds and frozen precipitation; form in the upper atmosphere

blood: [pg 52] liquid tissue in the circulatory system

boiling point: [pg 167] the temperature at which matter changes phase from a liquid to a gas

brain: [pg 56] primary center for the regulation and control of bodily activities, receiving and interpreting sensory impulses, and transmitting information to the muscles and body organs

bronchi: [pg 51] the large tubes that connect to your trachea (windpipe) and direct the air you breathe to your right and left lungs

budding: [pg 83] type of asexual reproduction; one-celled organism splits in half with an equal division of genetic material, unequal division of cell material

buoyancy: [pg 170] upward force of a fluid

butte: [pg 143] hill that rises abruptly from the surrounding area and has sloping sides and a flat top

-C-

Calorie: [pg 96] (with a capital "C") a kilo-calorie; unit used to measure the energy in foods

calorie: [pgs 96, 190] (with a small "c") amount of heat energy needed to raise the temperature of one gram of water by one degree Celsius

cancer: [pg 85] disease that results in abnormal cell division

capillary: [pg 51, 52] small blood vessel where exchange of materials between blood and cells occurs

carbohydrates: [pgs 95, 96] substances used for quick energy; examples; cereal, bread

carbon dioxide: [pgs 92, 93] waste gas released by plants and animals during respiration; used by plants during photosynthesis

carbon dioxide/oxygen cycle: [pg 102] recycling of carbon dioxide and oxygen on our planet

cardiac muscle: [pg 56] specialized striated muscle tissue of the heart; the myocardium

carotid artery: [pg 53] either of the two major arteries, one on each side of the neck, that carry blood to the head

carnivore: [pgs 5, 93] animal that gets its energy by eating other animals

cast: [pg 147] a type of fossil where sediment falls into an imprint

cavern: [pg 141] large underground chamber, as in a cave

Celsius: [pgs 220, 222] metric temperature scale; 0°C = freezing and 100° C = boiling point of water

cells: [pg 41] basic unit of structure for most living things; performs all physiological activities

cell membrane: [pg 43] surrounds the cell and controls what leaves and enters; also called plasma membrane

cell theory: [pg 42] well founded idea that all living things are composed of cells; all cells come from cells

cell wall: [pg 43] rigid structure composed of cellulose that surrounds, protects, and supports some types of cells, such as plant cells

centimeter: [pg 220] a metric unit of measurement used for measuring the length of an object. It is written as cm. A meter has 100 centimeters. 10 millimeters make one centimeter.

centrifuge: [pg 172] an apparatus consisting of a compartment spun about a central axis to separate materials of different specific gravities, or to separate particles suspended in a liquid

checklist: [pg 2] a list of items to be noted, checked, or remembered

chemical bond: [pg 176] forces or mechanisms, especially the ionic bond, covalent bond, and metallic bond, by which atoms or ions are bound in a molecule or crystal

chemical change: [pg 172] a change in matter where both physical and chemical properties are changed

chemical energy: [pg 185] stored energy found in all compounds; associated with chemical bonds

chemical equation: [pg 173] expression summarizing a chemical reaction

chemical properties: [pg 167] way a mineral changes, reacts, or behaves

chemical weathering: [pg 141] breakdown of rock by changing its chemical composition (make up)

chlorofluorocarbons (CFCs): [pgs 117, 139] any of various halocarbon compounds consisting of carbon, hydrogen, chlorine, and fluorine, once used widely as aerosol propellants and refrigerants. Chlorofluorocarbons are believed to cause depletion of the atmospheric ozone layer

chlorophyll: [pgs 93, 101] protein molecule used during the process of photosynthesis to help convert light energy into chemical energy

chloroplasts: [pgs 43, 46] cell structures (organelles) that contain chlorophyll, location of photosynthesis

chop: [pg 172] cut into small pieces

chromosomes: [pgs 43, 63] usually found in the nucleus of a cell, contain genetic information; composed of DNA, the chemical that directs heredity

chromatography paper: [pg 172] filter paper that separates mixtures by differences in their solubility and how they adhere to the paper

circuit: [pg 196] path followed by electrons as they travel through conductors

circulatory system: [pgs 45, 52] group of organs that includes heart, arteries, veins, lymph vessels, capillaries of body, and blood and lymph

cirrus clouds: [pg 158] high-altitude cloud composed of narrow bands or patches of thin, generally white, fleecy parts

class: [pg 48] a taxonomic category ranking below a phylum or division and above an order

classification (classifying): [pg 48, 226] organizing things based on characteristics

cleavage: [pg 147] tendency of a mineral to break along regular surfaces in one or more specific directions

climate: [pgs 112, 154] average weather conditions over a large geographic area that extends over a long period of time

climax community: [pg 115] a relatively stable group of populations that usually occurs at the end of ecological succession

circuit: [pg 196] a closed path followed or capable of being followed by an electric current

cloning: [pg 84] asexual reproduction; usually associated with creating a new organism from cells of another

coal deposits: [pg 188] fossil fuel formed form ancient fern forests

cold front: [pg 158] forms when a cold air mass of air meets a warm air mass

collision theory: [pg 176] idea that atoms must collide in order to react

color: [pg 167] appearance of objects or light sources described in terms of the individual's perception of them, involving hue, lightness, and saturation for objects and hue, brightness, and saturation for light sources

comets: [pg 129] celestial objects, rock, ice, dust, and gas, with a very elliptical orbit around the Sun

compaction: [pg 147] method of formation of sedimentary rocks

complete metamorphosis: [pg 87] a life cycle that contains four distinct stages; egg, larva, pupa, and adult

community: [pg 112] all of the populations of living things interacting in a given area

complex machine: [pg 214] made up of more than one simple machine

compound: [pg 176] two or more elements chemically combined

compound microscope: [pg 229] a microscope consisting of an objective and an eyepiece at opposite ends of an adjustable tube

compressional wave: [pg 193] (also longitudinal) travels parallel to its direction of motion; example, sound waves

concentrated solution: [pg 168] much solute dissolved in a solvent (see saturated solution)

conclusion: [pg 7] a judgment or decision reached after deliberation

condensation: [pgs 102, 177] phase change by which matter changes from a gas to a solid

conduction: [pg 191] energy transfer by direct contact

conductivity: [pg 169] ability of matter to conduct heat or electrical energy

conductor: [pg 169] substance or medium that conducts heat, light, sound, or especially an electric charge

conservation: [pg 117] the preservation or careful use of Earth's resources

constellations: [pg 125] an arbitrary formation of stars perceived as a figure or design, especially one of 88 recognized groups named after characters from classical mythology and various common animals and objects

consumer: [pgs 93, 105] organism that feeds on other living or dead matter

contour farming: [pg 143] following the contour lines of uneven terrain to limit erosion of topsoil

contour lines: [pg 234] a line on a map that joins points of equal elevation

control: [pg 11] condition of an experiment that does not change; method of regulating a system; used to provide a comparison

convection: [pg 191] heat energy transfer by differences in density; from high to low areas

convection cell: [pg 191] formed by a circular motion of air due to differences in densities

convection currents: [pgs 150, 156, 191] movement of one material through another by energy differences

Cook, James: [pgs 96, 196] 1728-1779. British navigator and explorer who commanded three major voyages of discovery, charting and naming many islands of the Pacific Ocean. He also sailed along the coast of North America as far north as the Bering Strait

coordinate system: [pg 130, 235] method of representing points in a space of given dimensions by coordinates

Copernicus, Nicolaus: [pg 122] 1473-1543 Polish astronomer who advanced the theory that the Earth and other planets revolve around the Sun, heliocentric view, disrupting the Ptolemaic system of astronomy

Coriolis Effect: [pgs 124, 125] observed effect of the Coriolis force, especially the deflection of an object moving above the Earth, rightward in the northern hemisphere and leftward in the southern hemisphere

Correns, Karl: [pg 67] German botanist that found that in some traits both genes of the pair for a trait can influence the trait in an individual

crater: [pg 129] bowl-shaped depression seen on volcanoes, geysers, and the moon.

Crick, Francis: [pg 62] Born 1916. British biologist who with James D.Watson proposed a spiral model, the double helix, for the molecular structure of DNA. He shared a 1962 Nobel Prize for advances in the study of genetics

crust: [pgs 139, 140] upper region of the solid part of Earth

crystallize: [pg 168] to cause to form crystals or assume a crystalline structure

cumulus clouds: [pg 158] towering thick white clouds that characterize cold fronts

current electricity: [pgs 196-197] electric charge that travels through a conductor

cytoplasm: [pg 43] watery material inside the cell.

-D-

DNA (Deoxyribonucleic Acid): [pg 62] substance that carries the code for hereditary information

Darwin, Charles: [pg 73] 1809-1882. British naturalist who revolutionized the study of biology with his theory of evolution based on natural selection. His most famous works include Origin of Species (1859) and The Descent of Man (1871)

data: [pg 14] information from an investigation that is important to the study (singular – datum)

daughter cell: [pg 64] either of the two identical cells that form when a cell divides

decantation: [pg 172] pouring off a liquid without disturbing the sediment

deceleration: [pg 208] to decrease the velocity

decimal system: [pg 219] system of measurement in which all derived units are multiples of 10 of the fundamental units

decomposer: [pgs 93, 105] an organism that gets it energy by breaking down wastes or dead organisms

density: [pgs 156, 169] amount of matter per unit of volume

density currents: [pg 191] heat energy transfer in a fluid due to density differences

dependent variable: [pg 11] in an experiment, the factor you measure to determine the effect of independent variable (also responding variable)

deposition: [pgs 143, 145, 177] agents of weathering and erosion settle out particles they can no longer transport; phase change from a gas to solid

desert: [pg 155] dry, often sandy region of little rainfall, extreme temperatures, and sparse vegetation

deuterium: [pg 174] hydrogen with one neutron

dew point: [pg 159] temperature at which air becomes saturated and produces dew

diaphragm: [pg 51] curved sheet of muscle between the chest and the abdomen, responsible for breathing

dichotomous key: [pg 226] system using pairs of contradictory statements to identify characteristics for classification

diffraction: [pg 195] change in direction and intensity of a wave after passing by an obstacle or through an aperture

digestive system: [pg 49] group of organs that break down food so it can be absorbed and transported to the cells

dilute solution: [pg 168] small amount of a solute dissolved in a solvent (see solution)

disease: [pg 57] a breakdown in the structure or function of an organism

dissolve: [pg 172] to pass into solution

distillation: [pg 172] process of purifying a liquid by evaporation and condensation

domain: [pg 197] any of numerous contiguous regions in a ferromagnetic material in which the direction of spontaneous magnetization is uniform and different from that in neighboring regions

dominant: [pg 66] a trait that hides the effect of another

double helix: [pg 62] coiled structure of double-stranded DNA in which strands linked by hydrogen bonds form a spiral configuration

ductility: [pg 171] ability of matter to be easily molded or shaped

-E-

Earth: [pg 130] third planet from the Sun, having a sidereal period of revolution about the Sun of 365.26 days at a mean distance of approximately 149 million kilometers (92.96 million miles)

earthquake: [pgs 152, 153] violent shaking of Earth's crust due to the motion of the plates; fault

eclipse: [pg 128] partial or total blocking out of a heavenly body by another heavenly body caused by the motion of Earth and Moon as they relate to the Sun

ecological succession: [pg 115] changes in a community that occurs over a long period of time as a community reaches a state of balance

ecology: [pg 111] the study of how all living things interact with one another and with their physical environment

ecosystem: [pgs 47, 113] the relationship between the community (plants and animals) and the nonliving factors with which they interact

efficiency: [pg 214] ratio of the energy delivered by a machine to the energy supplied for its operation

effort: [pgs 212-214] work put into a machine

effort distance: [pgs 213-214] measure of an object's travel using a machine

effort force (Fe): [pg 213] amount of push or pull applied to an object by a machine

egg: [pgs 58, 65, 84] female reproductive cell with half the normal number of chromosomes found in a body cell; gamete

Einstein, Albert: [pg 190] 1879-1955. German-born American theoretical physicist whose theories showed matter and energy can be changed into one another began the nuclear age

elasticity: [pg 171] ability of matter to be stretched and return to its original shape

electrical energy: [pg 185] produced from the motion of charges in an electric field

electricity: [pgs 195-198] flow of electrons through a conductor; energy that involves the motion of charges caused by the attraction of particles with opposite charges and the repulsion of particles with the same charge

electrolyte: [pgs 169, 192] chemical compound that can produce an electrically conductive medium; conducts electricity

electromagnet: [pg 197] magnet that uses electricity to create magnetism

electromagnetic field: [pg 197] area in which electromagnetic energy acts

electromagnetic spectrum: [pg 192] summary of all of the electromagnetic forms of energy; from lower frequencies such as radio-waves, short wave, AM, FM, TV, and radar to infrared rays, visible light and ultraviolet, and finally, the shorter wavelengths of x-rays and gamma rays

electron: [pg 173] negatively charged particle in an atom that is found in the electron cloud

electron cloud: [pg 173] location of electrons in an atom

electroscope: [pg 195] device that detects static electricity

element: [pg 146] pure substance made up of only one type of atom

elliptical: [pg 127] oval; shape of all orbits

embryo: [pg 86] organism at any time before full development, birth, or hatching

endocrine system: [pg 56] group of organs working together to regulate growth, development and reproduction; hormone control system

energy: [pg 185] ability to do work or cause a change

energy pyramid: [pgs 5, 105] movement of energy from the Sun as the original source through produces, consumers, and decomposers (see food chain)

environment: [pg 107] combination of external physical conditions that affect and influence the growth, development, and survival of organisms

epicenter: [pg 153] point above the focus of an earthquake

epidermis: [pg 46] outermost layer of cells covering the leaves and young parts of a plant

equator: [pg 235] imaginary great circle around the Earth's surface, equidistant from the poles and perpendicular to the Earth's axis of rotation

erosion: [pg 141] wearing away material from the Earth's surface by water, wind, or ice through natural processes, including weathering, dissolution, abrasion, corrosion, and transportation

esophagus: [pg 49] muscular, membranous tube for the passage of food from the pharynx to the stomach; the gullet

evaporation: [pgs 144, 172, 177] phase change from a liquid to gas

evolution: [pg 72] a process that results in changes in the genetic material of a population over time. Evolution reflects the adaptations of organisms to their changing environments and can result in altered genes, novel traits, and new species.

excretory system: [pg 54] a group of organs (liver, kidney, skin, lungs) that work together to remove wastes from the body

extinct: [pg 76] something that does not exist now, but once did

extinction: [pg 76] when all members of a species die off

-F-

fahrenheit: [pgs 220, 222, 229] English temperature scale 32° F = freezing and 212° F = boiling of water

family: [pg 176] group of elements with similar properties; found in the vertical columns of the periodic table

fats: [pgs 95-96] substance used to store energy

fertilization: [pg 84] the joining of an egg and sperm

fertilize, as in egg: [pg 58] the union of two sex cells, usually one sperm (male) and egg (female), the union produces a zygote

fertilize, as in soil [pg 145] addition of nutrients to the earth

field map: [pgs 157, 234] graphic description of the physical characteristics of a region

filtration: [pg 172] Separation of a liquid and a solid or two solids with different particle size

first cross: [pgs 67, 68] see Punnett Square method of solving genetic problems

First Law of Motion (Principle of Inertia): [pg 210] an object at rest will stay at rest while an objectin motion will stay in motion unless acted upon by an outside force

first quarter: [pg 128] Moon appears as a half-moon after new crescent; right side is completely illuminated, left side is dark

fission: [pgs 83, 189] act or process of splitting into parts; nuclear reaction in which an atom splits into fragments releasing millions of electron volts of energy; asexual reproductive where one-cell organism splits in half resulting in equal division of genetic material and an equal division of cell material

flammability: [pg 171] ease of matter to be ignited and capable of burning

flexibility: [pg 171] ability of matter to bend without breaking

flux lines: [pg 197] lines of magnetic force

focus: [pg 153] location of the crack caused by an earthquake; directly below the epicenter

food: [pgs 94-95] materials needed by an organism for maintenance, growth and repair

food chain: [pg 104] the path of energy between organisms

food web: [pg 104] path of energy between all of the organisms in a community

force: [pgs 206, 210] push or pull

fossils: [pgs 78, 147] the remains or imprints of an organism that once lived; found in rock

fossil fuels: [pg 188] a nonrenewable energy resource formed from ancient forms of life; oil, natural gas, coal

Foucault, Jean: [pg 123] 1819-1868. French physicist who performed an experiment to prove Earth rotates on its axis

frame of reference: [pg 205] point from which motion is detected

freezing: [pg 177] phase change from a liquid to solid

freezing point: [pg 167] temperature at which liquid changes to a solid; 32° F, 0° C

friction: [pg 213] force that opposes motion of two objects in contact

front symbols: [pg 159] The front marks the leading edge of the cold air. The blue triangles always point in the direction that the front (and the cold air) is going. A red line with half-circles on one side signifies a warm front. A warm front shows the leading edge of warmer air trying to replace a colder air mass.

frost action: [pg 141] form of erosion through the expanding action of frozen water

fulcrum: [pg 213] fixed point on a level around which the bar (lever) rotates

full Moon: [pg 128] phase when the Moon is behind Earth; entire face of the Moon is illuminated

fusion: [pg 189] nuclear reaction in which small nuclei join together releasing a lot of energy; all stars produce energy this way

-G-

gametophyte: [pg 88] gamete-producing phase in a plant characterized by alternation of generations

gamma rays: [pg 193] very high energy wave (ten thousand to ten million electron volts); part of the electromagnetic spectrum

gas: [pg 177] phase of matter in which the particles are moving the fastest and are the furthest away from one another

gene: [pg 63] the unit of genetic information that contains the code for a particular trait

genetic engineering: [pg 75] alteration of an organism's genetic material to eliminate undesirable traits or produce desirable ones

genus: [pg 48] a taxonomic category ranking below a family and above a species and generally consisting of a group of species exhibiting similar characteristics, used either alone or followed by a Latin adjective or epithet, to form the name of a species

geocentric model: [pg 122] early model of the universe in which Earth is the center of the universe

geothermal energy: [pg 188] energy from the inside of Earth

glaciers: [pg 143] a moving mass of ice formed over many years in which snowfall exceeds melting

gland: [pg 56] cell or organ that produces a secretion for use elsewhere in the body or in a body cavity or for elimination from the body

global warming: [pg 117] increase in the Earth's temperature due to pollution from the burning of fossil fuels

gradient: [pg 234] rate of ascending or descending change between values

graduated cylinder: [pg 221] device used to measure volume

gram: [pg 220] unit of mass

granite: [pg 148] common, coarse-grained, light-colored, hard igneous rock consisting chiefly of quartz, orthoclase or microcline, and mica, used in monuments and for building

grasslands: [pg 155] ecosystem with large, flat areas of grasses

great circle: [pg 235] segment of a circle representing the shortest distance between two terrestrial points

gravity: [pg 142] natural force of attraction between any two massive bodies, which is directly proportional to the product of their masses and inversely proportional to the square of the distance between them

gravitational potential energy: [pg 187] energy of position

Greenhouse Effect: [pgs 117, 138] the trapping of infrared radiation by pollution in Earth's atmosphere

greenhouse gases: [pg 138] gases that trap solar energy

groundwater: [pgs 140, 146] water beneath the Earth's surface, often between saturated soil and rock, that supplies wells and springs

Gulf Stream: [pgs 155-156] warm ocean current that originates from the Straits of Florida and travels in a general NE direction up along the coast of Europe

-H-

Habitat: [pg 47] area or type of environment in which an organism or ecological community normally lives

Halley's Comet: [pg 129] a comet named after Edmund Halley, with a period of approximately 76 years, the first one for which a return was successfully predicted, last appeared in 1986

hardness: [pgs 147, 167] resistance of a mineral to being scratched

Harvey, Dr.William: [pg 53] 1578-1657 English physician, anatomist, and physiologist who discovered the circulation of blood in the human body (1628)

heart: [pgs 52, 53] chambered, muscular organ in vertebrates that pumps blood received from the veins into the arteries, thereby maintaining the flow of blood through the entire circulatory system

heat: [pgs 185, 190, 191] energy that moves from areas of higher temperature to lower temperature

heliocentric model: [pg 122] model of the universe in which the Sun is the center of the universe

herbivore: [pg 93] an organism that gets its energy by eating plants

high pressure: [pg 157] in weather, an area of relatively high isobaric pressure when compared to a related low pressure area, often bounded by a front

horizon: [pg 234] the apparent intersection of the Earth and sky as seen by an observer

horizontal force: [pg 212] push or pull parallel to or in the plane of the horizon

Hooke, Robert [pg 41] 1635-1703 English physicist, inventor, and mathematician who formulated the theory of planetary movement

hormones [pg 56] substance, usually a peptide or steroid, produced by one tissue and conveyed by the bloodstream to another to effect physiological activity, such as growth or metabolism

hurricane: [pg 160] violent large storm; form over warm oceans; many convection cells merge into huge convection cells

hybrid: [pg 67] offspring of genetically dissimilar parents or stock, especially the offspring produced by breeding plants or animals of different varieties, species, or races; containing two different genes for a trait

hydroelectric energy: [pg 188] renewable energy that uses moving water to generate electricity

hydrosphere: [pg 140] liquid portion of Earth

hypothesis: [pg 4] a tentative, testable explanation of the observed relationship between variables

-I-

ice storm: [pg 160] forms when upper atmosphere temperature is warmer than that of lower atmosphere

Ideal Mechanical Advantage: [pg 213] calculated without including the force of friction

igneous rock: [pg 148] type of rock formed from liquid rock by solidification from a molten state (magma); intrusive and extrusive

immune system: [pg 57] a group of specialized cells working together to protect the body from infectious diseases

imprint: [pg 147] to produce a mark or pattern on a surface by pressure

inclined plane: [pg 214] simple machine that forms a slanted surface

incomplete dominance: [pg 67] heterozygous condition in which both alleles at a gene locus are partially expressed, often producing an intermediate phenotype

incomplete metamorphosis: [pg 87] a life cycle that contains three distinct stages; egg, nymph, and adult

independent variable: [pg 11] factor whose effect you are testing; manipulated variable

indicator: [pg 223] chemical used to gather information about a sample of matter

inert: [pg 176] not readily reactive with other elements; forming few or no chemical compounds

infectious diseases: [pg 57] caused by a pathogenic microorganism or agent

inference: [pg 7] conclude from evidence or premises; linking something you already know to an observation

infrared light: [pg 193] light energy that is transferred to heat energy

inner core: [pg 140] innermost zone within the Earth; iron/nickel solid

insolation: [pg 155] solar radiation striking Earth or another planet

insulator: [pg 169] matter that does not conduct heat or electricity easily

interdependently: [pg 44] in such a way that two or more things depend on each other.

International Date Line: [pg 235] an imaginary line through the Pacific Ocean roughly corresponding to 180° longitude, to the east of which, by international agreement, the calendar date is one day earlier than to the west

interpolating: [pg 26] estimate a value of a function or series between two known values

ion: [pg 177] a chemical species which holds a positive or negative charge of some magnitude

isobar: [pgs 157, 234] type of isoline on a weather map used to connects points of equal barometric pressure

isoline: [pgs 157, 234] lines on a map or model (such as contours, isotherms, isobars) that connect points of equal value

isotherm: [pgs 157, 234] type of isoline on a weather map that connects points of equal temperature

isotope: [pg 174] atoms of the same element with different atomic masses

-J-

jet stream: [pg 158] large stream of air moving in the upper atmosphere from west to east

joule: [pg 219] unit of work

Joule, James Prescott: [pg 190] 1818-1889. British physicist who established the mechanical theory of heat and discovered the first law of thermodynamics, showing that heat is related to the motion of the particles in matter

Jupiter: [pg 130] fifth planet from the Sun, the largest and most massive in the solar system, having a sidereal period of revolution about the Sun of 11.86 years at a mean distance of 777 million kilometers (483 million miles), a mean diameter of approximately 138,000 kilometers (86,000 miles), and a mass approximately 318 times that of Earth

-K-

Kelvin: [pg 220] temperature scale that begins with absolute zero; temperature at which no heat can be transferred from matter

Kepler, Johannes: [pg 127] 1571-1630 German astronomer and mathematician. Considered the founder of modern astronomy, he formulated three laws to clarify the theory that the planets revolve around the Sun

kidney: [pg 54] either one of a pair of organs in the dorsal region of the vertebrate abdominal cavity, functioning to maintain proper water and electrolyte balance, regulate acid-base concentration, and filter the blood of metabolic wastes, which are then excreted as urine

kilogram: [pg 220] (kg) metric unit of mass

kilometer: [pg 220] (km) metric unit of length equal to 1,000 meters

kinetic energy: [pg 186] energy at work or in motion

kingdom: [pg 48] largest grouping in the classification system

knowledge: [pg 1] familiarity, awareness, or understanding gained through experience or study

-L-

Lamarck, Jean Baptiste: [pg 73] 1744-1829. French naturalist whose ideas about evolution influenced Darwin's theory

land breeze: [pg 161] forms when air over land is cooler than air over water; air moves out toward the water

large intestine: [pgs 49, 54] portion of the intestine that extends from the ileum to the anus, forming an arch around the convolutions of the small intestine

last quarter: [pg 128] phase of the Moon that follows the old gibbous

latitude: [pgs 130, 155, 235] imaginary east-west parallel lines circling the Earth north and south of the equator; 0° parallel is the equator; used with meridians of longitude to locate places on the Earth

lava: [pg 148] rock formed by the cooling and solidifying of molten rock

lava plateau: [pg 153] build up of molten volcanic material in the formation of mountains

law: [pgs 8, 42] an idea or concept that is so well founded that it is almost equal to a fact

Law of Charges: [pg 196] like charges repel and unlike charges attract

Law of Conservation of Energy: [pgs 186, 197] energy cannot be created or destroyed just changed to other forms of energy

Law of Conservation of Mass: [pg 173] matter cannot be created or destroyed

Law of Gravity: [pg 187] any particle of matter in the universe attracts any other with a force varying directly as the product of the masses and inversely as the square of the distance between them

Law of Universal Gravitation: [pg 211] Newton's law determined that the strength of the gravitational force is dependent on the amount of mass, and on the square of the distance between the masses

Law of Poles: [pg 196] like poles repel and unlike poles attract

leaves: [pg 46] plant organ where photosynthesis occurs

lever: [pg 213] bar that is free to rotate around a fixed point called the fulcrum

light: [pgs 112, 120] visible form of the electromagnetic spectrum which provides energy for food and oxygen production by photosynthesis

lightning: [pg 160] abrupt, natural electric discharge in the atmosphere and accompanying visible flash of light

limiting factor: [pg 114] variables that affect how large a population can grow

Lind, James: [pg 96] Scottish physician known for his work in the prevention of scurvy by using citrus fruit

line graph: [pgs 21-26] visual representation of a relationship between two factors; used when the relationship is continuous

Linnaeus, Carolus: [pg 48] 1707-1778 Swedish botanist and founder of the modern classification system for plants and animals

liquid: [pg 177] phase of matter in which the particles are free to move over, under and around one another; the particles are further apart than in a solid

liter: [pg 220] basic unit of volume

lithosphere: [pg 139] solid portion of Earth

load: [pg 196] a device or the resistance of a device to which power is delivered

locomotion system: [pg 55] a group of organs working together to allow an organism to move

lodestone: [pg 196] rock that forms a natural magnet; magnetite

longitude: [pg 130] measures distance east or west of the prime meridian. Lines of longitude, also called meridians, are imaginary lines that divide the Earth. They run north to south from pole to pole, but they measure the distance east or west.

low pressure: [pg 157] places where the atmosphere is relatively thin.

lunar eclipse: [pg 128] occurs when the Sun, Earth, and Moon align so that the Moon passes into Earth's shadow. In a total lunar eclipse, the entire Moon falls within the darkest part of Earth's shadow, called the umbra.

lunar month: [pg 127] a month measured between successive new moons (roughly 29.5 days).

lungs: [pg 52] a pair of breathing organs located with the chest which remove carbon dioxide from and bring oxygen to the blood.

luster: [pgs 147, 167] the appearance of a mineral surface in terms of its light-reflective qualities.

lysosome: [pg 43] a membrane-bound cell organelle that contains digestive enzymes that break down protein

-M-

machine: [pg 213] system or device for doing work

magma: [pgs 142, 153] molten rock material under the Earth's crust, from which igneous rock is formed by cooling

magnet: [pgs 172, 196] an object that is surrounded by a magnetic field and that has the property, either natural or induced, of attracting iron or steel

magnetic compass: [pg 222] an instrument that uses a magnetized steel bar to indicate direction relative to the Earth's magnetic poles

magnetic field: [pg 197] the region around a magnet or an electric current where electromagnetic energy acts

magnetism: [pg 196] force exerted by creating fields of attraction and repulsion

malleability: [pg 171] ability of matter to be changed in shape; (example, hammered into sheets)

Malpighi, Marcello: [pg 53] 1628-1694 Italian anatomist who was the first to use a microscope in the study of anatomy and discovered the capillary system

manipulated variable: [pg 11] factor whose effect is being tested; independent variable

mantle: [pgs 139, 140] layer of the Earth between the crust and the core

Mars: [pgs 130, 240] fourth planet from the Sun, having a sidereal period of revolution about the Sun of 687 days at a mean distance of 227.8 million kilometers (141.6 million miles) and a mean diameter of approximately 6,726 kilometers (4,180 miles)

mass: [pgs 169, 210] the measure of the quantity of matter that a body or an object contains

mathematics: [pg 231] study of the measurement, properties, and relationships of quantities, using numbers and symbols

matter: [pgs 167, 170, 171, 185] anything that has mass and occupies space

mean: [pg 225] average of all scores

meander: [pg 144] follow a winding and turning course

measurement: [pgs 2, 219] the dimension, quantity, or capacity determined by measuring

mechanical energy: [pg 185] associated with the motion of matter; potential or kinetic energy

mechanical advantage: [pg 213] number of times a machine multiplies the effort force benefit realized by using a machine

median: [pg 225] half-way between the highest and lowest score

meiotic cell division: [pg 85] cell division where each new cells receives one-half of a complete set of chromosomes; occurs only in gonads and produces sex cells (gametes)

melting: [pg 177] phase change from a solid to a liquid

Mendeleev, Dmitri: [pg 174] 1834-190. Russian chemist who first devised and published the periodic table of the elements (1869).

meniscus: [pg 221] curve on liquid surface in a graduated cylinder; point from which volume is measured

Mendel, Gregor: [pg 66] 1822-1884. Austrian botanist and founder of the science of genetics. Through years of experiments with plants, chiefly garden peas, he discovered the principle of the inheritance of characteristics through the combination of genes from parent cells

Mercalli Scale: [pg 153] assigns a numerical value to the damage associated with an earthquake

Mercury: [pgs 130, 240] smallest of the planets and the one nearest the Sun, having a sidereal period of revolution about the Sun of 88.0 days at a mean distance of 58.3 million kilometers (36.2 million miles) and a mean radius of approximately 2,414 kilometers (1,500 miles)

meridian: [pg 130] imaginary great circle on the Earth's surface passing through the north and south geographic poles

mesa: [pg 143] broad, flat-topped elevation with one or more cliff-like sides, common in the southwest United States

metabolism: [pgs 46, 95, 192] the sum of all the chemical reactions in an organism

metalloid: [pg 176] element with both metal and nonmetal properties

metamorphic rock: [pg 148] type of rock formed from exposure to extreme pressure and temperature

metaphase: [pg 64] stage of mitosis and meiosis, following prophase and preceding anaphase, during which the chromosomes are aligned along the metaphase plate

meteor: [pg 129] rock fragments formed from the beginning of the universe

meteor shower: [pg 129] light show that forms from the friction of meteors moving through Earth's atmosphere

meteorologist: [pgs 157, 160] one who reports and forecasts weather conditions, as on television

meter: [pg 220] (m) metric standard for line measure; unit of distance

methane: [pg 138] odorless, colorless, flammable gas, CH_4, the major constituent of natural gas, that is used as a fuel and is an important source of hydrogen and a wide variety of organic compounds

metric ruler: [pg 222] the standard instrument for measurement in the scientific laboratory

metric system: [pg 219] decimal system of units based on the meter as a unit length, the kilogram as a unit mass, and the second as a unit time

microgravity: [pg 212] not experiencing the effects of gravity; free fall

micro-lightning: [pg 187] small electrical charge and this electrical charge is manifested in the form of light energy.

microscope: [pgs 41, 229] devise used to examine objects too small to be observed well with the unaided eye

Mid-Atlantic Ridge: [pg 152] a series of mountain ranges on the ocean floor, more than 84,000 kilometers (52,000 miles) in length, extending through the North and South Atlantic, the Indian Ocean, and the South Pacific. According to the plate tectonics theory, volcanic rock is added to the sea floor as the mid-ocean ridge spreads apart

Milky Way Galaxy: [pg 122] galaxy containing the solar system, visible as a broad band of faint light in the night-sky

mineral: [pg 146] pure substance made up of only one type of matter

mitochondria: [pg 43] organelle that is the center for cellular respiration

mitotic cell division: [pg 85] cell division where both new cells receive a complete set of chromosomes identical to that of the parent cell

mixture: [pg 172] two or more substances physically combined; both keep their own characteristics

mode: [pg 225] average that identifies the most frequently occurring score

model: [pg 5] representation of an idea made to better understand the idea

Moh Scale: [pg 147] a scale used to determine the relative hardness of a mineral

Mohorovicic (mo-ho-Ro-vi-chich), Andrija: [pg 140] associated with earthquake study, found way to determine the depth of the crust using seismic waves

molecule: [pg 176] smallest particle into which an element or a compound can be divided without changing its chemical and physical properties; a group of like or different atoms held together by chemical forces

momentum: [pg 207] measure of the motion of a body equal to the product of its mass and velocity

Moon: [pg 211] natural satellite of Earth

Mosely, Henry: [pg 174] 1887-1915. British physicist who organized the modern-day periodic table, after his discovery that an atomic number of an element can be deduced from the element's x-ray spectrum

motion: [pg 186] change in an object's location in space over a period of time

mountain building: [pg 151] result of collision of continental plates

mouth: [pg 49] body opening through which an animal takes in food

multicellular: [pg 44] organism consisting of many cells

mutation: [pg 75] a change in the structure of the DNA of an organism

-N-

natural barriers: [pg 155] Earth's physical features that act as limiting agents in weather, weathering, and migrations

natural gas: [pg 188] fossil fuel; usually found in association with petroleum because it is formed in the same way

natural selection (theory of): [pg 73] Darwin's theory on the basis for evolution; only organisms that best fit their environment will survive

natural satellite: [pgs 127, 130] a celestial body that orbits a planet; a moon

natural variation: [pg 17] differences due to chance

nerves: [pg 56] cord-like bundles of fibers made up of neurons through which sensory stimuli and motor impulses pass between the brain or other parts of the central nervous system and the eyes, glands, muscles, and other parts of the body

nervous system: [pgs 56, 57] a group of organs working together to control and coordinate the bodies responses to changes in its environment; brain, spinal cord, nerves

neurons: [pg 56] impulse-conducting cells that constitute the brain, spinal column, and nerves, consisting of a nucleated cell body with one or more dendrites and a single axon

neutralization reaction: [pg 223] a reaction between an acid and a base that yields a salt and water

neutron: [pg 173] neutral particle found in the nucleus of an atom

new crescent: [pg 128] Moon phase that follows when the Moon is between the Sun and Earth; only the back side of the Moon is illuminated (side not visible to a viewer on Earth)

new gibbous: [pg 128] Moon phase following first quarter; appears that part of the full Moon is missing.

new Moon: [pg 128] Moon is between the Sun and Earth; only the back side of the Moon is illuminated (the side not visible to a viewer on Earth)

newton: [pgs 170, 220] (N) basic metric unit of force (weight)

Newton, Sir Isaac: [pgs 8, 210] 1642-1727. English mathematician and scientist who developed the Laws of Motion and the Law of Universal Gravitation; his treatise on gravitation was supposedly inspired by the sight of a falling apple

nitrogen bases: [pg 62] A molecule that contains nitrogen and has the chemical properties of a base. The nitrogenous bases in DNA are adenine (A), guanine (G), thymine (T), and cytosine (C). The nitrogenous bases in RNA are the same, with one exception: adenine (A), guanine (G), uracil (U), and cytosine (C)t

nitrogen cycle: [pg 103] flow of nitrogen between the living and non-living environment

nitrogen oxide: [pg 138] product of combustion of fossil fuels; greenhouse gas

noble gases: [pgs 173, 176] any of a group of rare gases that include helium, neon, argon, krypton, xenon, and sometimes radon and that exhibit great stability and extremely low reaction rates.

nonrenewable energy resource [pg 188] a substance that is used up more quickly than it can replace itself. The supply of a nonrenewable resource is finite, which means it cannot easily be replenished. Nonrenewable resources are extracted directly from the Earth.

nose: [pg 51] the northernmost point on Earth

North Atlantic Drift: [pg 156] the broad, north-ward flow of surface waters that replaces the sinking waters in the N. Atlantic polar seas.

north pole: [pg 125] the prominent structure between the eyes that serves as the entrance to the respiratory tract and contains the olfactory organ

nuclear age: [pg 190] sometimes called the Atomic Age or the Atomic Era, refers to the period of time following the first atomic bomb during the Second World War.

nuclear energy: [pg 185] a form of energy released from the nucleus, the core of atoms, made up of protons and neutrons

nucleus: [pgs 43, 173] information center for the cell; the membrane-enclosed organelle within a cell that contains the chromosomes.

-O-

oblate sphere: [pg 126] shape of Earth; almost perfectly round but bulges at the equator

observing: [pgs 3, 229] using one or more senses to examine something carefully

obsidian: [pg 149] usually black or banded, hard volcanic glass that displays shiny, curved surfaces when fractured and is formed by rapid cooling of lava

occluded front: [pg 158] fast moving cold front overtakes a warm front, lifting it quickly

ocean currents: [pgs 143, 151, 155, 163] (see Gulf Stream)

ocean ridges: [pg 152] underwater mountains (see Mid-Atlantic Ridge)

ocean trench: [pg 151] deep furrow or ditch cause by collision of plates when denser one slides under another

Oersted, Hans Christian: [pg 197] 1777-1851. Danish physicist, demonstrated that current electricity gives rise to a magnetic field

old crescent: [pg 128] This is the last illuminated phase before a New Moon

old gibbous: [pg 128] Moon phase occurs as the amount of illumination decreases on the full Moon

omnivore: [pgs 5, 93] organism that gets energy by eating plants and/or animals

operational definition: [pg 14] a description that allows people to agree or recognize the thing being described associated with scientific experimentation

orbital motion: [pgs 127, 211] circular motion that occurs when the force of gravity and the forward inertia of an object are in balance

order: [pg 48] a taxonomic category of organisms ranking above a family and below a class

organs: [pg 45] group of tissues working together to perform certain functions; example - stomach has muscle and nervous tissue working together to help digest food

organ system: [pg 45] a group of organs working together to perform certain functions; example - the mouth, esophagus, stomach, and intestines all work to make food available for use

organelle: [pg 43] differentiated structure within a cell, such as a mitochondrion, vacuole, or chloroplast, that performs a specific function

origin: [pg 23] point at which something comes into existence or from which it derives or is derived

outer core: [pg 140] liquid Earth zone between the inner core and the mantle; like the inner core, composed of iron and nickel

oxygen: [pg 141] nonmetallic element constituting 21 percent of the atmosphere by volume that occurs as a diatomic gas, O_2, and in many compounds such as water and iron ore. It combines with most elements, is essential for plant and animal respiration, and is required for nearly all combustion

oxygen cycle: [pg 102] movement and recycling of oxygen through the ecosystem through various forms and compounds

ozone: [pgs 117, 137, 138] gas located in the stratosphere that protects Earth from UV radiation

ozone depletion: [pg 139] loss of atmospheric ozone associated with destruction by CFCs and human health concerns

Pacific Rim: [pg 151] countries and landmasses surrounding the Pacific Ocean

Pangaea: [pg 150] super-continent that existed millions of years ago when all of the continents were joined

parallel: [pgs 130, 235] lines of latitude run east-west and are parallel to one another

parallel circuit: [pg 196] each load has its own path back to the source of electricity; example, if a lamp light goes out in your home, it does not open the circuit for all of the other loads

parasitism: [pg 114] relationship whereby one organism benefits (parasite) and one is harmed (host)

percentage of error: [pgs 16, 224] compares data to a standard value

pedigree chart: [pg 69] shows relationships between generations; a representation of pairings and offspring that helps show patterns of inheritance

period: [pg 176] row on the periodic table; properties of the elements in a period change in a predictable manner

Periodic Table of Elements: [pgs 174, 175] organization of all elements based on similar and repeating properties

petrification: [pg 147] formation of fossils when minerals replace living tissue

petroleum: [pgs 171, 188] fossil fuel; formed mostly from remains of plants and animals living in shallow oceans

pH scale: [pg 223] unit of measure for strength of an acid or base; the range goes from 0 - 14, with 7 being neutral. pHs of less than 7 indicate acidity, whereas a pH of greater than 7 indicates a base

phase: [pg 177] state of material; solid, liquid, or gas

phases of the Moon: [pg 128] new Moon, new-crescent, first quarter, new gibbous, full Moon, old gibbous, last quarter, old crescent

phase change: [pgs 172, 177] change from one material state to another

phloem: [pg 45] tissue that carries food products down the stem to the roots from the plant's leaves

photovoltaic cell: [pg 188] device that converts solar energy directly into electrical energy

photosynthesis: [pgs 46, 93, 101] process by which carbon dioxide, water, and light energy are converted into chemical energy (simple sugar)

phylum: [pg 48] a primary division of a kingdom, as of the animal kingdom, ranking next above a class in size

physical factors: [pg 112] Any environmental factor not caused by living organisms

physical properties: [pg 167] phase, shape or size of matter

physical weathering: [pg 141] breakdown size and shape of rock without change in chemical composition

planet: [pg 130] nonluminous celestial body larger than an asteroid or a comet, illuminated by light from a star, such as the Sun, around which it revolves and it must be big enough that its gravity clears away any objects of similar size.

plasma: [pg 177] a phase of matter achieved under very high temperature such as that which exists on our Sun and other stars

plates: [pg 177] gigantic pieces of the Earth's crust and uppermost mantle. They are made up of oceanic crust and continental crust. Earthquakes occur around mid-ocean ridges and the large faults which mark the edges of the plates.

plotting a graph: [pg 25] placing the data points at the correct positions on the graphing grid

Pluto: [pg 130] ninth and usually farthest planet from the Sun, having a sidereal period of revolution about the Sun of 248.4 years, 4.5 billion kilometers or 2.8 billion miles distant at perihelion and 7.4 billion kilometers or 4.6 billion miles at aphelion, and a diameter less than half that of Earth

pollution: [pg 113] substances that are harmful to living things

population: [pg 112] all members of the same type of living things in a given area

potential energy: [pg 186] stored energy; energy at rest as a result of its state or position; energy of position such as gravitational potential energy

power: [pg 212] amount of work done per unit of time

power plant: [pg 197] structures, machinery, and associated equipment for generating electric energy from another source of energy, such as nuclear reactions or a hydroelectric dam

precipitation: [pg 145] any form of water, such as rain, snow, sleet, or hail, that falls to the Earth's surface

predation: [pg 114] when one organism (predator) eats another organism (prey)

predicting: [pg 7] making an educated guess as to what might happen; forming a hypothesis

pressure: [pgs 139, 168] force applied uniformly over a surface, measured as force per unit of area

primary wave (P-wave): [pg 194] compressional wave of energy transmitted through Earth during earthquake

Prime Meridian: [pg 130, 235] zero meridian (0°), used as a reference line from which longitude east and west is measured, passes through Greenwich, England

principle of dominance and recessiveness: [pg 66] used to describe the functional relationship between two alleles of one gene in a heterozygote.

principle of inertia: [pgs 127, 210] Newton's first Law of Motion

principle of segregation: [pgs 66] the individual has two alleles for each particular characteristic, and during the development of gametes, these alleles become segregated.

producer: [pgs 5. 93, 105] organism that can make their own food from nonliving materials; example, green plants

product: [pg 173] substance created as a result of a reaction

projectile motion: [pg 212] curved motion due to the forward inertia of an object and the pull of gravity

prophase: [pg 64] first stage of mitosis, during which the chromosomes condense and become visible, the nuclear membrane breaks down, and the spindle apparatus forms at opposite poles of the cell

protein: [pgs 95, 96] nutrient needed for growth and repair of cells and tissues

protium: [pg 174] hydrogen with no neutrons

proton: [pg 173] positively charged particle found in the nucleus of an atom

Ptolemy, Claudius: [pg 122] proposed geocentric view of the universe, placing the Earth as the center of the universe

pulley: [pg 213] grooved wheel that rotates around (not with) an axle; machine based on the lever

pulse: [pg 53] surge of blood through the arteries as the heart pumps blood

pulse point: [pg 53] a place on the body, such as the neck or wrist, where the pulse can be easily felt

pumice: [pg 148] light, porous, glassy lava, used in solid form as an abrasive and in powdered form as a polish and an abrasive

Punnett, R.C. [pg 67] discovered a method of statistically, mathematically determining the genetic characteristics of offspring (see Punnett Square)

Punnett Square: [pg 67] grid used to help determine the genes in offspring containing two similar genes for a trait (see first cross)

-Q-

qualitative data: [pg 17] information that is relevant to an investigation that cannot be put directly into numeric form

quantitative data: [pg 17] numeric information that is relevant to an investigation

-R-

radial artery: [pg 54] a blood vessel that supplies blood to the forearm (lower part of the arm) and hand

radiant energy: [pg 185] type of energy that makes up the electromagnetic spectrum

radiation: [pg 191] heat energy transfer that can occur across space

radio wave: [pg 193] relatively low energy wave that is part of the electromagnetic spectrum

reactant: [pg 173] substance participating in a chemical reaction, especially a directly reacting substance present at the initiation of the reaction

reaction to acid: [pg 147] carbon dioxide being released by a chemical reaction between the carbonate material and acid

reactivity: [pg 171] how readily matter combines with other matter

recessive: [pg 67] trait whose effect may be hidden by a dominant trait

recording: [pg 2] information or facts, set down especially in writing as a means of preserving knowledge

reflection: [pg 195] when a light or sound wave bounces off a surface

refraction: [pg 195] when a wave bends because it is slowing down or speeding up

relative humidity: [pg 159] water vapor in a given sample of air compared to the maximum amount of water vapor that air sample can hold

renewable energy resources: [pg 188] alternative energy resources that will be around for millions of years such as wind and solar energy

reproductive system: [pgs 57, 58] male and female organs, glands; necessary to produce offspring

repulsion: [pg 196] tendency of particles with like charges and like poles to push away from one another

resistance distance: [pg 213] actual distance the machine has to travel

resistance force: [pg 213] weight of the object

respiratory system: [pgs 47, 51] group of organs that bring oxygen to and remove carbon dioxide from the blood

responding variable: [pg 11] the factor you measure to determine the effect of manipulated variable; dependent variable

revolution: [pg 127] orbital motion about a point, especially as distinguished from axial rotation

rib: [pg 51] one of a series of long, curved bones occurring in 12 pairs in human beings and extending from the spine toward the sternum

ribosomes: [pg 43] cell structure that assists in building proteins

Richter Scale: [pg 153] measures the relative energy of an earthquake

Ring of Fire: [pg 151] circular region in the Pacific Ocean where the plates are so active that many volcanoes occur

rock: [pg 150] solid Earth made up of more than one mineral

rock cycle: [pg 150] cycle by which rocks change to other types of rocks

roots: [pg 46] plant organs that help a plant take in materials from the soil

rotation: [pg 127] act or process of turning around a center or an axis

runoff: [pgs 142, 146] water that moves on the surface of Earth

-S-

saliva: [pg 49] watery mixture of secretions from the salivary and oral mucous glands that lubricates chewed food, moistens the oral walls, and contains ptyalin

salt: [pg 173] A salt in which no hydrogen or hydroxyl (OH) ion is replaced by a metallic ion. Sodium chloride (NaCl) is a simple salt.

San Andreas fault: [pg 151] a major zone of fractures in the Earth's crust extending along the coastline of California from the northwest part of the state to the Gulf of California

saturated solution: [pg 168] solution that holds the maximum solute for a given temperature

Saturn: [pg 130] sixth planet from the Sun and the second largest in the solar system, having a sidereal period of revolution about the Sun of 29.5 years at a mean distance of about 1,425,000,000 kilometers (886,000,000 miles), a mean diameter of approximately 119,000 kilometers (74,000 miles), and a mass 95 times that of Earth

scaling a graph: [pg 23] setting up values for the graphing grid to best display the data

Schleiden, Matthias: [pg 41] German botanist responsible for the discovery that all plants are composed of cells

Schwann, Theodor, [pg 42] 1810-1882 German physiologist and pioneer histologist who described the cell as the basic structure of animal tissue

science: [pg 1] exploring and explaining the natural world

scientific inquiry: [pg 1] scientific study of the natural world

scientific notation: [pg 223] a method of writing or displaying numbers in terms of a decimal number between 1 and 10 multiplied by a power of 10

screw: [pg 214] simple machine that forms a spiral inclined plane

scurvy: [pg 96] disease caused by deficiency of vitamin C, characterized by spongy and bleeding gums, bleeding under the skin, and extreme weakness

sea breeze: [pg 161] forms when air over the water is cooler than air over land; denser, cooler air moves toward the land

season: [pg 125] one of the four natural divisions of the year, spring, summer, fall, and winter, in the North and South Temperate zones

second cross [pg 68] see Punnett Square method of solving genetic problems

Second Law of Motion: [pg 210] force is directly related to mass and acceleration

secondary wave (S-wave): [pg 194] transverse wave of energy transmitted through Earth during earthquake

sedimentary rock: [pg 147] formed from weathered rocks, through the process of erosion; may contain fossils

sedimentation: [pg 143] deposit of particles by water, wind, and glaciers

seed: [pg 88] reproductive embryo for young plant surrounded by stored food and a protective coat

segregation: [pg 66] separation of paired alleles especially during meiosis, so that the members of each pair of alleles appear in different gametes

seismic waves: [pg 153] wave that radiates from the point of origin of at earthquake, moving in all directions through solid rock

seismograph: [pg 153] very delicate instrument that detects and records passing earthquake waves

selective breeding: [pg 75] a form of genetic engineering in breeding in order to choose special traits for the offspring

senses: [pg 167] any of the faculties by which stimuli from outside or inside the body are received and felt, as the faculties of hearing, sight, smell, touch, taste, and equilibrium

series circuit: [pg 196] electricity travels through all of its loads before it travels back to its source of electrical energy; example - if one light in a string goes out, it acts to open the circuit, blocking the flow of electricity

sex cell: [pgs 65, 85] germ cell or gamete

sexual reproduction: [pgs 65, 73, 84, 92] union of sperm and egg producing an offspring where half of the genetic information for the new individual comes from each parent

shooting star: [pg 129] meteor, a bright trail or streak that appears in the sky when a meteoroid is heated to incandescence by friction with the Earth's atmosphere

sidereal day: [pg 124] measured or determined by means of the apparent daily motion of the stars; uses star as a frame of reference

sidereal time: [pg 124] measured or determined by means of the apparent motion of the stars; uses distant star as a frame of reference

skeletal muscles: [pg 55] attached to the internal structure composed of bone and cartilage that protects and supports the soft organs, tissues, and other parts of a vertebrate organism

skin: [pg 54] membranous tissue forming the external covering or integument of an animal and consisting in vertebrates of the epidermis and dermis

slope: [pg 29] a numerical value that describes the steepness of a line

small intestine: [pg 49] narrow, winding, upper part of the intestine where digestion is completed and nutrients are absorbed by the blood

smell: [pgs 56, 167] quality of something that may be perceived by the olfactory sense

smooth muscle: [pg 55] tissue that contracts without conscious control, having the form of thin layers or sheets made up of spindle-shaped, unstriated cells with single nuclei and found in the walls of the internal organs, such as the stomach, intestine, bladder, and blood vessels, excluding the heart

soil: [pgs 112, 144] organic material mixed with rock fragments; formed from weathered rock

solar day: [pg 124] measured or determined by means of the apparent daily motion of the Sun; uses Sun as a frame of reference

solar collector: [pg 188] a physical collection unit for solar energy

solar eclipse: [pg 128] partial or total blocking of the Sun; Moon is between Earth and Sun; total solar eclipse lasts only a few minutes

solar energy: [pg 188] energy from sunlight

solar radiation: [pg 126] Sunlight

solid: [pg 177] phase of matter in which the particles are closest together

solubility: [pgs 167, 168] how well a substance can dissolve in a given amount of solvent

solubility graphs: [pg 168] summary of solubility of various substances in given mass of water

solute: [pg 167] substance that dissolves in a solvent; component of a solution present in the lesser amount

solution: [pg 167] homogeneous (similar) mixture of two or more substances, which may be solids, liquids, gases, or a combination of these

solvent: [pg 167] substance in which a solute is dissolved

sound energy: [pg 185] form of vibrational energy

south pole: [pg. 125] The southern end of Earth's axis of rotation, a point in Antarctica

specialize: [pg 44] to develop so as to become adapted to a specific function or environment

species: [pg 48] organisms with similar characteristics; smallest grouping in the classification system; ability to reproduce

specific heat: [pg 190] amount of heat needed to raise the temperature of one gram of matter by one degree Celsius

speed: [pg 207] rate or a measure of the rate of motion

sperm: [pgs 58, 84] male fertilizing cell, contains one half the number of chromosomes found in a body cell

sphere: [pg 126] a celestial body, such as a planet or star

spinal cord: [pg 56] thick, whitish cord of nerve tissue that extends from the medulla oblongata down through the spinal column and from which the spinal nerves branch off to various parts of the body

spring scale: [pg 221] instrument that measures force

spores (sporophyte): [pg 88] small, usually single celled reproductive body that is highly resistant to desiccation and heat and is capable of growing into a new organism, produced especially by certain bacteria, fungi, algae, and nonflowering plants

staining: [pg 233] adding a chemical to a specimen to make it easier to observe

star trails: [pg 123] time-lapse photograph created because of Earth's rotation

stars: [pg 129] celestial body that can produce its own light

statistics: [pg 224] collection, organization, and interpretation of numerical data, especially the analysis of population characteristics by inference from sampling

static electricity: [pg 195] electrical charge "at rest," charges builds up on matter and are then discharged

stationery front: [pg 158] meeting of a cold and warm air mass; move very slowly relative to one another; produces weather closely resembling warm front weather

stem: [pg 46] plant organ that helps to support the plant

stomach: [pg 49] one of the principal organs of digestion, located in vertebrates between the esophagus and the small intestine

stomates: [pg 46] minute pores in the epidermis of a leaf or stem through which gases and water vapor pass

stop watch: [pg 222] a watch that can be instantly started and stopped by pushing a button and used to measure an exact duration of time

stored energy: [pg 186] energy held in reserve - potential energy

stratus cloud: [pg 158] low-altitude cloud formation consisting of a horizontal layer of gray clouds

stratosphere: [pgs 137, 139] upper portion of the atmosphere; above the troposphere, below the mesosphere

streak: [pg 146] color of the powdered mineral which can easily be seen by rubbing the mineral on a white or clear smooth surface such as ceramic

strip-mining: [pg 142] to mine (ore) from the surface of the Earth, by removing layers exposed to the atmospheric environment

subduction: [pg 151] sliding of a denser plate under a less dense plate

sublimation: [pgs 145, 177] phase change of matter from solid to gas

summer solstice: [pg 126] beginning of summer (about June 21st); longest period of daylight in one day

Sun: [pgs 121-130] star closest to Earth

supersaturated solution: [pg 168] This is a solution that holds more solute than it normally can for a given temperature

synoptic weather map: [pg 157] weather map that summarizes a number of weather conditions

-T-

taiga: [pg 155] subarctic, evergreen coniferous forest of northern Eurasia located just south of the tundra and dominated by firs and spruces

teeth: [pg 49] hard, bone-like structures rooted in sockets in the jaws of vertebrates used for mechanical digestion

tectonic plate: [pg 150] section into which the Earth's crust is divided and that is in constant motion relative to other plates, which are also in motion

telophase: [pg 64] final stage of mitosis or meiosis during which the chromosomes of daughter cells are grouped in new nuclei

temperate deciduous forest: [pg 155] shedding or losing foliage at the end of the growing season

temperate rain forest: [pg 155] dense evergreen forest

temperature: [pgs 141, 155, 168, 176] measures the average kinetic energy in a sample of matter

texture: [pg 167] appearance and feel of a surface

theory: [pgs 1, 8, 42] carefully thought out, logically reasoned, well founded idea about the natural world

Theory of Continental Drift: [pg 150] continents were once all joined into one large continent and over millions of years separated into the present continents

Theory of Plate Tectonics: [pg 151] Earth consists of a number of plates which make up the lithosphere; plates can move

thermometer: [pg 222] an instrument for measuring temperature, especially one having a graduated glass tube with a bulb containing a liquid, typically mercury or colored alcohol, that expands and rises in the tube as the temperature increases

Third Law of Motion: [pg 210] for every action there is an equal and opposite reaction

Thompson, Benjamin (Count Rumford): [pg 190] 1753-1814. American-born British public official and physicist, first scientist credited with showing that heat is a form of energy

thunderstorms: [pg 160] hazardous storms normally form after severe heating of Earth's surface; results in numerous powerful convection cells

thunder: [pg 160] crashing or booming sound produced by rapidly expanding air along the path of the electrical discharge of lightning.

tissues: [pg 45] group of similar cells that perform the same function; example- cells in muscle tissue work together to create motion

topographic map: [pg 234] graphic representation of the surface features of a place or region on a map, indicating their relative positions and elevations

top soil: [pg 145] surface soil usually including the rich upper layer in which plants have most of their roots and which the farmer turns over in plowing

tornadoes: [pg 160] rotating column of air usually accompanied by a funnel-shaped downward extension of a cumulonimbus cloud, having a vortex several hundred yards in diameter whirling destructively at speeds of up to 500 miles (800 kilometers) per hour

trachea: [pg 51] tube that carries air from the nose to the bronchi on the way to the lungs. Also called the windpipe

transpiration: [pg 145] to give off vapor containing waste products, as through animal or plant pores

transverse wave: [pgs 193, 201] wave that travels perpendicular to its direction of motion; radiant energy travels as transverse waves

trait: [pg 66] genetically determined characteristic or condition

trench: [pg 151] formation that results when one plate slides under another plate

triple beam balance: [pg 221] device used to measure mass

tritium: [pg 174] hydrogen with two neutrons

troposphere: [pg 137] lower portion of the atmosphere where weather takes place

tsunamis: [pg 153] giant waves caused by underwater earthquakes

tundra: [pg 155] treeless area between the icecap and the tree line of Arctic regions, having a permanently frozen subsoil and supporting low growing vegetation such as lichens, mosses, and stunted shrubs

-U-

ultraviolet light: [pg 193] a type of electromagnetic radiation that makes black-light posters glow, and is responsible for suntans

ultraviolet radiation: [pg 117] high energy part of the electric magnetic spectrum that is associated with Sun tanning in humans and causing skin cancer

unicellular: [pg 44] one-celled organism

universal solvent: [pg 141] water

universe: [pg 185] all matter and energy, including Earth, the galaxies and all therein, and the contents of intergalactic space, regarded as a whole

Universal Law of Gravitation: [pgs 126, 211] Newton's Law which states that all matter in the universe is attracted to one another

unknown: [pg 224] a quantity of unknown numerical value

unsaturated solution: [pg 168] solution that can hold more solute for a given temperature

Uranus: [pg 130] seventh planet from the Sun, revolving about it every 84.07 years at a distance of approximately 2,869 million kilometers (1,790 million miles), having a mean equatorial diameter of 52,290 kilometers (32,480 miles) and a mass 14.6 times that of Earth

urinary structures: [pg 54] regulatory organs involved in the formation and excretion of urine and the removal of metabolic wastes

U-shaped valleys: [pg 143] a valley formed by a process called glaciation. Glaciation happens when a glacier carves into a valley and scours it into a distinctive U shape with high, straight sides and a rounded or flat bottom.

-V-

vacuole: [pg 43] cell structure for storage

vaporization: [pg 177] phase change from a liquid to gas

variable: [pgs 5, 11] factor whose change might influence a relationship

variations: [pg 73] the differences or deviations from the recognized norm or standard

vegetation: [pgs 143, 154] plants of an area or a region; plant life

vegetative propagation: [pg 84] asexual reproduction where a part of an organism can be separated and grown into another individual

vein: [pg 52] blood vessels that carry blood and wastes to the heart

velocity: [pg 207] vector quantity measures speed with direction; represents time rate of change in displacement

Venn diagram: [pg 5] a diagram using circles to represent an operation in set theory, with the position and overlap of the circles indicating the relationships between the sets

Venus: [pg 130] second planet from the Sun, having an average radius of 6,052 kilometers (3,760 miles), a mass 0.815 times that of Earth, and a sidereal period of revolution about the Sun of 224.7 days at a mean distance of approximately 108.1 million kilometers (67.2 million miles)

Vents: [pg 153] Any opening at the Earth's surface through which magma erupts or volcanic gases are emitted

vernal equinox: [pg 126] marks the beginning of spring, about March 21; point at which the Sun passes the celestial equator (on its way north) causing equal amounts of daylight and darkness; see autumnal equinox

vertical force: [pg 212] upright push or pull at right angles to the horizon

Virchow, Rudolf: [pg 42] 1821-1902 German physician and pathologist known for his contributions to cell theory and the study of disease

virus: [pg 42] considered the "exception" to the Cell Theory, extremely small, often associated with harmful diseases, only capable of reproducing when inside an organism

visible light: [pg 192] part of the electromagnetic spectrum that you can see

vitamins: [pg 95] a group of substances needed in small quantities for the health of an organism

volcanic islands: [pg 152] islands formed from volcanoes

volcano: [pgs 152, 153] ridge or mountain that releases liquid rock from an opening near the top

voltmeter: [pg 222] electrical device to measure voltage drop across a resistor; always attached in parallel; galvanometer with a high resistance in series

volume: [pgs 167, 169] amount of space occupied by a three-dimensional object or region of space

von Baer, Karl Ernst: [pg 72] 1792-1876. Estonian born German naturalist and pioneer embryologist who discovered the mammalian egg (1827)

-W-

warm front: [pg 158] leading edge of an air mass which has warmer temperatures than the preceding air mass; usually associated with steady precipitation

water: [pg 95, 112, 141] clear, colorless, odorless, and tasteless liquid, H_2O, essential for most plant and animal life

water budget: [pg 154] method of comparing the inflow and outflow of water within the Earth's hydrosphere

water cycle: [pg 145] phase change and movement of water between the living and nonliving environment; atmosphere, hydrosphere, and lithosphere; hydrologic cycles

water pressure: [pg 141] pressure exerted by the liquid part of Earth

water vapor: [pgs 138, 145] water in a gaseous state, especially when diffused as a vapor in the atmosphere and at a temperature below boiling point

Watson, James: [pg 62] Born in 1928. American biologist who with Francis Crick proposed a spiral model, the double helix, for the molecular structure of DNA. He shared a 1962 Nobel Prize for advances in the study of genetics

watt: [pg 219] unit of power

wavelength: [pg 193] distance from top (crest) of one wave through the bottom (trough), and back to the top of another wave

weather: [pgs 154, 156] local or short-term changes determined by variables in temperature, wind, moisture, and pressure

weather front: [pg 158] area by which two air masses meet

weather map: [pg 157] graphic representation showing temperature, barometric pressure, wind speed, and relative humidity

weather patterns: [pg 157] (see weather)

weather satellite: [pg 157] space craft orbiting the Earth to relay weather data

weather station: [pg 157] a facility or location where meteorological data are gathered, recorded, and released

weathering: [pg 140] when rock is changed physically or chemically as it interacts with air, water and living things

wedge: [pg 214] simple machine; a moving inclined plane

Wegener, Alfred: [pg 150] 1880-1930. German geophysicist, meteorologist, and explorer who proposed the theory of continental drift

weight: [pgs 169, 206] force created by the response of mass to the pull of gravity; measure of the heaviness of an object

weightlessness: [pg 212] not experiencing the effects of gravity

wet mount: [pg 233] when an object is prepared for view under a microscope by adding water

wheel and axle: [pg 214] lever that rotates in a circle

white blood cells: [pg 57] also called leukocytes. They protect you against illness and disease. Think of white blood cells as your immunity cells. In a sense, they are always at war. They flow through your bloodstream to fight viruses, bacteria, and other foreign invaders that threaten your health.

wind: [pgs 143, 188] a natural and perceptible movement of air parallel to or along the ground

wind belts: [pg 155] associated with the pressure belts of the Earth, responsible for the movement of massive weather systems

wind energy: [pg 188] renewable energy; can turn turbines in order to generate electricity

winter solstice: [pg 126] marks the beginning of winter; occurs about December 21; shortest day of the year; when the Sun is over the tropic of Capricorn

wood: [pg 188] form of renewable energy

work: [pgs 190, 212] force applied through a distance

-X-

x-axis: [pg 21] horizontal axis of a two-dimensional Cartesian coordinate system

x-ray: [pg 193] high energy wave that is part of the electromagnetic spectrum

xylem: [pg 45] woody tissue that carries water and minerals up through the plant

-Y-

y-axis: [pg 21] vertical axis of a two-dimensional Cartesian coordinate system

yield: [pg 173] the amount of product resulting from a chemical reaction

-Z-

zenith: [pg 235] highest point above the observer's horizon attained by a celestial body

zygote: [pgs 58, 84] cell produced after a sperm and egg cell unites; contains the full number of chromosomes